T0245421

CAMBRIDGE LIBRARY COLLECTION

Books of enduring scholarly value

Technology

The focus of this series is engineering, broadly construed. It covers technological innovation from a range of periods and cultures, but centres on the technological achievements of the industrial era in the West, particularly in the nineteenth century, as understood by their contemporaries. Infrastructure is one major focus, covering the building of railways and canals, bridges and tunnels, land drainage, the laying of submarine cables, and the construction of docks and lighthouses. Other key topics include developments in industrial and manufacturing fields such as mining technology, the production of iron and steel, the use of steam power, and chemical processes such as photography and textile dyes.

Principles of Mechanism

Robert Willis (1800–75) was an scientist, inventor and architectural historian of international repute. As Jacksonian Professor of Natural and Experimental Philosophy at Cambridge, he demonstrated specially made mechanical devices to huge audiences. First published in 1841, *Principles of Mechanism* provided the theory behind the demonstrations. He defined mechanism as the means by which any relations of motion could be realised. The book was extremely influential, with all books in English, French and German on the subject for the next generation adopting Willis's classification and nomenclature. He worked closely with William Whewell, whose *Mechanics of Engineering* was published in the same year. These two books established the science of mechanism, and provided study materials for the rapidly growing engineering profession. The work became a standard textbook for engineering and mathematics students, with a second edition issued in 1870.

Cambridge University Press has long been a pioneer in the reissuing of out-of-print titles from its own backlist, producing digital reprints of books that are still sought after by scholars and students but could not be reprinted economically using traditional technology. The Cambridge Library Collection extends this activity to a wider range of books which are still of importance to researchers and professionals, either for the source material they contain, or as landmarks in the history of their academic discipline.

Drawing from the world-renowned collections in the Cambridge University Library, and guided by the advice of experts in each subject area, Cambridge University Press is using state-of-the-art scanning machines in its own Printing House to capture the content of each book selected for inclusion. The files are processed to give a consistently clear, crisp image, and the books finished to the high quality standard for which the Press is recognised around the world. The latest print-on-demand technology ensures that the books will remain available indefinitely, and that orders for single or multiple copies can quickly be supplied.

The Cambridge Library Collection will bring back to life books of enduring scholarly value (including out-of-copyright works originally issued by other publishers) across a wide range of disciplines in the humanities and social sciences and in science and technology.

Principles of Mechanism

Designed for the Use of Students in the Universities, and for Engineering Students Generally

Robert Willis

CAMBRIDGE UNIVERSITY PRESS

Cambridge, New York, Melbourne, Madrid, Cape Town, Singapore,
São Paolo, Delhi, Dubai, Tokyo, Mexico City

Published in the United States of America by Cambridge University Press, New York

www.cambridge.org
Information on this title: www.cambridge.org/9781108023092

This edition first published 1841
This digitally printed version 2010

ISBN 978-1-108-02309-2 Paperback

PRINCIPLES

OF

MECHANISM,

DESIGNED FOR THE USE OF STUDENTS
IN THE UNIVERSITIES,

AND FOR

ENGINEERING STUDENTS GENERALLY.

BY

ROBERT WILLIS, M.A., F.R.S., &c.

JACKSONIAN PROFESSOR OF NATURAL AND EXPERIMENTAL PHILOSOPHY
IN THE UNIVERSITY OF CAMBRIDGE.

LONDON:
JOHN W. PARKER, WEST STRAND.
CAMBRIDGE: J. & J. J. DEIGHTON.

M.DCCC.XLI.

PREFACE.

In the present work I have employed the term Mechanism as applying to combinations of machinery solely when considered as governing the relations of motion. Machinery as a modifier of force, has in the science of Mechanics occupied the attention of nearly every mathematician of eminence who has arisen in the world; but, by some strange chance, very few have attempted to give a scientific form to the attractive and valuable results of mechanism; for it cannot be said that the few and simple machines which form the examples in books of mechanics, are to be regarded as even forming a foundation for the principles upon which is to be based a science that will enable us either to reduce the movements and actions of a complex engine to system, or to give answers to the questions that naturally arise upon considering such engines;—for example, are the means by which the results are obtained the best that might have been employed? or what are the various methods that might have been substituted for them? Yet there appears no reason why the construction of a machine for a given purpose should not, like any usual problem, be so reduced to the dominion of the mathematician, as to enable him to obtain, by direct and certain methods, all the forms and arrangements that are applicable to the desired purpose, from which he may select at pleasure. At present, questions of this kind can only be solved by that species of intuition which long familiarity with a subject usually confers upon experienced persons, but which they are totally unable to communicate to others.

When the mind of a mechanician is occupied with the contrivance of a machine, he must wait until, in the midst of his meditations, some happy combination presents itself to his mind which may answer his purpose. Yet upon analysing the mental operations by which the nascent contrivance is gradually made to assume form and consistency, it will generally be observed, that the motions of the machine are the principal subject of contemplation, rather than the forces applied to it, or the work it has to do. For every machine will be found to consist of a train of pieces connected together in various ways, so that if one be made to move they all receive a motion, the relation of which to that of the first is governed by the nature of the connexion. The work which the machine has to do will require that the pieces appropriated to this work shall move with respect to each other in some given manner, and the forces applied to the machine to set it in motion must also move the piece which receives them in some other manner. Thus the question of contriving a machine by which a given kind of power may be made to perform given work, is reduced to a problem of mere motion,—to a question of connecting the pieces which receive the power and those which do the work; so that when the first move according to the law required by the economy of the power, the last shall necessarily receive the motion which will enable them to do the work. There are, of course, many essential considerations of force and arrangement which must be entered into before the machine can be completed, but they admit of being abstracted in the first instance; and it is only by so doing that we can hope to create a science of mechanism. Yet this view seems to have presented itself but lately, with due clearness, to the minds of writers on this subject; and it may be interesting to trace the history of its rise and progress.

Apart from the writings on the science of Mechanics, the history of which is well known, a number of books have been produced from time to time, having for their subject *Machinery*. At first, however, the leading principle of classification in these is derived from the purpose for which each machine is designed, and accordingly these books are either confined to machines destined for one particular kind of work, as in the early treatises of Valturius (1472) and Agricola (1550) on warlike and mining machinery respectively; or else they are collections of machines classed and described with reference to the objects for which they are constructed; divided, for example, into machines for raising water, for grinding flour, sawing timber, and so on. The earliest of these collections are the treatises of Besson (1569), Ramelli (1580), Strada (1618), Zonca (1621), Branca (1629), Bockler (1662); and the list might be continued without interruption to the present day*. The voluminous "Theatrum Machinarum" (1724) of Leupold, although it falls under the same description, yet in its first volume contains the first attempt to consider the parts of machinery separated from their work, and referred to the modifications of motion. And although these parts are made to follow the usual mechanical powers, and are mixed up with considerations of force, yet we find chapters on the *crank*, on *cams*, on machines for *converting a circular motion into a rectilinear*, or a *back and forwards motion*, and for converting a *back and forwards motion into a continued circular motion*; and so on. This must, in fact, be considered as the first attempt to produce a systematic treatise on Mechanism. The next step appears to have been made in 1794, by Monge, who, in planning the organization of the Ecole Polytechnique, proposed to devote two months of the first year of study to the *elements of machines*. "By these *elements* are to be understood the means by which the directions of motion are changed; those by which progressive motion in a right line, rotative motion, and reciprocating motion, are made each to produce the others. The most complicated machines being merely the result of a combination of some of these elements, it is necessary that a complete enumeration of them should be drawn up†." This enumeration formed the subject of part of his lectures, and was the basis of the two similar systems of Hachette, and of Lanz and Betancourt. The latter was finally

* This list might be preceded by Vitruvius, Book x., the works of Hero and other Greek mechanists, &c. Vide Veterum Mathematicorum Opera. Par. 1693.

† Vide Essai sur la Composition des Machines, par MM. Lanz and Betancourt, Par. 1808. p. 1.

adopted for the Ecole Polytechnique, and printed in 1808, under the title of "An Essay on the Composition of Machines." It was subsequently translated into English. Postponing for the moment the discussion of the system, we may observe, that Monge, in the above programme, distinctly proposes to study machines by treating them merely as contrivances for changing one kind of motion into another, apart from any considerations of force. We shall see presently, however, that this plan did not extend beyond the mere enumeration and description of the elements, without containing a provision for the calculation of the laws of the motion, or changes of motion produced. Ampère, however, appears to have contemplated the formation of a system that would also include these latter objects; for in his Essay on the Philosophy of the Sciences, published in 1834, we find it distinctly asserted, "that there exist certain considerations which if sufficiently developed would constitute a complete science, but which have been hitherto neglected, or have formed only the subject of memoirs or special essays. This science, (which he terms *Kinematics*,) ought to include all that can be said with respect to *motion* in its different kinds, independently of the forces by which it is produced. It should treat in the first place of spaces passed over, and of times employed in different motions, and of the determination of velocities according to the different relations which may exist between those spaces and times.

"It ought then to develope the different instruments by the help of which one motion may be converted into another, so that, calling these instruments by the usual name of *machines*, this science will define a machine to be, not as usual, *an instrument by means of which we may change the direction and intensity of a given force*; but, *an instrument by means of which we may change the direction and velocity of a given motion.* The definition is thus freed from the consideration of the forces which act on the machine; a consideration which merely distracts the attention of those who endeavour to unravel the mechanism.

"To understand, for example, the wheel-work by means of which the minute-hand of a watch makes twelve turns while the hour-hand makes but one, why need we trouble ourselves with the force that sets the watch in motion? The effect of the wheel-work, so far as it governs the relative velocity of the hands, is the same, by whatever cause the motion may be produced, as, for example, when the minute-hand is turned by the finger.

"After these general considerations relating to motion and velocity, this new science might pass on to the determination of the ratios that exist between the velocities of the different points of a machine, or generally of any system of material points, in all the movements of which the machine or system is susceptible; in a word, to the determination, independently of the forces applied to the material points, of what are called *virtual velocities*; a determination which is infinitely more comprehensible when thus separated from considerations of Force*."

* Vide Ampère, Essai sur la Philosophie des Sciences, 1835, p. 50.

It is much to be regretted that this distinguished writer did not attempt to follow up this clear and able view of the subject, by actually developing the science in question.

A similar separation of the principles of motion and force formed the basis of the Lectures on Mechanism which I delivered for the first time to the University of Cambridge, in 1837; and the same views were subsequently sanctioned by the high authority of Professor Whewell, who, in his Philosophy of the Inductive Sciences, has assigned a chapter to the Doctrine of Motion*, in which, under the title of Pure Mechanism, he has defined this science nearly in the above words of Ampère, whom he quotes.

To make the plan of the following pages more intelligible, it will be necessary in the first place to take a short review of the system of MM. Lanz and Betancourt, which, as we have seen, is founded upon the views of Monge. Their system is thus detailed at the opening of their work:

"The motions of the parts of machines are either (1) *Rectilinear*, (2) *circular*, (3) or *curvilinear;* and each of these may be *continuous* in direction or *alternate*, that is, *back and forward.* These six motions admit of being combined two and two in twenty-one different ways, each motion being supposed to be also combined with itself. The object of every simple machine being to counterchange or communicate these motions, the following system will include them all.

* Whewell, Philosophy of the Inductive Sciences, 1840, p. 144.

Continuous Rectilinear*, changed into	rectilinear	continuous†	1
		alternate†	2
	circular...	continuous†	3
		alternate†	4
	curvilinear	continuous†	5
		alternate†	6
Continuous Circular*, into	rectilinear	alternate†	7
	circular...	continuous†	8
		alternate†	9
	curvilinear	continuous†	10
		alternate†	11
Continuous Curvilinear*, into	rectilinear	alternate†	12
	circular...	alternate†	13
	curvilinear	continuous†	14
		alternate†	15
Alternate Rectilinear*, into	rectilinear	alternate†	16
	circular...	alternate†	17
	curvilinear	alternate†	18
Alternate Circular*, into............	circular...	alternate†	19
	curvilinear	alternate†	20
Alternate Curvilinear*, into.........	curvilinear	alternate†	21”

Of many of these combinations, however, no direct solution is given. Thus for (2) we are told to convert rectilinear motion into circular by one of the combinations in (3), and then to convert this into alternate rectilinear by one of those in (7). In this way also classes 5, 6, 11, 12, 13, 15, 16, 18, and 21, are disposed of; so that there remain only twelve, under which our authors proceed to arrange the elementary combinations into which, according to them, mechanism may be resolved.

This celebrated system, which has been pretty generally received, must however be considered as a merely popular arrangement, notwithstanding the apparently scientific sim-

* With velocity either uniform or varying according to a given law.

† With a velocity of the same nature as that which produces it, preserving a constant proportion to it or varying according to a given law. In the same or in different planes.

plicity of the scheme. In the first place, it is not confined to pure combinations of mechanism, but is embarrassed by the intrusion of several dynamical and even hydraulic contrivances. Thus, a water-wheel and a windmill-sail are considered to be a means of converting continuous rectilinear motion into continuous circular; and a ferry-boat attached to one end of a long rope, of which the other is fixed to the bank, is admitted into Class 4, as a means of converting continuous rectilinear motion into alternate circular. Fly-wheels, pendulums with their escapements, parallel motions, are all placed in one class or other of this scheme. No attempt is made to subject the motions to calculation, or to reduce these laws to general formulæ, for which indeed the system is totally unfitted.

The plan of the great work of Borgnis, published in 1818, is much more comprehensive and complete, really embracing the whole subject of machinery, instead of being confined by its plan to elementary combinations for the modification of motion. Borgnis, in the volume on the *Composition of Machines*, divides mechanical organs into six orders, each of which have subordinate classes. His orders are*; (1) Receivers of power; (2) Communicators; (3) Modifiers; (4) Frame-work, fixed and moveable: (5) Regulators; (6) Working parts.

For the mere purposes of descriptive mechanism this system is much better adapted than that of MM. Lanz

* In the original, (1) Récepteurs, (2) Communicateurs, (3) Modificateurs, (4) Supports, (5) Régulateurs, (6) Opérateurs.

and Betancourt, but still does not provide for the investigation of the laws of the modifications of motion, which is an especial object of the proposed science of Kinematics. Many essays, however, have been from time to time written concerning various detached portions of this science. The teeth of wheels is the most remarkable of these, from having occupied the attention of so many of the best mathematicians. But in fact, the description of all the mechanical curves, as epicycloids and conchoids, may be held to belong to this science, which would thus be made to include a great mass of matter that has hitherto been classed with geometry. The calculation of trains of wheel-work is also a branch of it, to which the first contribution was made by Huyghens, who employed continued fractions, for the purpose of obtaining approximate numbers for the trains of his Planetarium*.

The following pages must not however be considered as an attempt to carry out the able and comprehensive views of Ampère, being confined to machinery alone, and not passing from it to the more abstract generalities of motion, which he seems to have contemplated.

My object has been to form a system that would embrace all the elementary combinations of mechanism, and at the same time admit of a mathematical investigation of the laws by which their modifications of motion are governed. I have confined myself to the Elements of Pure

* Vide also Young's Nat. Philosophy, vol. II. p. 55. Arts. 365, 366, the substance of which will be found in this work. Arts. 34 and 237.

Mechanism, that is, to those contrivances by which motion is communicated purely by the connexion of parts, without requiring the essential intermixture of dynamical effects.

I have taken a different course from the one hitherto followed, in respect that instead of considering a machine to be an instrument by means of which we may change the direction and velocity of a *given motion*, I have treated it as an instrument by means of which we may produce any *relations of motion* between two pieces.

For Monge and his followers began by dividing motion into rectilinear and rotative, continuous and reciprocating, and so based their system upon the *actual motion* of the parts; and Ampère defines his machine in the words quoted above as modifying a *given motion*. But a little consideration will shew that any given element of machinery can only govern the *relations of velocity and direction* of the pieces it serves to connect; and that this connexion and the law of its action are for the most part independent of the *actual velocities*. By establishing a system upon the relations of motion instead of upon the actual motions, it will be found that many of the redundancies and difficulties that have hitherto obscured the subject are got rid of.

Thus, to follow up the example given by Ampère of the hands of a watch, it is clear that the connexion governs the relation of their angular velocities, which at every instant is in the proportion of twelve to one; and also provides that they shall both revolve the same way, whether that be to the right or to the left. If then the one be made to revolve through a small angle back and forwards, the other will also revolve back and forwards through an angle of one twelfth of that described by the first. Now in the usual system this identical contrivance, which in its ordinary employment belongs to the class of conversion from continuous circular into continuous circular, is thus also thrown into the class of alternate circular changed into alternate circular. In the system which I propose, this contrivance at once finds its place as a combination in which the velocity ratio and directional relation are constant.

I have also dismissed, or given a subordinate place, to the distinction between circular and rectilinear motion, and have introduced a new distinction between those motions which are capable of being from the nature of the contrivance continued indefinitely in either direction, and those of which the extent is limited by the nature of the contrivance.

The first ground of my classification is the effect of the combination upon the Velocity Ratio of the pieces, and upon the relation of their directions of motion, or Directional Relation; from which considerations I have divided all the Elementary Combinations into three classes.

The second ground of the classification, and the one by means of which the calculation of the law of communication of the velocities and directions is effected, is the *mode in which the motion is transmitted;* a part of the subject which appears wholly neglected by the writers already referred to. These modes I have divided into Rolling and Sliding Contact, Link-work, Wrapping Connexion, and Reduplication. The relative motions produced by each of these methods will be found to be governed by a different geometrical principle, and every possible mode of communication may be placed under one or other of these divisions. Many combinations, however, derive their principle of action from a mixture of two or more of these methods of communication. In this case their place in the system is always determined by that method which has the greatest influence; besides which, each combination is reduced to its equivalent simple form, and its position determined by that alone; for the object of the system is to reduce the motions to calculation; and for this purpose the equivalent simple form of every combination must be employed.

For example, the action of combinations in which rows of teeth are used depends partly upon rolling contact and partly upon sliding contact; for the action of the individual teeth is of the latter kind, but the total action of them is equivalent to the rolling contact of their pitch-lines, and the pitch-lines only need be considered in calculating the motion. Accordingly, all combinations in which rows of teeth are employed will be found under the head of Rolling Contact. Again, when cam-plates or curves are used a friction roller is often employed for these plates to act against. At first sight this would appear to convert the action of the combination into rolling contact. But besides that this contrivance merely transfers the sliding action to the axis of the roller, and that our definition of rolling contact supposes the two axes of motion of the rolling curves to be fixed in position, the calculation of the motion of all such combinations is effected by supposing the roller reduced to a point, and the curve thus obtained upon the principles of pure sliding contact, is afterwards adapted to the roller by tracing a second curve within it at a normal distance equal to the radius of the roller. All combinations of this kind are therefore placed under Sliding Contact, notwithstanding the employment of friction-rollers. Either of these considerations, the Velocity Ratio and Directional Relation, or the modes of communication, might have been made the primary ground of the classification; and some advantages might result from adopting the second for this purpose. I was induced to select the first, because it enabled me to separate from the others all that most important class of combinations in which the Velocity Ratio and Directional Relation remain constant, and which are also the foundation of most of those contained in the subsequent classes. The Synoptical Table, which immediately follows this Preface, will shew the general arrangement of the Elementary Combinations under the proposed system.

In the Second Part of the work I have assembled a number of contrivances which appear to me to be connected by a general principle which has not hitherto been defined; these I have ventured to term Aggregate Motions. One portion only of these contrivances has usually been treated as a separate class under the name of Differential Motions.

The Third Part contains several problems relating to the calculation and arrangement of mechanism in which it is necessary to have the power of altering the relations of motion at pleasure. This part, for the reasons which will be found in its first chapter, is not to be considered as a complete essay on this branch of the subject.

I have, in the course of the work, endeavoured in every case to acknowledge the sources from whence I have derived any portion of its contents by references at the foot of the page. But so little of the subject has been hitherto treated mathematically, that I must hold myself answerable for the greatest portion of it. The teeth of wheels, which occupies the first part of Chapter III. is the only branch of mechanism in which the original papers have been already wrought into a system, and published in a collected form. This was first done by Camus, and has been subsequently effected by Buchanan in his Essays, and by Hachette.

These works being so well known I have not so constantly referred to my authorities in this chapter, but it will be found that I have incorporated into it extracts from the valuable paper of Professor Airy, as well as the entire contents of my own paper from the Transactions of the Society of Civil Engineers, and have added several original investigations relating to the proportions of the teeth, and their least numbers. Some of these questions have been discussed by Kæstner*, but not in a manner

* Commentat. Gott. 1781, 1782.

adapted to practice. Tredgold has also given some results*, but has unfortunately vitiated them by the coarseness of his approximations. It will be found that I have calculated all the results that are required in practice, and have arranged them in tables for reference.

On the whole, it will be seen that the present volume is limited to a very small portion of the important subject of machinery. The object of it is, as has been already stated, to systematize the subject, and to free it from the considerations of force, with which it has been usually mixed up.

To complete the plan, therefore, it will be necessary, in the next place, to apply these considerations of force to the combinations thus obtained, as well as to describe and investigate those parts of machinery in the action of which forces are essential; a task which I shall probably undertake at some future time.

In carrying out this branch of the subject, great assistance will be derived from the works of modern French writers—Navier, Poncelet, Morin, &c., who have with so much success and originality applied themselves to this purpose. Their works are now beginning to attract the attention of our own writers; and in the present year Professor Whewell has introduced many of their results into his Mechanics of Engineering, generalizing them with his usual ability; and in the same work he has flattered me by the

* Vide Buchanan's Essays.

adoption of my own views upon the classification of the modes in which motion is communicated from one piece to another of a machine, adding to them the investigation of the effects of force and resistance; which may be considered as carrying out a portion of the plan above alluded to, as necessary to complete this arrangement of the science of Machinery.

I am not without hopes, that in addition to its principal object of giving a scientific and systematic form to its subject, the results of the volume which I now venture to present to the world, may be found a useful addition to mathematical studies in general, by affording simple illustrations of the application and interpretation of formulæ, and by suggesting new subjects for problems, and for farther investigation.

CAMBRIDGE,
Sept. 28, 1841.

SYNOPTICAL TABLE

OF THE

ELEMENTARY COMBINATIONS OF PURE MECHANISM.

	DIRECTIONAL RELATION CONSTANT.		DIRECTIONAL RELATION CHANGING PERIODICALLY.
	Velocity-Ratio Constant.	*Velocity-Ratio Varying.*	*Velocity-Ratio Constant or Varying.*
	CLASS A.	CLASS B.	CLASS C.
DIVISION A. By Rolling Contact.	Rolling cylinders, cones, and hyperboloids. General arrangement and form of toothed wheels. Pitch.	Rolling curves and rolling curve-wheels. Roëmer's & Huyghens' wheels, &c. Wheels with intermitted teeth. Rolling-curve levers.	Mangle-wheels. Mangle-racks. Escaping geerings.
DIVISION B. By Sliding Contact.	Forms of the individual teeth of wheels. Cams. Screws. Endless screws or worms and their wheels.	Pin and slit lever. Cams. Unequal worm. Geneva stop and other intermittent motions.	Pin and slit lever. Cams in general. Swash plate. Double screw. Spiral and solid cams. Escapements.
DIVISION C. By Wrapping Connectors.	Arrangement and material of bands. Form of their pullies. Guide pullies. Geering chains. Arrangements for limited motions.	Curvilinear pully. Fusees.	Curvilinear pully and lever.
DIVISION D. By Link-work.	Cranks and link-work for equal rotations. Cranks for limited motions. Bell crank-work.	Link-work. Hooke's joints.	Cranks, excentrics, and other link-work. Ratchet wheels and clicks. Intermittent link-work.
DIVISION E. By Reduplication.	Tackle of all kinds, with parallel cords and in trains.	Tackle with unparallel cords.	

GENERAL TABLE OF CONTENTS.

CONTENTS OF THE SEPARATE ARTICLES.

INTRODUCTION.

PART THE FIRST.

On Trains of Mechanism in General.

Class A. { Directional Relation Constant. Velocity Ratio Constant. } Division A. By Rolling Contact.

By pure Rolling Contact.

To apply these Solutions to practice.

Class A. Division B. By Sliding Contact.

Axes Parallel.

On Combinations for producing Aggregate Paths.

On parallel Motions.

PART THE THIRD.

On Adjustments.

To alter the Velocity Ratio by determinate changes.

To alter the Velocity Ratio by gradual changes.

LIST OF TECHNICAL AND NEW TERMS,

WHICH ARE DEFINED OR EXPLAINED IN THE FOLLOWING ARTICLES.

As this example is rather curious, I have thought it worth while to give the complete solution of it. Thus: Since V and A_1 are constant, they are in the proportion of the spaces described by the reciprocating piece and the point whose radius is unity upon the first axis; and as one revolution of the latter corresponds to a complete double oscillation of the former, we have $\frac{V}{A_1} = \frac{2\rho}{\pi} = k$, whence $r_2 = \frac{c\rho\,.\sin\theta}{\rho\sin\theta + k} = c\frac{\pi\sin\theta}{\pi\,.\sin\theta + 2}$, whence the follower curve may be laid down. Again, by Art. 260, if θ_1 be the corresponding value of θ in the driving curve, we have

$$\theta_1 = \int \frac{r_2 d\theta}{c - r_2} = \frac{\pi}{2}\int \sin\theta d\theta = C - \frac{\pi}{2}\cos\theta,$$

and when $\theta = 0$, and $\frac{\pi}{2}$, $\theta_1 = 0$, and $\frac{\pi}{2}$, respectively, whence $C = \frac{\pi}{2}$,

$$\text{also } \theta_1 = \frac{\pi}{2}\,.\,\text{versin}\,\theta, \text{ and } r_1 = c - r_2,$$

will give the driving curve. In the following Table a sufficient number of values are computed to enable these two curves to be laid down by points.

The radius of the follower, however, vanishes at two points of the circumference, the form of its curve resembling that of the figure ∞. These points correspond to the passage of the crank over the dead points, where, as it communicates for the moment no velocity to the reciprocating piece, the velocity of the crank must become infinite to maintain the conditions of the problem, which requires a constant velocity in the reciprocating piece, and therefore no loss of time in the change of direction. All which being practically impossible, it is necessary to alter the figure of the curve at these points, and reduce it to the form ∞, shortening the points of the driver accordingly; teeth may then be added to these curves in the usual manner.

FOLLOWER.		DRIVER.	
θ	$\frac{r_2}{c}$	θ_1	$\frac{r_1}{c}$
0°	0	0°	1.0000
5°	.1204	0° 20′	.8796
10°	.2143	1° 22′	.7857
15°	.2890	3° 4′	.7110
20°	.3495	5° 25′	.6505
30°	.4399	12° 4′	.5601
40°	.5025	21° 3′	.4975
50°	.5461	32° 9′	.4539
60°	.5763	45°	.4237
70°	.5963	59° 13′	.4037
80°	.6075	74° 23′	.3925
90°	.6109	90°	.3891

NOTE *to page* 361.

THE following mode of communicating an aggregate velocity to a worm-wheel, ought to have been inserted at page 361, as a mixture of sliding and rolling contact.

In fig. 198, let the axis of motion of the worm-wheel B be supposed fixed in position. Then, if the endless screw or long worm Aa revolve, it will communicate a rotation to the wheel B in the usual manner, at the rate of one tooth of the latter for each turn of the former. Again, if an endlong travelling motion without rotation be communicated to Aa, it will now act as a rack upon the teeth of B. If, therefore, the two motions of rotation and travelling be communicated to the endless screw, which can be done in various ways from two sources, the wheel B will receive the aggregate motion, and its angular velocity be affected accordingly. For example, let the screw revolve uniformly, and at the same time travel back and forwards through a small space endlong, the wheel will then revolve with a hobbling motion, making a short trip in one direction and a long trip in the other direction continually.

ERRATA.

22, line 6, *for* ...we should find in like manner the velocity of Q triple that of P. And...
read ...M being now the moving point and P the fixed point, we should find in like manner the velocity of Q triple that of M. And M being again the fixed point... (Vide Chap. VI.)

24, line 4, *for* line of centers, *read* link.

127, line 3 from bottom, *for* (fig. 6), *read* (fig. 61).

215, last line, *for* Chap. IX, *read* Chap. X.

225, lines 11 and 23; 236, last line, *for* $\frac{S_1}{S_m}$, *read* $\frac{L_1}{L_m}$,

278, line 2, *for* Chapter, *read* Class.

284, line 11, *for* DIVISION D, *read* DIVISION E.

393, line 5 from bottom, *for* velocity, *read* velocity-ratio.

397, line 3 from bottom, *for* not, *read* not necessarily.

PRINCIPLES OF MECHANISM.

INTRODUCTION.

1. EVERY machine is constructed for the purpose of performing certain mechanical operations, each of which supposes the existence of two other things beside the machine in question, namely, a moving power, and an object subjected to the operation, which may be termed the work to be done.

Machines, in fact, are interposed between the power and the work, for the purpose of adapting the one to the other.

2. As an example of a machine whose construction is familiar to all, the grinding machine so commonly seen in our streets may be cited, in which the grindstone is made to revolve by the application of the foot to a treadle.

Here the *moving power* is derived from muscular action. The *operation* is carried on by pressing the edge of the cutting instrument, which is the subject of it, against the surface of the grindstone, which is caused to travel rapidly under it.

The arrangement and form of this surface, and its connexion with the foot in such a manner that the pressure of the latter shall communicate the required motion to the former, is the office and object of the machine.

Two portions of the machine are given, the one by the nature of the power, and the other by that of the work. The first is a treadle placed at a proper level to receive the pressure of the foot, by the action of which it may be made to perform, without unnatural exertion, about eighty or ninety vertical oscillations in a minute. The second part of the machine is the cylindrical grindstone, which is mounted on a horizontal axis at the upper part of the frame, and at a convenient height to allow the tool to be pressed upon its revolving surface. The surface should pass under the edge of the tool at the rate of about 500 feet in a minute, and therefore supposing the diameter of the grindstone to be eight inches, it must revolve at the rate of 250 turns in a minute. The remainder of the mechanism serves to connect the treadle and grindstone, and may consist of any contrivance that will compel the latter to revolve when the former is made to oscillate, and in the proportion of 250 revolutions to 80 oscillations, or about three to one.

3. It appears, then, that this machine consists of a series of connected pieces, beginning with the treadle whose construction, position, and motion, are determined by the nature of the moving power, and ending with the grindstone, which in like manner is peculiar and adapted to the work. But this is, in fact, the description of every machine. There is always one or more series of connected pieces, at one end of each of which is a part especially adapted to receive the action of the power, such as a water-wheel, a windmill-sail or a horse-lever, a handle or a treadle. At the other end of each series will be a set of parts determined in form, position, and motion, by the nature of the work they have to do, and which may be called the working pieces. Between them are placed trains of mechanism connecting them

so that when the first parts move according to the law assigned them by the action of the power, the second must necessarily move according to the law required by the nature of the work.

4. These three classes of mechanical organs are so far independent of each other, that any given set of working parts may be supplied with power from any source: thus a grindstone may be turned either by the foot or by the hand of an assistant, by water or by a horse. Again, a given water-wheel or other receiver of power may be employed to give motion to any required set of working parts for whatever purpose. Also between a given receiver of power and set of working parts the interposed mechanism may be varied in many ways. Moreover the principles upon which the construction and arrangement of these three classes are founded are different. The *receivers of power* derive their form from a combination of mechanical principles with the physical laws which govern the respective sources of power. The *working parts* from a combination of mechanical principles, with considerations derived from the processes or objects in view. But the principles of the *interposed mechanism* admit of being developed without reference to the powers employed or transmitted, or to the resistances or work to be done, or, in fact, to the objects for which machinery is constructed. By defining mechanism in the abstract to be a combination of parts for the purpose of connecting two or more pieces, so that when one moves according to a given law, the others must move according to certain other given laws, this branch of the subject may be reduced to geometrical principles alone: whereas by considering mechanism as usual, as a modifier of force, the subject becomes embarrassed by a condition foreign to the connexion of parts by which the modification is pro-

duced; and which condition and its consequences admit more conveniently of subsequent consideration and separate investigation.

5. The hour-hand of a clock, for example, is connected with the minute-hand by a mechanism which compels the former to perform one revolution while the latter completes twelve; or generally, the angular velocity of the first is always one twelfth of that of the second. The connexion is independent of the force which puts the minute-hand in motion, and also of the actual velocity of the minute-hand. If this be turned by hand quickly or slowly, uniformly or variably, back or forwards, the hour-hand will still follow these motions at an angular rate of one twelfth of the original. The constant relation of the angular velocities depends in this as in other similar cases only upon the proportion between the diameters or number of teeth of the wheel-work that connects the two hands—a purely geometrical relation, the comprehension of which is rather obscured than assisted by the introduction of statical principles, of which the connexion is independent, but which find their proper place, when it becomes necessary to investigate the proportion between the forces and resistances in any given case, and the strains thrown upon the different parts of the mechanism by their application, and thus to find the requisite strength of each part.

6. The term *mechanism*, then, must be understood to be in this work confined to those mechanical combinations which govern the relations of motion only, and which therefore admit of being entirely separated from the consideration of force. This, of course, excludes not only those mechanical organs which have been already alluded to, as receivers of power and working parts, but also those which

are employed to govern the motions of machinery; such as the escapements of clocks, and contrivances by which machinery is made self-acting and self-regulating; all of which are derived from combinations of pure mechanism with statical or dynamical principles, but from which they do not admit of separation. The exposition of such contrivances will naturally and easily follow from the principles of the present work, but are excluded from it by its plan, which is, to reduce the various combinations of *Pure Mechanism* to system, and to investigate them upon geometrical principles alone.

7. Neither is it my purpose to enter into minute details of the actual construction of machinery, of the different forms which each combination may assume, or of the infinitely varied methods of framing and putting them together; for, in the first place, the choice of these forms in every particular case is mainly determined by the strains to which the machinery is to be exposed; and, in the next place, this branch of the subject is sufficiently important and extensive to admit of separation from the others, under the name of *Constructive Mechanism.* Although some details of this kind are unavoidable in the present work, I have carefully avoided them when possible, and for this purpose have excluded from the drawings all unnecessary and extraneous framing or connexions that tend to individualize the combinations, and thus to oppose the very object which I have proposed to myself, namely, to introduce such a degree of generalization and system, as would give to *Pure Mechanism* a claim for admission into the ranks of the Sciences.

8. I must here recapitulate the ordinary definitions and measures of motion and velocity, for the purpose of

introducing certain modifications which they require to adapt them to our present purposes.

A body is absolutely at *rest* when it remains in the same position in space, and at rest relatively to another body when it continues in the same relative position to that body, as it is usually said to be at rest when it remains in the same relative position to the earth. Thus, too, a body which remains in the same place in a boat or a carriage, is at rest with respect to that boat or carriage, although these may be in motion; and so a wheel or other portion of a machine may be carried into different positions relatively to the fixed frame, and yet remain at rest with respect to the arm or carriage upon which it is mounted.

A body is in *motion* when it occupies successively different positions in space; motion being relative as well as rest. Two bodies moving with respect to a third will be at rest with respect to each other, if they retain in their motions the same relative positions; or a body absolutely at rest may be said to move with respect to another moving body, if the latter be assumed as the standard to which the motion is to be referred.

9. Motion is essentially continuous; that is to say, a point cannot pass from one position to another without going through a series of intermediate positions. Thus the motion of a point describes in space a line necessarily continuous, which line is termed its *path*. The path of a solid body must be understood as the line described by some principal point in that body, such as the center of a sphere.

The path of a body being assigned, there are only two *directions* in which it can move. Direction of motion being relative, may be indicated by naming some fixed point which the body is approaching or retiring from: as, for example, the points of the compass, the zenith or nadir, or by personal or other relations, such as right and left, larboard and starboard, windward and leeward, upwards and downwards, &c.; otherwise its direction of motion may be defined by comparing it with that of the sun or of the hands of a watch; the latter is an exceedingly convenient standard for rotative motion. By supposing the path of the sun projected upon the plane of motion, it may be employed as a standard for rotative direction in every case but that of motion in a plane perpendicular to its orbit.

The path and direction being assigned, the body may move in its given path and direction quickly or slowly, with a greater or less *velocity;* and this velocity is estimated by comparing the space passed over with the time occupied in describing it.

10. When a body describes equal portions of its path in equal successive times, the motion is said to be uniform, and the velocity measured by the space (that is, the length of path) described in the unit of time. The units usually employed are feet and seconds. Thus a body is said to move at the rate of 3 feet per second.

Since the same space is described in every unit of time, the entire space described is proportional to the time employed in describing it, and the measure of velocity is obtained by dividing the number of feet passed over by that of the seconds employed.

If V be the velocity, S the space in feet, T the time in seconds, $V = \frac{S}{T}$. The direction is indicated analytically by the sign of the velocity for a given path; if the velocity in one direction be assumed positive, that in the opposite direction will be negative.

11. The motion of a revolving body may be measured by the linear velocity of a point whose radial distance is equal to the unit of space. This is termed the angular velocity of the body, which is said to revolve uniformly when its angular velocity is uniform.

In uniform angular velocity the angles described by a given radius, are manifestly proportional to the times; and since the linear velocity of every point is the arc described in the unit of time, which arc is proportional to the radius, so the linear velocity of every point is proportional to its radius. If A be the angular velocity, R the radius of the point in feet, the linear velocity $V = RA$.

The motion of a uniformly revolving body may also be conveniently measured by the number of rotations performed in a given time. In uniform rotation the angles described are proportional to the times, and any given point describes its own circle with uniform linear velocity. Let T be the time of performing k revolutions, where k may be a whole number or a fraction. Then, since 2π is the circumference whose radius is unity; $2\pi k$ will be the space described in k revolutions by the point whose radius is unity, but A is the space described by the same point in the unit of time;

$$\therefore\ A : 2\pi k :: 1 : T; \quad \therefore\ T = \frac{2\pi k}{A}\ (1); \quad k = \frac{TA}{2\pi}\ (2);$$

Hence the number of turns in a given time varies as the angular velocity.

Let R be the radius of a wheel and V its perimetral velocity;

$$\therefore\ V = RA. \quad \text{And } k = \frac{TV}{2\pi R}\ (3);$$

whence the number of turns in a given time varies directly

as the perimetral velocity, and inversely as the radius or diameter of the wheel*.

Let the time in which a wheel performs one complete revolution be termed its Period ($= P$); $\therefore P = \frac{2\pi}{A}$ {putting $k = 1$ in (1)}; and the period varies inversely as the angular velocity.

Also from (2) $k = \frac{T}{P}$; whence the period varies inversely as the number of turns in a given time. When the rotations of two wheels are to be compared, the number of turns they respectively make in a given time may be termed their synchronal rotations.

12. When the velocity is not uniform, these expressions can no longer be applied, because the velocity is different at different times. In this case, then, the velocity at every instant is measured by the space that would be described in the succeeding unit of time, were the velocity with which that unit is commenced continued uniformly throughout it.

If the velocity of a body increase, it is said to be accelerated, and if the velocity diminish, to be retarded.

13. Varied motion admits of convenient graphical representation, by which its characteristic points and general laws are rendered much more easy of comprehension than they are by the use of formulæ alone.

* In practice linear velocity is commonly referred to seconds, and angular velocity to minutes; thus a millwright will define the velocity of a given wheel by either saying that it performs twenty revolutions in a minute, or that its circumference moves at the rate of three feet per second. In the expression (3) if k and T be expressed in minutes, and V is to be expressed in seconds, we must put $60V$ for V;

$$\therefore k = \frac{60\,TV}{2\pi R} = \frac{10\,TV}{R} \text{ very nearly.}$$

Thus to represent the motion of a body of which the velocities at certain given intervals of time are known, take an indefinite straight line *AX*, and from *A* set off abscissæ *Ab*, *Ac*, *Ad*...... proportional to the given intervals of time as measured from the beginning of the motion.

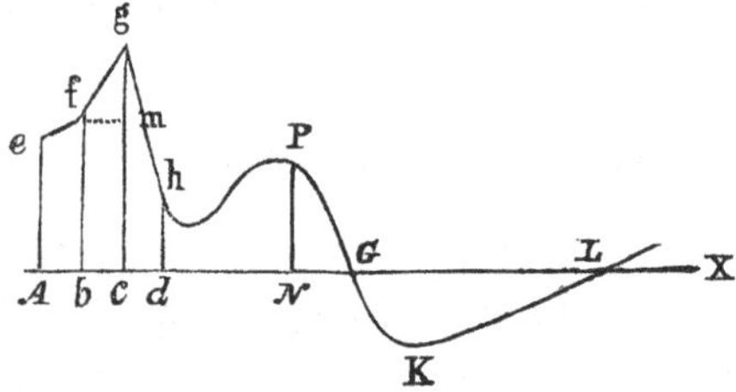

Upon *A*, *b*, *c*, *d*...... erect ordinates *Ae*, *bf*, *cg*, *dh*, respectively proportional to the velocity of the body at the beginning of the motion and after each interval of time. By joining the extremities of these ordinates, a polygon *efgh*...... is obtained, which if the intervals of time be taken with their differences sufficiently small, will become a curve as *hPGKL*, of which the abscissa *AN* at any given point *P*, will represent the time elapsed from the beginning, and the ordinate *NP* the corresponding velocity of the body.

If the motion of the body cease, its velocity becomes zero, and the curve meets the axis, as at *G* and *L*. If the body change its *direction* in its path, this is indicated by the change of sign in the velocity; for either direction being assumed positive, the other will be negative; and so in this curvilinear representation, the ordinates representing the velocity for one direction being set off upwards from the line, as from *e* to *G*, those of the opposite direction will be set off downwards as from *G* to *L*.

14. By another method a curve is constructed of which the abscissæ shall represent the time as before, but the ordinates the space described by the body. Thus, if the

last figure be supposed to be constructed on this second hypothesis, Ae will represent the distance of the body at the beginning of the motion from that point of its path whence the space is to be measured; bf its distance from the same point at the end of the time Ab; cg its distance after the time Ac; and so on. But the motion in one direction being accounted positive, that in the opposite direction will be negative. If then the body change its direction in the interval cd, the ordinates will decrease.

And, as in the former case, if the ordinates are taken in sufficient number, a continuous curve is obtained, as $pPGKL$, which will tend upwards when the body moves in one direction, and downwards when in the other direction.

Now since the space described in any interval of time is represented by the difference of the two ordinates corresponding to the beginning and end of that interval, so the velocity is proportional to that difference divided by the difference of the abscissæ. Thus in the interval $bc\,(=fm)$, gm is the space described, and $\frac{gm}{fm}$ the velocity, which is proportional to the tangent of gfm, or ultimately to the tangent of the angle which the curve makes with the axis Ax.

15. This method is better adapted for representing the motion of the parts of mechanism than the other, because the tendency of the sinuous line corresponds with the direction of the body, changing from upwards to downwards, and *vice versa*, as the direction changes; while its more or less rapid inclination indicates the change of velocity. Thus the line is a complete picture of the motion, as the line formed by the notes in music is a picture of the undulations of the melody; whereas by the first method where the ordinates represent the velocities, the

directions are indicated by the situation of the curve above or below the axis, which is a distinction of a different kind from the thing it represents, and requires an effort of thought for its comprehension.

Sometimes the axis Ax of the time is drawn vertically, and the ordinates consequently are horizontal.

16. The two methods are compared in the following figure, which represents the motion of the lower extremity of a pendulum, the continuous line upon the first hypothesis, and the dotted line upon the second.

The axis of the abscissæ Ak is vertical, AM is the interval of time corresponding to one oscillation from left to right, and MN to the returning oscillation from right to left.

In the continuous line the horizontal ordinates represent the velocities, which beginning from zero at the left extremity of the vibration at A, reach their maximum values in the middle of each oscillation at H and K, and vanish at the extremities of the oscillations at M and N. The right side of the axis is appropriated to the direction of motion from left to right, and the left side to the opposite direction.

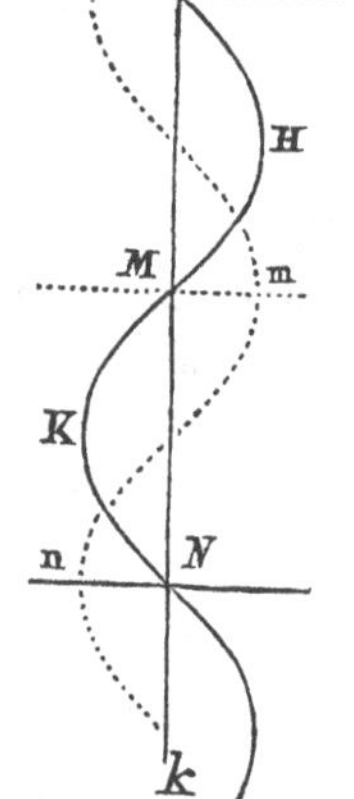

In the dotted line the ordinates represent the distances from the middle or lowest point, which are greatest at the beginnings and ends of the oscillations at a, m, n. But the curve in this case moves from right to left, and *vice versa*, as the pendulum moves.

17. In the varied motion of mechanical organs it generally happens that the changes of velocity recur per-

petually in the same order, in which case the movement is said to be *Periodic*. The *period* is the interval of time which includes in itself one complete succession of changes, and the motion is made up of a continual series of similar periods. But the changes of velocity in the different periods may be similar in the law of their succession only, and may differ either in the actual values, or in the interval of time required for each period. In most cases, however, the periods are precisely alike in the law and value of the successive velocities, as well as in the interval of time assigned to each. Such motion is termed a *Uniform Periodic Motion*; of which examples are the motion of pendulums, or of the saws in a saw-mill, supposing the prime mover to revolve uniformly.

The complete set of changes in velocity included in one period may be termed the *Cycle* of Velocities. This phrase is indeed generally applicable to any thing that is subject to recurring variations, whereas *Period* is applicable to time alone. The successive phenomena of motion in each period are sometimes termed its *Phases*, so that the periodic motion is thus a recurring series of phases. The choice of the phase in this series, which shall be reckoned as the beginning and end of the period, is arbitrary. Thus we may reckon the beginning of the periods of a pendulum, either from one of the extremities of its oscillation, or from the middle and lowest point.

PART THE FIRST.

CHAPTER I.

ON TRAINS OF MECHANISM IN GENERAL.

18. MECHANISM may be defined to be a combination of parts, connecting two or more pieces, so that the motion of one compels the motion of the others, according to a law of connexion depending on the nature of the combination. The motion of elementary combinations are *single* or *aggregate*.

Aggregate motions are produced by combining in a peculiar manner two or more *single* combinations, as will hereafter appear in Part II. All that follows in this Part relates to the *single* combinations alone.

19. The motion of every piece in a machine being defined, as in the Introduction, by path, direction and velocity, it will be found, that its path is assigned to it by its connexion with the frame-work of the machine; but its direction and velocity are determined by its connexion with some other moving piece in the train. Thus a wheel describes circles, because its axis is supported by holes in the frame; but it describes them swiftly or slowly, backwards or forwards, by virtue of its connexion with the next wheel in the train, which lies between it and the moving power.

This connexion affects the *ratio* of the velocities, and the *relative* direction of motion of the two pieces in question, but its action is independent of the *actual* velocities or directions of either piece*, as in the familiar example already quoted of the two hands of a clock, where the connexion by wheel-work is so contrived, that while one hand revolves uniformly in an hour, the other shall revolve uniformly in twelve. But this connexion has this more general property, that it will also compel the latter to revolve with an angular velocity of one twelfth of the former, whatever be the actual velocity communicated to either; as, for example, when we set the clock by moving the minute-hand rapidly to a new place on the dial, and similarly with respect to direction, the two hands will always revolve the same way, whether we turn one of them backwards or forwards.

Since Mechanism is a connexion between two or more bodies, governing their proportional velocities and relative directions, and not affected by their actual velocities or directions; it follows that a systematic arrangement of the principles of mechanism must be based upon the proportions and relations between the velocities and directions of the pieces, and not upon their actual and separate motions.

20. Proportional velocities may be divided into those in which the ratio is constant, and those in which it varies.

Let V and v be the velocities of two bodies, then $\frac{V}{v}$ is the velocity ratio; and if the velocities are uniform, let S, s be the spaces described in the same time T; $\therefore \frac{V}{v} = \frac{S}{s}$ a constant ratio; consequently between uniform velocities the velocity ratio is constant, which indeed is sufficiently obvious.

If however the velocities be not uniform, and yet the velocity ratio constant, let the bodies in any successive

* We shall find a few contrivances in which this is not strictly true with respect to the direction, but they are not of a nature to vitiate the generality of the principle.

intervals of time T, $T_{\prime}$, $T_{\prime\prime}$ ... move with velocities V, $V_{\prime}$, $V_{\prime\prime}$ ... and v, $v_{\prime}$, $v_{\prime\prime}$ respectively, of any different magnitudes, but so that the two velocities at the same instant always preserve the same ratio;

$$\therefore \frac{V}{v} = \frac{V_{\prime}}{v_{\prime}} = \frac{V_{\prime\prime}}{v_{\prime\prime}}, \text{ \&c.} \ldots = c.$$

Hence if S, $S_{\prime}$, $S_{\prime\prime}$ ... and s, $s_{\prime}$, $s_{\prime\prime}$ be the spaces described with these velocities by the two bodies in the intervals T, $T_{\prime}$, $T_{\prime\prime}$ respectively, we have

$$c = \frac{S}{s} = \frac{S_{\prime}}{s_{\prime}} = \frac{S_{\prime\prime}}{s_{\prime\prime}} \ldots\ldots = \frac{S + S_{\prime} + S_{\prime\prime} + \ldots}{s + s_{\prime} + s_{\prime\prime} + \ldots}.$$

And as this is true whatever be the magnitude of the intervals of time, it is also true when they are taken so small that the changes of velocity become continuous, and therefore *when the velocity ratio is constant it is obtained by comparing the entire spaces described in the same interval of time, whatever changes the actual velocities of the bodies may have undergone during that time.*

And in the same manner it may be shewn that in revolving bodies the angular velocity ratio, if constant, is equal to the ratio of the synchronal rotations, notwithstanding the velocities of rotation may vary, and also to the inverse ratio of the periods if the angular velocities be uniform.

When the velocity ratio varies, the relations of motion between two pieces may often be more simply defined by means of the law of their corresponding positions than by the ratio of their velocities.

21. With respect to actual direction we have seen that it has only two values, but the relation of direction between two bodies moving in given paths may be conveniently divided into two classes. In the first, while one continues to move in the same direction, the other shall also

persevere in its own direction; but if one change the other shall change. To this class belongs the clock-hands; and in this instance both hands move the same way round the circle. But this is not necessary; it may be that when one piece revolves to the right the other may revolve to the left, and *vice versa*, as in a pair of flatting rollers; or again in the old simple mangle, so long as the handle is turned in one direction, the bed of the mangle will travel forwards, but when the motion of the handle is reversed, the bed of the mangle also returns. In all these cases *the directional relation is constant*. In another class the connexion is of this nature, that while one body perseveres in the same direction, the other shall change its direction; as, for example, in a saw-mill. The saw-frame moves up and down, changing its direction periodically, but the piece from which it derives this motion revolves continually in the same direction. In cases of this kind *the directional relation changes*.

22. We have thus two kinds of directional relation, and two of the velocity ratio, by means of which it will appear, that all the simple combinations of mechanism, for the modification of motion, may be distributed into three classes:—

Class A. *Directional relation and Velocity ratio constant.*

Class B. *Directional relation constant—Velocity ratio varying.*

Class C. *Directional relation changing periodically—Velocity ratio either constant or varying.*

This latter class might have been divided into two, by arranging the constant and variable velocities under separate heads; but it will be found that the contrivances for effecting these two conditions are so much alike, that this division would only introduce needless complication.

23. In those classes of combinations in which either the velocity ratio or the directional relations change, it will generally happen, from the very nature of mechanism, that the changes will recur in cycles. But, since these changes are independent of the actual velocities of the bodies, the cycles cannot be periodic in time, but will recur with reference to the path of one of the moving bodies, the same velocity ratio and directional relation generally corresponding to the arrival of this body at the same point of its path, and so on in succession for the different phases. The true *argument**, as it is called, of the change being in fact the *path* of one of the bodies, and not the *time* of its motion.

24. A *train* of mechanism is composed of a series of moveable pieces, each of which is so connected with the frame-work of the machine, that when in motion every point of it is constrained to move in a certain path, in which, however, if considered separately from the other pieces, it is at liberty to move in the two opposite directions, and with any velocity. Thus wheels, pullies, shafts, and revolving pieces generally, are so connected with the frame of the machine, that any given point is compelled when in motion to describe a circle round the axis. Sliding pieces are compelled by fixed guides to describe straight lines, and so on.

25. These pieces are connected in successive order, either by contact or by intermediate pieces, so that when

* Vide Whewell's *Philosophy of the Inductive Sciences*, Vol. II. p. 542.

the first piece in the series is moved from any external cause, it compels the second to move, which again gives motion to the third, and so on.

26. The act of giving motion to a piece is termed *driving* it, and that of receiving motion from a piece is termed *following* it. The piece or part of a piece which is appropriated to transmitting motion to the next is the *driver*, and the part which receives motion is the *follower*.

27. The law of motion of one piece in a train may differ in any way from the law of motion of the next piece in the series, and the change is effected by the mode of connexion. The systematic examination of the different cases under which these changes may be arranged, constitutes the principles of mechanism.

One piece may drive another either by immediate contact or by an intermediate or connecting piece. The different modes of doing it will be best explained by taking an example of each in its most elementary and general shape.

28. *Communication of Motion by Contact.* Let *AC*, *BD* be two successive pieces of a train of mechanism, moving on centers *A* and *B* respectively, and let *BD* be the driver, and *AC* the follower, the curved edge of the first touching that of the second. If the driver be moved into a new position near the first, as shewn by dotted lines, its edge will press that of the follower, and move it also into a new position. Let *m* be the point of contact in the first position, and let *n* and *p* be the respective points of the edges that come into contact in the second position as at *r*. Now, during the motion every

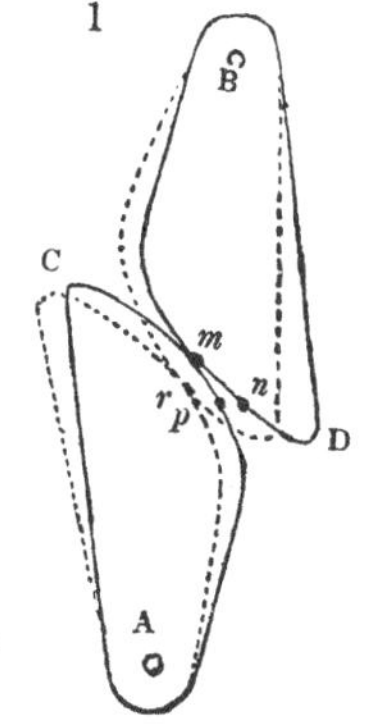

point between p and m in one curve has been successively in contact with some other point between n and m in the other; and if from the nature of the curves nm is not equal to pm, sliding must have taken place between the edges through a space equal to the difference. But if nm be equal to pm no sliding will have happened. In the first case the communication of motion is said to be by sliding contact, and in the second by rolling contact.

This mode of action supposes either that the curves are both convex; or should the curvature lie in the same direction, that the convex edge has a greater curvature than that of the concave edge at the point of contact. If this be not the case, successive contacts may take place at discontinuous points.

29. *Communication of Motion by Intermediate Pieces.* Let AP, BQ be a driver and follower, moving on centers at A and B respectively, and let a rod or link, PQ, be jointed at its extremities to the driver and follower at P and Q. Then, if the driver be moved into a new position Ap, it will by means of the link place the follower in a position Bq. If the driver push the follower before it, the link *must* be rigid, but if the driver drag the follower after it, the link may be flexible, the principle of linkwork only requiring that the connexion between the link and its pieces shall be at constant points, and the distance between the two points of attachment invariable.

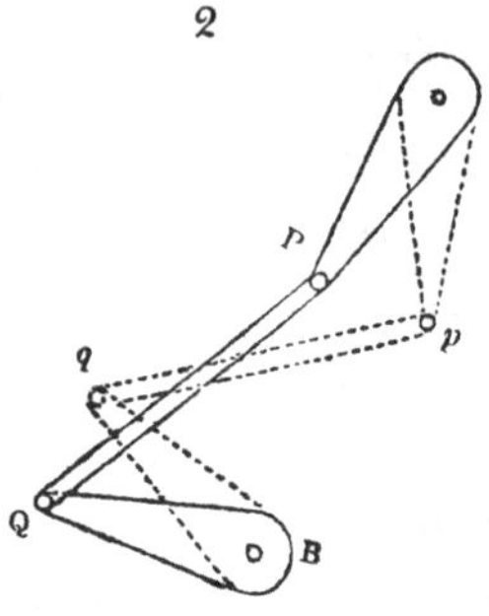

Let ACE be a driver, BDF a follower whose centers of motion are A and B, and whose edges CE, DF, are curved and connected by a flexible band, which is attached at C and D to the curves, and wraps round them, but lies between them in a state of tension in the direction of the common tangent of the curves.

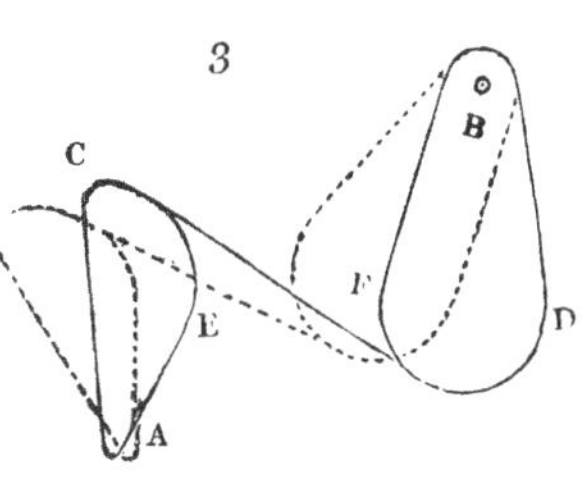

If the driver be moved, it will through this connexion drag the follower after it, and the connector will wrap and unwrap itself from the edges respectively, so as always to lie in the direction of the common tangent.

A connector may be jointed at one end, and wrap at the other, as in Figure 4.

Under the first of these modes of connexion will be arranged all kinds of jointed-work, cranks, and so on; and the second includes all manner of endless bands and belts, wrapping cords, &c.

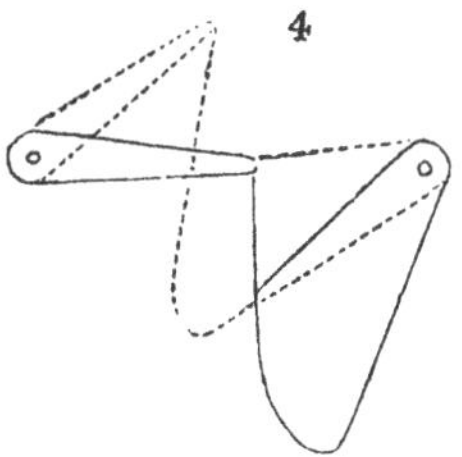

There remains yet another method of modifying motion, which depends on a totally different principle, which I shall denominate Reduplication.

30. *Communication of Motion by Reduplication.* Let P be a pin or piece capable of sliding in the direction Pp, and let a cord be attached to a fixed point M, doubled over P, brought back in a direction parallel to

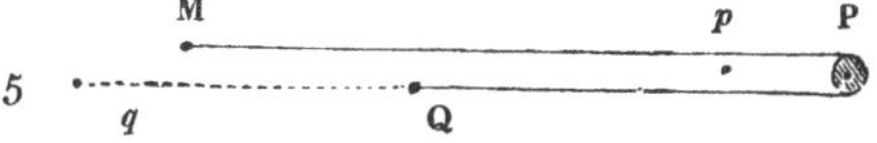

MP, and attached to a point Q. If the point Q be moved into a new position q in the line PQ produced, the point P will be drawn into a position p. And as the length of the cord is invariable, $MP + PQ = Mp + pq$;

that is, $(Mp + pP) + (Pp + pQ) = Mp + (pQ + Qq)$;

$\therefore 2pP = Qq$, or the velocity of the point Q is double that of the point P.

If the cord were again doubled over a pin at M, and returned back to Q, Q then moving in the opposite direction, we should find in like manner the velocity of Q triple that of P. And if the cord were again doubled over P and brought back to Q, the velocity would be quadrupled. Generally if there be n threads between M and P, the velocity of Q will be $n \times$ velocity of P.

In practice, pullies are employed at the points P and M to reduce the friction, but they do not affect the modifications of velocity.

31. Every elementary combination of mechanism, with the exception of those that fall under the head of Aggregates, may now be distributed into one or other of these five divisions:

Immediate Contact....... { *Rolling Contact,* / *Sliding Contact.*

Intermediate Connexion.. { *Links,* / *Wrapping Connectors,* / *Reduplication.*

The velocity ratio has been already determined for the last class, and may be found for the other classes as follows:

32. *To find the velocity ratio in Link-work.* Let AP, BQ be two arms moving on fixed centers A and B respectively, and let them be connected by a link PQ, jointed to their extremities at P and Q. Let AR, BS be perpendiculars from the centers upon the direction of the link produced if necessary, and let AP, BQ, PQ be moved into the new positions Ap, Bq, pq, very near to the first. Draw pm and Qn perpendicular to PQ, then in the right-angled triangles Ppm, APR, Pp is perpendicular to AP.

and Pm to AR; therefore the angle at P in the small triangle is equal to the angle at A in the large triangle,

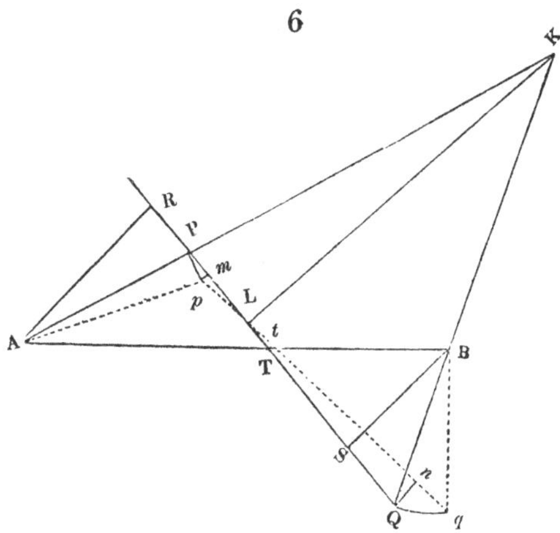

and the triangles are similar. In like manner the small triangle qnQ is similar to BSQ; whence

$$Pp : Pm :: PA : AR,$$

$$qn : Qq :: BS : BQ,$$

$$\text{also } AT : BT :: AR : BS \ldots\ldots(1),$$

by similar triangles ART, BTS. Compounding these proportions and arranging the terms, remarking that $qn = Pm$ ultimately since the length of the link $PQ = pq$, we finally obtain

$$\frac{Pp}{AP} : \frac{Qq}{BQ} :: BT : AT \ldots\ldots(2),$$

that is to say, *the angular velocities of the arms* AP, BQ *are to each other inversely as the segments into which the link divides the line* AB, *which joins the centers of motion, and which is technically termed the line of centers.*

COR. 1. $\frac{Pp}{AP} : \frac{Qq}{BQ} :: BS : AR$, by compounding (1) and (2); therefore *the angular velocities of the arms* AP, BQ *are inversely as the perpendiculars from their centers of motion upon the line of centers.*

COR. 2. Produce AP and BQ to meet in K, and drop KL perpendicular to PQ,

$$\text{then } pm : Pm :: PL : KL,$$
$$\text{and } qn : Qn :: KL : QL;$$

whence, compounding, $pm : Qn :: PL : QL$, which shews that L is the intersection of the two positions of the link.

COR. 3. If the path of the pieces be rectilinear, or any other curve than a circle, let Pp, Qq be the elements of the paths;

then since $Pm = qn$, $Pp \cdot \cos pPm = Qq \cdot \cos Qqn$;

$$\therefore \frac{Pp}{Qq} = \frac{\cos Qqn}{\cos pPm},$$

where the angles are those made by the link with the respective directions of motion; and hence

The linear velocities are to each other inversely as the cosines of the angles which the link makes with the respective paths.

33. *To find the Velocity Ratio in Contact Motions.* Let A, B be the centers of motion of two pieces connected by the contact of curved edges, and let M be the point of contact in a given position.

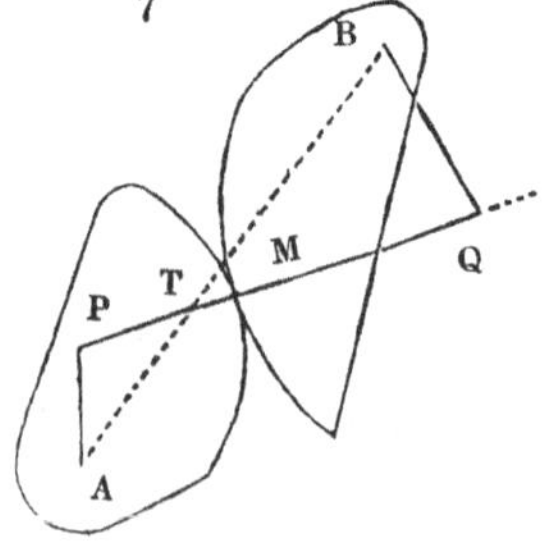

Let P, Q be the respective centers of curvature of the edges, corresponding to the point of contact M; join PQ; therefore this line will pass through the point of contact M. Now in considering the communication of motion through a small angle, the circles of curvature may be substituted for the curved edges. But the line PQ being thus equal to the sum of the radii of two circles, will be constant during that small motion, and hence the motion be the same if a pair of rods AP, BQ, connected by a link PQ, be substituted. Join AB, meeting PQ in T, then by the last proposition, the angular motions of the arms AP, BQ are to each other as the segments BT, AT, and PQ is the common normal to the two curves; whence *in the communication of motion by contact, the angular motions of the pieces are inversely as the segments into which the common normal divides the line of centers.*

34. *To find the quantity of sliding in Contact Motions.* Let A and B be the two centers, M the point of contact, MD the common normal; then,

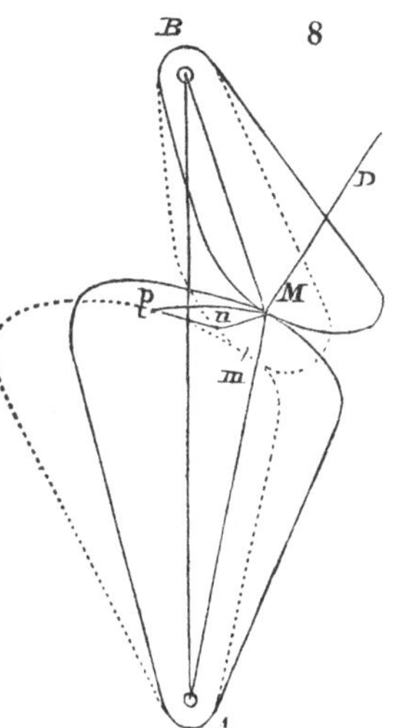

Suppose the curves to move into the new positions, shewn by the dotted lines, and very near the first, and let m be the new point of contact, and p and n the new positions of the points which were in contact at M.

Now since every point of mn must have necessarily touched some point or other of mp, during the change from the first to the second position, a sliding or shifting of the surfaces must have taken place equal to the difference between mp and mn. Join pn, which will ultimately represent this difference, and become a right line perpendicular to the normal MD. Also Mp, Mn are ultimately perpendicular to AM, BM.

In the small triangle Mpn, the sides Mp, Mn, pn are respectively perpendicular to AM, BM, MD, and consequently make mutually the same angles with each other as these latter lines;

$$\text{therefore } \frac{pn}{pM} = \frac{\sin pMn}{\sin pnM} = \frac{\sin BMA}{\sin DMB},$$

in which expression $\frac{pn}{pM}$ is the ratio of the sliding to the elementary quantity of motion of the point of contact in one of the pieces, DMB is the angle between the normal and the radius of contact of the other piece, and $\sin BMA = \sin(BAM + ABM) =$ the sine of the sum of the angular distances of the radii of contact from the line of centers.

$$\text{Similarly, } \frac{pn}{nM} = \frac{\sin BMA}{\sin DMA}.$$

35. From these expressions it appears that in the small triangle pnM, pn can only vanish with respect to nM or pM when $\sin BMA$ vanishes; that is, when the radii of contact coincide with the line of centers. But when pn vanishes the sliding vanishes, and the contact becomes *rolling contact.* Hence it appears that in rolling contact the curves must be so formed, that the point of contact shall always lie on the line of centers. Also the common normal will cut the line of centers at the point T, (Fig. 7.) which will be now the point of contact, and therefore *in rolling contact, the angular velocities are inversely as the segments into which the point of contact divides the line of centers.*

36. Let the curves be a pair of involutes of circles, and let BD be a perpendicular from B upon MD. But this perpendicular is constant in the involute;

$$\text{therefore } \sin DMB = \frac{BD}{BM} \propto \frac{1}{BM};$$

$\therefore \; \dfrac{pn}{pM} \propto BM \times \sin BMA$ varies as the perpendicular upon AM produced.

But if the curves be an epicycloid turning on the center A, in contact with a radial line which turns round B; then DMB is a right angle,

$$\text{and } \frac{pn}{pM} \propto \sin BMA.$$

37. *To find the Velocity Ratio in wrapping connexions.*

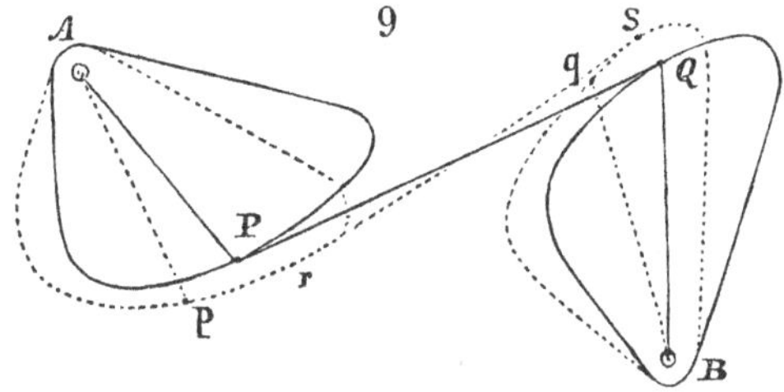

Let A, B be the centers of motion, PQ the wrapping connector touching the curves at P and Q, and let the point P be moved to p very near to its first position, then will Q be drawn to q, and the connector will touch the curves in two new points of contact, which may be r and s respectively. Now, in the action of wrapping or unwrapping, the connector touches the curves in a series of consecutive points between q and S or p and r, and ultimately q coincides with S and p with r. The extremities of the connector may therefore be considered at any given moment as if jointed to the two curves at the points of contact, and turning upon these points in the manner of a link. The relative velocities of the curves are therefore momentarily the same as if AP, BQ were a pair of rods connected by a link PQ. *Hence the angular velocities of the pieces are to each other inversely as the segments into which the connector divides the line of centers.*

38. If the line of direction of the link in link-work, of the common normal to the curves in contact motion, and of the connector in wrapping motion, be severally termed the line of action, we can express the separate propositions which relate to the Velocity Ratio, by saying that the angular velocities of the two pieces are to each other inversely as the segments into which the *line of action* divides the line of centers, or inversely as the perpendiculars from the centers of motion upon the line of action.

I have confined these investigations, for the present, to motions in the same plane. The cases of motions in different planes, are more simply examined as the individual combinations which require them occur.

39. It has been shewn that the principal pieces which constitute a train of mechanism are compelled to move each in a given path. Now it is generally sufficient to consider this path as a circle, for in fact the pieces always either revolve or move in right lines; and the contrivances by which motion is communicated in a rectilinear path, are the same as those by which it is given to a revolving piece, and derived from the latter by that familiar geometrical artifice by which a right line is considered as the arc of a circle whose radius is infinite. It will presently appear that, in this way, much complication of classes will be avoided. Thus, for example, a pinion driving a rack is plainly the same contrivance as a pinion driving a toothed wheel, the rack being considered as a portion of a toothed wheel whose radius is infinite.

It is true, that pieces in a train may be found which describe other paths, such as elliptical, epicycloidal, or sinuous lines; but these are always produced by combining together various circular motions, and therefore the motion of the piece is actually produced by pieces that travel in circular paths. Cases of this kind fall under the head of Aggregate Motions, to which a separate Part of this work has been assigned.

40. The path of a revolving piece may be considered as *unlimited* in extent in either direction, since the piece may go on performing any number of rotations in the same direction. But a piece that travels in a right line is necessarily *limited* in its motion either way, to the length of that line.

Again, the method by which motion is communicated from one piece to another, may be of such a nature as to limit the motion of these pieces, although they may be capable of unlimited motion, considered apart from this connexion. For example, if the driver and follower be revolving cylinders, and therefore capable of unlimited motion, the communication of motion may be effected by a string whose ends are fixed one to each cylinder, and coiled round it, so that when the driver revolves it shall communicate motion to the follower by coiling the string round itself and uncoiling it from the follower; in which case the rotations of each cylinder are limited to the number of coils which its circumference contains when the other is empty.

It appears, then, that the motion of a pair of connected pieces may be limited either by the figure of one or both of their paths, or by the nature of their connexion; and a *limited* connexion may be formed between *unlimited* paths, or *vice versâ*, but if either the paths or the connexion be *limited*, the motion of the pieces will be limited.

In classifying the communication of motion, however, the union of *unlimited* connexions with *limited* paths, will require but little attention, as the modifications to which they lead are in general sufficiently obvious; but the distinction between limited and unlimited methods of communication is of more importance.

CHAPTER II.

ELEMENTARY COMBINATIONS.

CLASS A. { DIRECTIONAL RELATION CONSTANT.
VELOCITY RATIO CONSTANT. }

DIVISION A. COMMUNICATION OF MOTION BY ROLLING CONTACT.

---

41. IN rolling contact it has been shewn that the point of contact is always in the line of centers; and the angular velocities are inversely as the segments into which the point of contact divides that line. Therefore if the velocity ratio is constant, the segments must be constant, and the curves become circles, revolving round their centers, and whose radii are the segments, and no other curves will answer the purpose.

Let R be the radius of the driving circle, and r that of the following circle; L and l their synchronal rotations; then as they are (by § 20) in the ratio of the angular velocities:

$$\therefore \frac{L}{l} = \frac{r}{R}.$$

This ratio will be preserved, whatever be the absolute velocity of the driver, but when this is uniform, which is generally the case, let P and p be the respective periods of the driver and follower;

$$\therefore \text{(by § 20)} \ \frac{P}{p} = \frac{l}{L} = \frac{R}{r}.$$

The motions being supposed hitherto to be in the same plane, the axis of rotation of the circles will be parallel.

42. *When the axes are parallel.* Let Aa, Bb be two parallel axes, mounted in any kind of framework that will allow them to revolve freely, but retain them parallel to each other at a constant distance, and prevent endlong motion, and let two cylinders or rollers, E, F, be fixed opposite to each other, one on each axis, and concentric to it; the sum of their radii being equal to the distance of the axes. The cylinders will therefore be in contact in all positions, and if one of these axes, and consequently its attached cylinder, be made to revolve, its superficial motion will be communicated to the surface of the other cylinder by the adhesion of the parts which are brought successively into contact; and thus the second cylinder will be *driven* by the first by rolling contact, and their perimetral velocities will be equal.

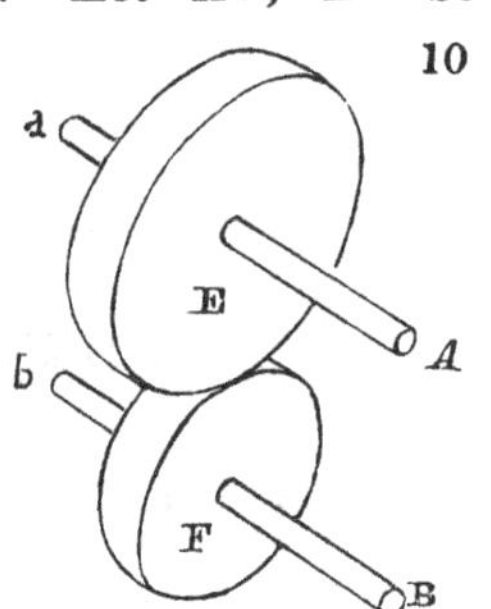

Let R be the radius of the driver, and r of the follower; then a section of the cylinders, made by a plane passing through them at any point at right angles to the axis, will present a pair of circles in contact, whose radii are R and r;

$$\text{and therefore, as before, } \frac{P}{p} = \frac{l}{L} = \frac{R}{r};$$

which is indeed manifest, for since the same length of circumference of the driver and of the follower passes the line of centers* in the same time, let M. circumferences of the driver, equal m. circumferences of the follower;

$$\therefore \; 2\pi RM = 2\pi rm, \text{ and } \frac{M}{m} = \frac{r}{R}.$$

* The *line of centers* is the right line which joins the centers of motion, as already stated, and, in the case of rolling circles, passes through their point of contact. The *plane of centers* is the plane which contains the two axes, whether they be parallel or intersect. These two phrases are of continual use.

But the number of circumferences that pass a given point measure the number of revolutions of the wheel;

$$\therefore \frac{M}{m} = \frac{L}{l} = \frac{r}{R}, \text{ as before.}$$

43. If the axes of rotation be not parallel, they may either meet in direction or not, and these cases must be considered separately.

Axes meeting. Let AB, AC be two axes of rotation intersecting in A,

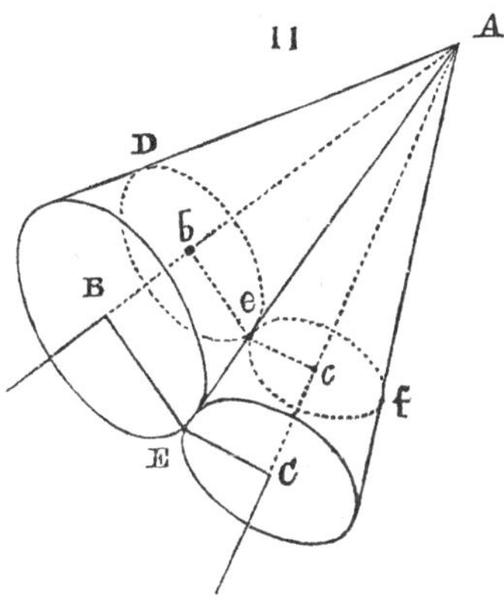

to which are attached cones ABE, AEC, whose apices coincide with A, and which have angles at their vertices of such a magnitude that their surfaces are in contact. Let AE be the line of contact, and Dbe, ecf sections of the cones at any point e of the line and respectively perpendicular to their axes, which sections are necessarily circles touching at e, whose radii are be, ce. If angular velocities A, a be given to the cones ABE, AEC, the perimetral velocities of these sections will be $A.be$ and $a.ce$, and if these are equal,

$$\frac{A}{a} = \frac{ce}{be} = \frac{CE}{BE}$$

a constant ratio. If then the perimetral velocities of any pair of corresponding sections be equal, those of every other such pair will be equal; therefore the cones will roll together as in the former case, and the ratio of the angular velocities be inversely as the radii of the bases of the cones.

44. In practice, a thin frustum only of each cone is employed. Let the position of the axes be given, and also the ratio of the angular velocities, it is required to describe the cones, or rather the frusta.

Let AB, AC be the axes intersecting in A. Through any point D in AB draw DF parallel to AC, and make $DF : AD$ in the ratio of the angular velocity of AB to that of AC. Join AF, then will AF be the line of contact of the two cones, by means of which the required frusta may be described at any convenient distance from A,

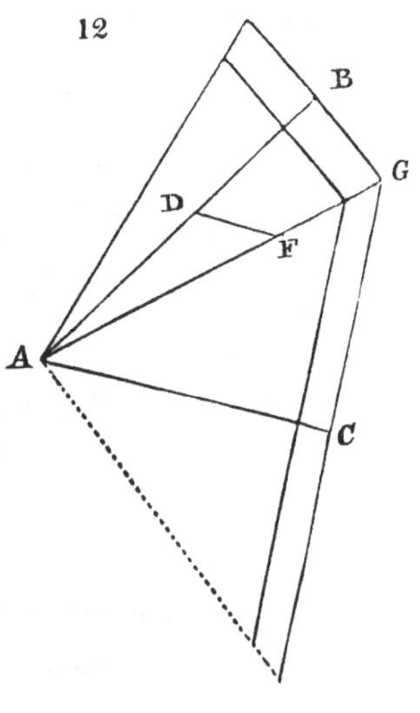

$$\text{for } \frac{DF}{AD} = \frac{\sin DAF}{\sin AFD}$$

$$= \frac{\sin DAF}{\sin FAC} = \frac{BG}{CG};$$

that is, the angular velocities are in the ratio required by last Article.

Cor. The angles at the vertices of the cones may be readily found thus:

Let θ be the angle BAC, κ the semiangle of the vertex of the cone of AB, $\frac{m}{n}$ the given ratio of the angular velocities;

$$\therefore \frac{\sin \kappa}{\sin \theta - \kappa} = \frac{m}{n}; \text{ (}m\text{ being the less)}$$

$$\text{whence } \tan\kappa = \frac{\sin\theta}{\frac{n}{m} + \cos\theta};$$

which may be adapted to logarithms by taking a subsidiary angle ϕ, so that $\cos\phi = \frac{m}{n}\cos\theta$;

$$\text{whence } \tan\kappa = \frac{m\sin\theta}{2n\cos^2\frac{\phi}{2}}$$

If θ be a right angle, which is generally the case, then

$$\tan\kappa = \frac{m}{n}.$$

45. *Axes neither parallel nor meeting.* The hyperboloid of revolution is a well known surface, generated by the revolution of a straight line about an axis to which it is not parallel and which it does not meet*. If two such hyperboloids *EF* be placed so that their generating lines

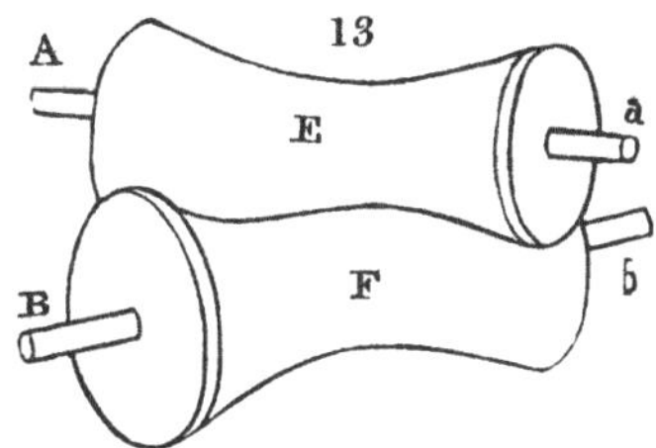

coincide, the solids will touch along this line, and their axes *Aa*, *Bb* will neither be parallel nor meeting in direction.

If the solids revolve about their axes, the contact will plainly continue throughout this line; and that they may be so proportioned as to roll together, can be shewn as follows.

Let *AB*, *CD* be the axes of rotation, *ab*, *cd* their respective projections upon a plane to which they are both

* Vide Newton's Universal Arithmetic, Prob. 33. Hymers' Analytical Geometry, p. 142; or any Treatise on Analytical Geometry.

parallel; gk their common perpendicular* produced to

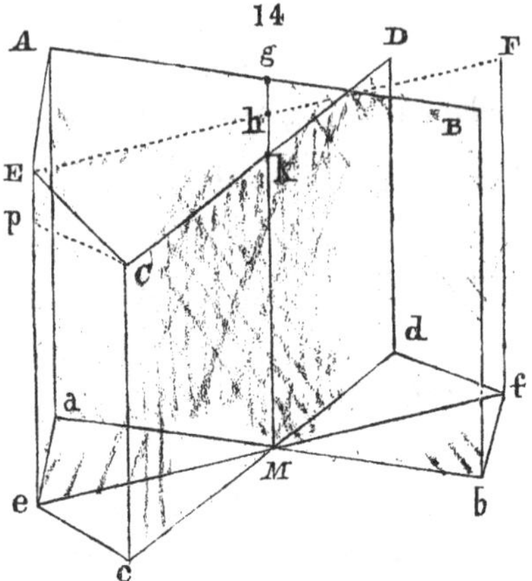

meet the plane in M; EF a line intersecting gk in h, and also parallel to the plane and projected in ef.

Let EF revolve round AB to generate one hyperboloid, and round CD to generate the other; then EF will be their line of contact. From any point E let fall EA, EC, perpendiculars upon AB and CD.

Now as the hyperboloids are solids of revolution, these lines AE, CE will be the radii of a pair of circles in contact at E.

Draw Cp parallel to ce, therefore $Ep = hk$ and $Cp = ce$; whence $EC^2 = hk^2 + ce^2$; and similarly $AE^2 = gh^2 + ae^2$.

If therefore $\frac{ce}{ae} = \frac{hk}{gh}$, we have $\frac{EC}{AE} = \frac{hk}{gh}$, a constant ratio for every corresponding pair of circles; whence it follows, that if the superficial velocities be equal at any point of contact, they will be equal at every other; and as in the cones, the angular velocity ratio $\frac{A}{a} = \frac{r}{R}$, where r and R are the corresponding radii of the hyperboloids.

* Vide Playfair's Geometry, Sup. B. II. Prop. xix.

46. But in rolling cones, the perimeters of every corresponding pair of circles move in the same direction at the point of contact, although the circles are not in the same plane; for the radii of contact are perpendicular to the line of intersection of the two planes which contain the circles, and consequently the tangents at the point of contact coincide with that line and with each other. On the contrary, this is not the case in rolling hyperboloids.

For the circles whose radii are CE, AE, lie in planes whose intersection is the line Ee; and the tangents to these circles at the common point E, plainly cannot coincide either with that line or with each other; so that the motion of the surfaces is not in the same direction, and the rolling action is imperfect, and the more so the greater the angle ECp, that is, the greater the distance gk between the axes; for as this distance diminishes, the hyperboloids approach to a pair of cones whose common apex is h.

47. In practice, as in the case of cones, a thin frustum only is required of each hyperboloid, and these frusta include so small a portion of the curve surface, that a frustum of the tangent cone at the mean point of contact may be substituted without sensible error; and this may be found as follows:

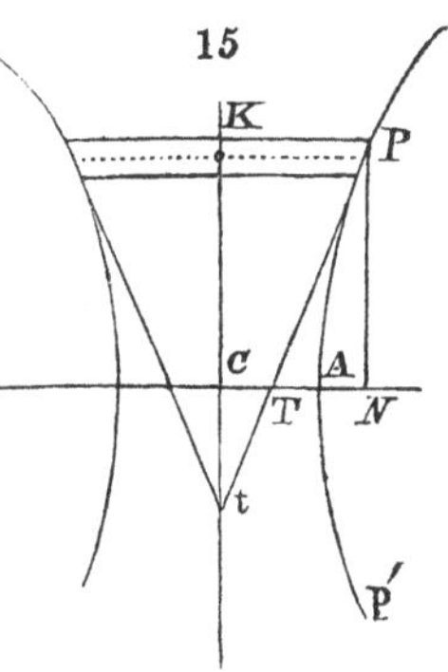

Let CK be the axis of the given hyperboloid; $CK = NP = y$ the mean distance of the required frustum from the center C, $KP = CN = k$ its radius; $CA = a$ the least radius of the hyperboloid and semi-axis-major of its generating hyperbola pAP; all which may be found when the position of the axes and ratio of the angular velocities are given in any particular case, by the last proposition; observing that in Fig. 14. AE, EC are the mean radii of the frusta; g, k, the centers of the generating hyperbolas, and gh, hk their semiaxes-major.

Let PTt be the tangent at P; then, by the known properties of the hyperbola,

$$y^2 = \frac{b^2}{a^2}(x^2 - a^2), \text{ and } Ct = \frac{b^2}{y^2};$$

$$\therefore \; Ct = \frac{a^2 y}{x^2 - a^2} \text{ gives the apex } t \text{ of the required cone.}$$

Or, take $CT = \frac{a^2}{y}$; join PT, and produce it to t.

48. This third case of axes, neither parallel nor meeting, admits of solution by means of the cones of the second case; thus*:

Let Aa, Bb be the two axes, take a third line intersecting the axes at any convenient points C and D respectively; and let a short axis be mounted so as to revolve in the direction of this third line between the other two axes.

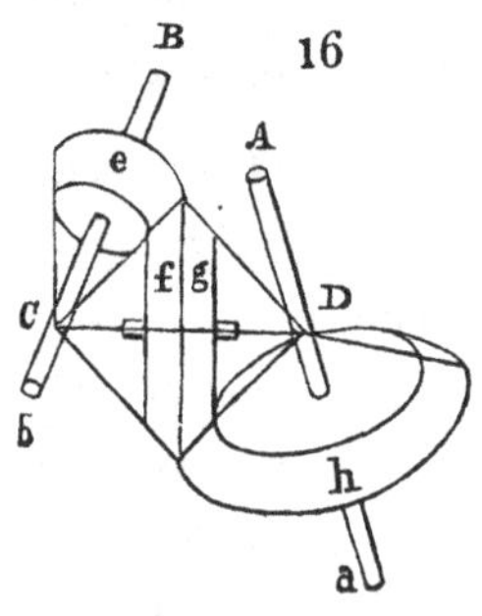

Now a pair of rolling cones, e, f, with a common apex at c, will communicate motion from the axis Bb to the intermediate axis; and another pair g, h, with a common apex at D, will communicate motion from the intermediate axis to Aa; and thus the rotation of Bb is communicated to Aa by pure rolling contact.

Let A, $A_{\prime}$, a, be the respective angular velocities of the axes Bb, CD, Aa; and R, $R_{\prime}$, r the radii of the bases of their cones, those of the cones, f, g, being the same;

* From Poncelet, Mecanique Industrielle, p. 300.

$$\therefore \quad \frac{A}{A_{,}} = \frac{R_{,}}{R}, \quad \text{and} \quad \frac{A_{,}}{a} = \frac{r}{R_{,}}, \quad \text{whence} \quad \frac{A}{a} = \frac{r}{R},$$

exactly as if the cones e, h, were in immediate contact.

To apply these Solutions to Practice.

49. Theoretically we have now the complete solution of the problem in all the three cases; having shewn how to find a pair of cylinders in the first case, and of conical frusta in the other cases, by which a given angular velocity ratio will be effected. If these solids could be formed with mathematical precision, then, the axes having been once adjusted in distance so that the surfaces should touch in one position, they would touch in every other position; but in practice this is impossible, and various artifices are employed to maintain the adhesion upon which the communication of motion depends.

The surface of one or both rollers may be covered with thick leather, which by giving elasticity to the surface enables it to maintain adhesional contact, notwithstanding any small errors of form.

One of the axes may be either made to run in slits at its extremities instead of round holes, or else it may be mounted in a swing frame. Both methods allowing of a little variation of distance between the two axes, the contact of the rollers will in this way also be maintained, notwithstanding small errors of form.

If the weight of the uppermost roller is not sufficient to produce the required adhesion, or if the rollers lie with their axes in the same horizontal plane, then weights or springs may be employed to press the axes together. The practical details of these methods belong rather to the department of Constructive Mechanism than to the plan of the present work.

50. But the most certain method of maintaining the action of the surfaces is to provide them with teeth. The plain cylindrical or conical surfaces of contact are exchanged for a series of projecting ridges with hollow spaces between. These ridges or *teeth* are distributed at equal distances from each other on the two surfaces, and generally in the direction of planes passing through the axis, so that when the driving wheel is turned, its teeth enter in succession the spaces between those of the follower. They are so adjusted that before one tooth has quitted its corresponding space the next in succession will have entered the next space, and so on continually; consequently, the surfaces cannot escape from each other, and there can be no slipping, notwithstanding slight errors of form.

The action of this contrivance falls partly under the head of rolling contact, and partly under that of sliding contact; for the teeth considered separately act against each other by sliding contact, and the forms of their acting surfaces must be determined, as we shall see, upon that principle.

On the other hand, the total action of a pair of toothed wheels upon each other is analogous to that of rolling contact. Equal lengths of the two circumferences contain equal numbers of teeth, and therefore equal lengths will pass the line of centers in the same time, if measured by the unit of the space occupied by one tooth and a hollow between. In fact, the adhesion which enables the surface of one plain roller to communicate motion to another arises from the roughness of the surfaces, the irregular projections of one indenting themselves between those of the other, or pressing against similar projections; and the contrivance of teeth is merely a more complete developement of this mode of action, by giving to these projections a regular form and arrangement. I shall proceed therefore to explain in this Section all that relates to the general action, arrangement, and construction of toothed wheels; leaving the exact form of the individual teeth to the next Section, and observing, that this arrangement corresponds to the ordinary practical view of the subject; for all that belongs to the complete action or construction of a pair of toothed wheels is always referred to a pair of corresponding plain rollers, or rolling circles, which are termed the *pitch circles*, or *geometrical circles*.

51. *Geering* is a general term applied to trains of toothed wheels. Two toothed wheels are said to be *in geer* when they have their teeth engaged together, and to be *out of geer* when they are separated so as to be put out of action; and generally speaking, a driver and follower, whatever be the nature of their connexion, are said to be in geer when the connexion is completely adjusted for action, and out of geer when the connexion is interrupted.

52. Toothed wheels with few teeth are termed *pinions*. This phrase is merely to be considered as the diminutive of *toothed wheel*; and there is no impropriety or ambiguity in calling a pinion a toothed wheel, if more convenient.

53. The teeth of wheels may be either made in one piece with the body or rim of the wheel, or they may be each made of a separate piece and framed into the rim of the wheel.

The first method is employed in cast-iron wheels of all sizes, from the largest to the smallest; also for brass or other metal wheels in smaller machinery, which are formed out of plain disks by cutting out a series of equidistant notches round the circumference, and thus leaving the teeth standing.

Figure 17, *A* and *C*, represents the form of the modern cast-iron wheels, in which, for the sake of uniting lightness and stiffness, a thin web or fin runs along the inner edge of the rim and on each side of the arms, so that the transverse section of the arm is a cross.

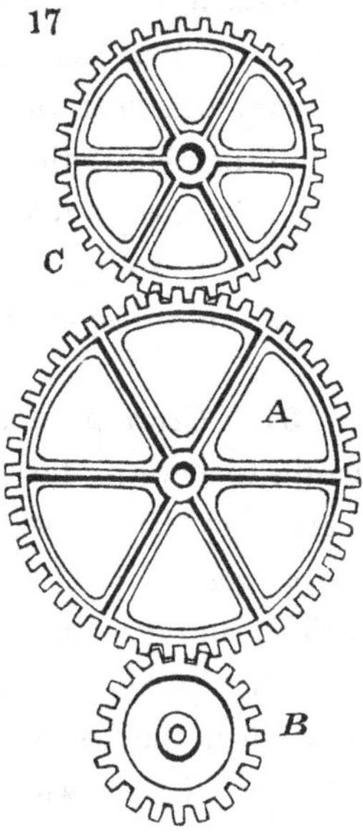

In smaller wheels the arms are omitted, as at *B*, and the rim of teeth united to the central boss by a thin continuous plate. These wheels are *plate* wheels, and when arms are employed, wheels are said to be *crossed out*; but this phrase rather belongs to clock-work. Wooden wheels in one piece with their teeth are too weak to be trusted beyond the construction of models, or wheel-work which transmit little pressure. The wheels of Dutch clocks of the coarser kind are constructed in this manner.

54. Figure 18 exemplifies the construction of mill-work, and larger machinery, previous to the introduction of cast-iron wheels by Messrs. Smeaton and Rennie, at the latter end of the last century*. The wheel *A* is framed of wood, not like carriage-wheels with radial spokes, but with two pair of parallel bars set at right angles, so as

* Mr. Smeaton was the first who began to use cast-iron in mill-work at the Carron Iron-works, in 1769. It was first employed for the large axes of water-wheels, and soon afterwards for large cog-wheels; but the complete introduction of it is due to Mr. Rennie.—Vide Farey on the Steam Engine, p. 443.

to leave a square opening in the midst for the reception of the shaft, which is also of wood, and square, and the opening being purposely left larger than the section of the shaft, the wheel is secured upon it by driving wedges in the intermediate space. This frame carries the rim of the wheel, which is made truly cylindrical on the outer surface, and annular in front. Equidistant mortises are pierced through the rim in number equal to those of the teeth or *cogs*, as they are called when made in separate pieces.

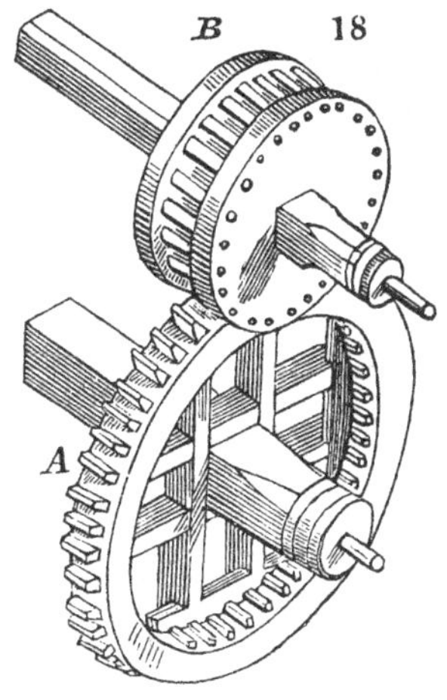

18

The cogs are made of well-seasoned hard wood, such as mountain-beech, hornbeam, or hickory; the grain is laid in the direction of the length, which being the radial direction, gives them the greatest transverse strength. A cog consists of a head *a*, and a shank *b*, of which the head is the acting part or actual tooth which projects beyond the rim, and the shank or tenon is made to fit its mortise exceedingly tight, and is left long enough to project on the inside of the rim. When the cog is driven into its mortise up to its shoulders a pin *c* is inserted in a hole bored close under the rim of the wheel, by which it is secured in its place.

19

a *b* *c*

55. This construction of a toothed wheel has been partly imitated in modern mill-work, for it is found that if in a pair of wheels the teeth of one be of cast-iron, and in the other of wood, that the pair work together with much less vibration and consequent noise, and that the teeth abrade each other less than if both wheels of the pair had iron

teeth. Hence in the best engines one wheel of every large sized pair has wooden cogs fitted to it exactly in the manner just described; only that instead of employing a wooden framed wheel to receive them, a cast-iron wheel with mortises in its circumference is employed. Such a wheel is termed a *Mortise wheel.*

Wheels of the kind hitherto described, in which the teeth are placed radially on the circumference, whether the teeth be in one piece with the wheel, or separate, are termed *spur-wheels;* and when the term *pinion* is applied to a wheel its teeth are usually called *leaves.*

56. The pinions in large wooden machinery are commonly formed by inserting the extremities of wooden cylinders into equidistant holes, in two parallel disks attached to the axis or shaft*, as at *B*, (fig. 18.) thus forming a kind of cage, which is termed a *lantern*, *trundle*, or *wallower*; the cylindrical teeth being named its *staves*, *spindles*, or *rounds.* This construction is very strong, and the circular section of its teeth or *staves* gives it the advantage of a very smooth motion, when the *lantern* is *driven*, as will be shewn in its proper place. In Dutch clock-work this plan is imitated on a small scale, and small wire used for the staves.

57. A similar system to this is of great antiquity, for in early machinery the toothed wheels were often cut out of thin metal plates; and it would be obviously impossible to make a pair of such thin wheels work together, as in fig. 17; for the smallest deviation of one of the wheels from the plane of rotation of the pair, would cause the teeth to lose hold of each other sideways. For this reason one of the wheels of a pair were always made either in the lantern form as just

* *Axis* is the general and scientific word, *shaft* the millwright's general term, and *spindle* his term for smaller shafts; *axle* is the wheelwright's word, and *arbor* the watchmaker's.

described, or with pins inserted at one end only into a disk, as at *A* fig. 20, or else the teeth of one of the wheels were cut out of a hoop, as at *C*, forming what is termed a crown wheel, or contrate wheel.

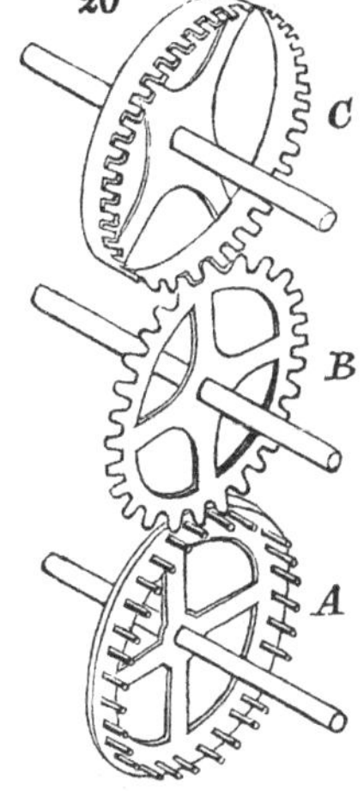

In this figure it is evident that the thin wheel *B* would retain hold of the pins of *A*, or of the teeth of *C*, notwithstanding a little deviation from the plane of rotation, or a little end-play in the axes.

58. *Annular wheels* have their teeth cut on the inside edge of an annulus, so that the pinion which works with them shall lie within the pitch circle. Hence *the two axes revolve in the same direction.* The arms of an annular wheel necessarily lie behind the annulus, in order to make room for the pinion, and the latter must be fixed at the extremity of its axis, otherwise this will stop the wheel by passing between the arms. Annular wheels are more difficult to execute than common spur-wheels, but it will be shewn that the action of their teeth is smoother. A pin-wheel like *A*, fig. 20, may be employed as an annular wheel, and is much easier to construct.

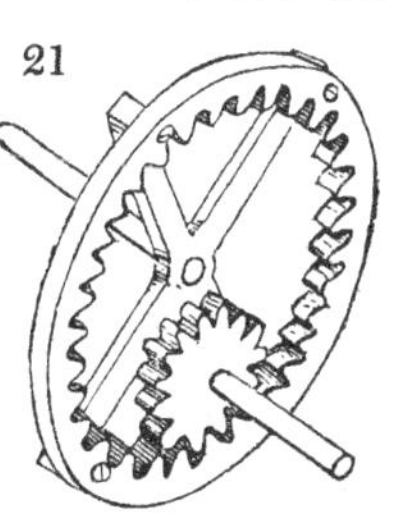

59. When the path of one of the pieces is rectilinear, or, in other words, if it be a sliding piece, then the teeth are cut on the edge of a bar attached to this piece, so that the teeth may work with those of the wheel or pinion, which is to drive or

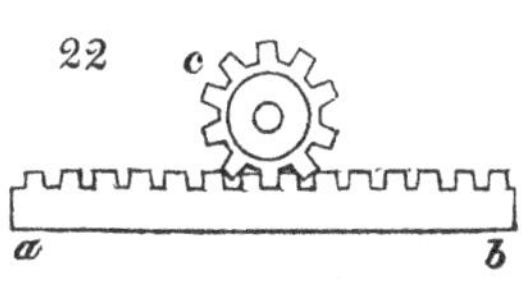

follow it, as in this figure, where the bar *ab* is supposed to be confined by proper guides, so as to move only in the direction of its length, and the pinion *c* to geer with it either as a driver or a follower.

Such a toothed bar is termed a *rack*. The teeth admit of all the different forms and arrangements of which the teeth of wheels in general are susceptible; the rack being merely a toothed wheel whose radius is infinite. Similarly, an annular wheel may be considered as a toothed wheel whose radius is negative.

60. If the space through which the bar moves is less than the circumference of the wheel, the latter may assume the form of a sector, as in this figure.

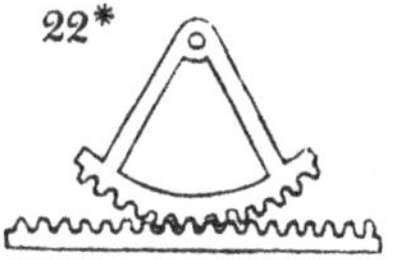

61. All these examples belong to the first case of position in the axes, that is, when they are parallel; but the second case, in which their directions meet, presents itself also very early in the history of mechanism.

A water-wheel, for example, has its axis necessarily horizontal, and near the surface of the water. The axis of a mill-stone, on the other hand, is vertical, and it is convenient to place the latter in an upper floor of the building. This is the disposition of the water-mill of Vitruvius, and is in fact universal.

But the exact method of deriving the form of the toothed wheels from a pair of rolling cones, was not introduced until the middle of the last century, when its mathematical principles were completely laid down by Camus, in 1766*.

* Camus, Cours de Mathématique, Par. 1766. The part relating to toothed wheels has been printed separately in England, and is well known. The principle of rolling cones was first published in England by Imison. In his treatise of the Mechanical Powers, 1787, he uses the term *bevel geer*, and speaks of such wheels as well known.

Previously to this it was thought sufficient to dispose the teeth of the wheels, as in this figure, upon the face

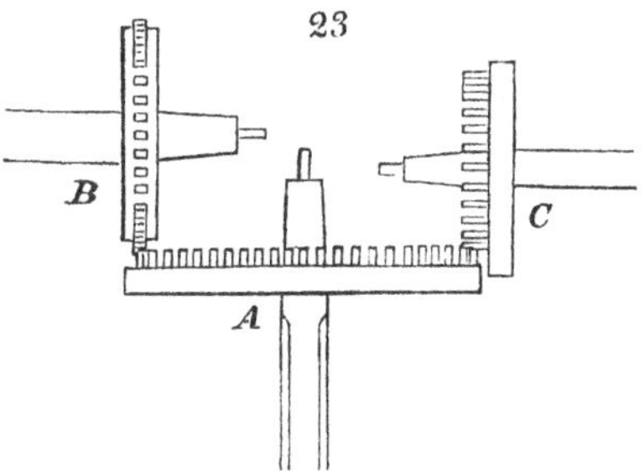

of one of the wheels as *A*, so as to catch those of an ordinary spur-wheel *B* with teeth on the circumference; or else to place the teeth of both wheels on the face, as in those of *A* and *C*. Sometimes the teeth of both wheels were placed on the circumference, as in the ordinary spur-wheels; with this difference, that the teeth require to be much longer, to enable them to lay hold of each other in this relative position. For the forms of the individual teeth no certain principles were followed, and for the arrangements in question the only principle appears to have been to place the teeth so that on passing the line or rather plane of centers*, the teeth should present themselves in the same relative position as if they belonged to a pair of wheels with parallel axes.

A similar principle is, indeed, clearly stated by De la Hire, in the extract which follows the next paragraph.

62. When the axes intersected each other at right angles, and one of them revolved much quicker than the other, a cylindrical lantern was universally given to the latter, and the teeth of the former placed on its face,

* Vide Note, p. 32.

as in this figure, at *A* and *B*. This form and arrange-

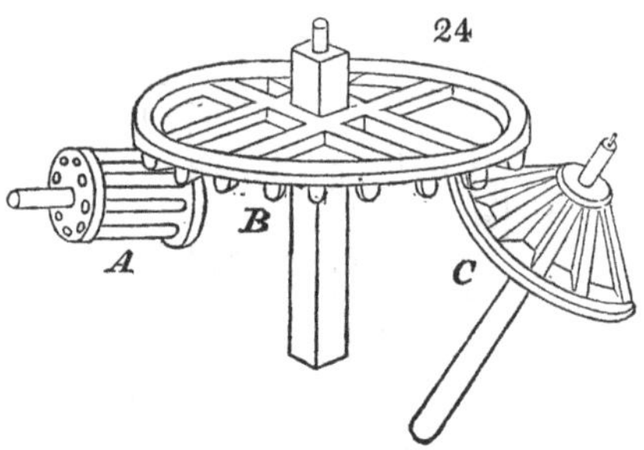

ment is found in mills of all kinds, from the earliest known printed figures to the wooden mill-work of the last century.

The wheel *B* is termed a face-wheel; it generally revolved in a vertical plane. This figure is copied from one in De la Hire's Mechanics*, in a chapter where he proposes to shew how the direction of motion may be changed by toothed wheels; and after giving the cylindrical lantern *A* for the case of axes at right angles, he proceeds to axes inclined at any other angle, thus:—"If a lantern *C* be constructed having staves inclined to the axis at any given angle, then will the horizontal motion of the power be changed into a motion inclined to it at any angle we please, provided only that the staves of this lantern *C* must be so arranged that they come successively into the horizontal position at the moment of meeting the teeth of the wheel *B*, in order that they may apply themselves to the teeth in the same manner as if this lantern was like the other *B*. These changes of direction in motions may be of great use in machinery."

* De la Hire's Treatise on Mechanics, Par. 1695. Prop. LXVI. This was early translated into English, in part, by Mandey, in his Mechanical Powers, 1709, p. 304.

It is rather singular, that upon the authority of this conical lantern the invention of *bevil geer* has been attributed to De la Hire, when it is plain that the principle of rolling cones, which is essential to them, has nothing whatever to do with this arrangement; which is solely founded upon the notion of presenting the teeth to each other at the plane of centers, in the same relative position as in spur or face-wheels. The apex of the cone is turned in the wrong direction for bevil-wheels, and the cylindrical lantern is employed for the axes at right angles.

63. But the necessity of changing the direction of motion through other angles than right angles had arisen long before the time of De la Hire; suggested, as I believe, by the use of the Archimedean screw for raising water, which appears to have been a great favourite with the early mechanists. Figure 25, for example, is part of a complex piece of mill-work extracted from one of the early printed collections of machinery*. The object of the mechanism in question is to enable a water-wheel to give motion to a series of three Archimedean screws placed one above the other. A face-wheel, carried by the axis of the water-wheel, geers with a trundle (Art. 56) at the lower extremity of a vertical axis, which extends to the top of the building, and of which *A* is a portion.

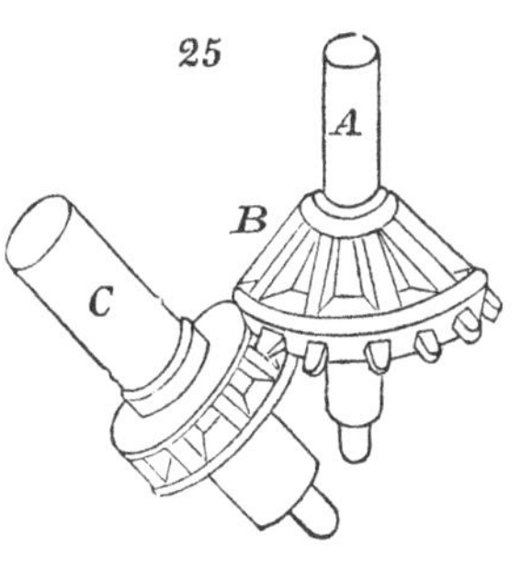

Three conical wheels, similar to *B*, are placed one opposite to the lower end of each screw, as *C*, which it turns by geering with a square-staved trundle, as shewn in the figure.

* Le Diverse et Artificiose Machine del Capitano A. Ramelli. Par. 1580, ch. XLVIII.

4

These conical wheels are derived from the common spur-wheel, by the same principle of placing the teeth so that they shall, in crossing the line of centers, lie in the same relative position as if the axis of the wheel had been parallel to that of the trundle; which principle it was, in this case, oddly enough, thought necessary to extend also to the spokes or arms of the wheel.

64. The common *crown-wheel* and pinion, Fig. 26, which is used in clock and watch-work, in cases where axes meet at right angles, is another example of the same principle. The axis *A*, which carries the pinion, is at right angles to *B*, which carries the crown-wheel.

The teeth are cut on the edge of a hoop, and the action

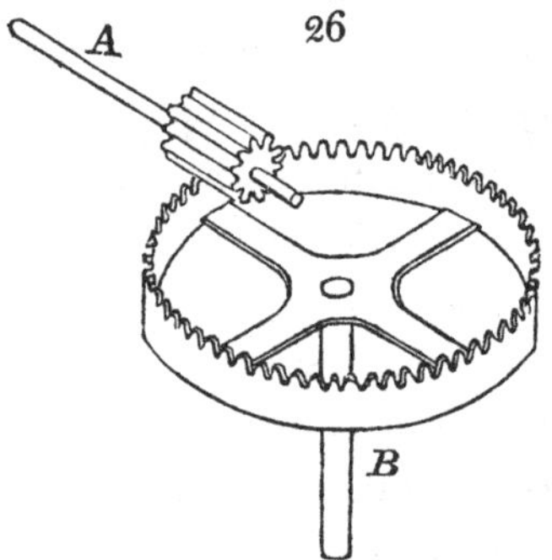

of the pinion upon them is nearly the same as if it worked with a rack; the combination being made on the presumption, that the curvature of that portion of the hoop whose teeth are engaged is so small, that it may be neglected; in which case, the hoop coincides with a rack which is tangent to it, along its line of intersection with the plane of centers, and which travels in a direction perpendicular to that plane.

The *crown-wheel* is often termed a *contrate wheel.*

65. *To form a pair of bevil-wheels,* a pair of conical frusta having been described (by Art. 44) to suit the required

angular positions of the axes and the given velocity ratio,

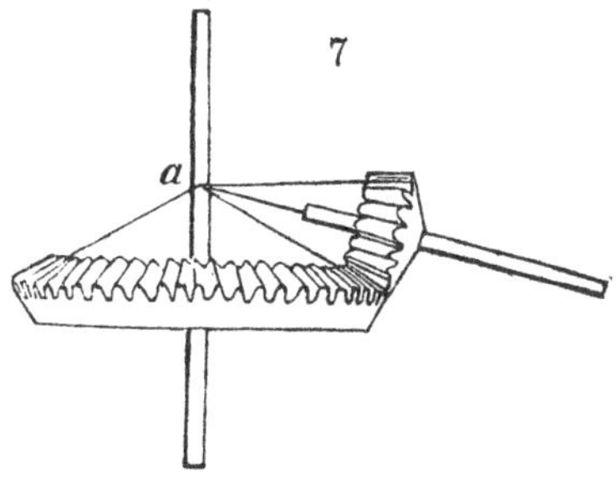

the smooth surface of these cones must be exchanged for a regular series of equidistant teeth, projecting nearly as much beyond the surface as the intermediate hollows lie below it, and directed to the apex of the cone, so that a line passing through this apex shall, if brought into contact with any part of the side of a tooth, touch it along its whole length. Thus the contact of one tooth with another will also take place along the line; whereas in face geering the contact of the teeth is between two convex surfaces at a point only.

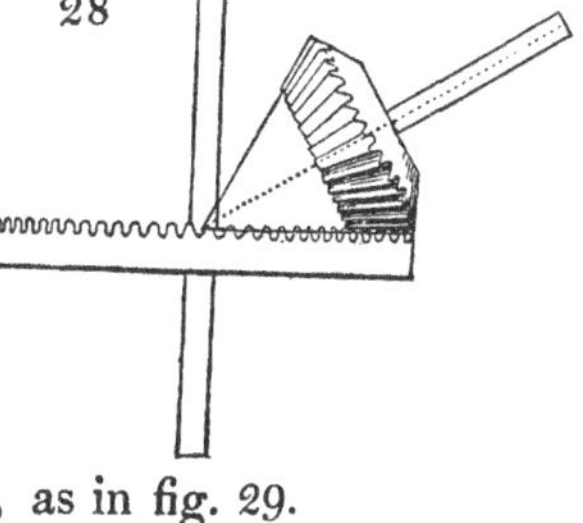

66. It may happen that the common apex of the two cones shall lie so that one of them becomes a plane surface, as in fig. 28; in which case the teeth become radial. Also one of the cones may even be hollow, as in fig. 29.

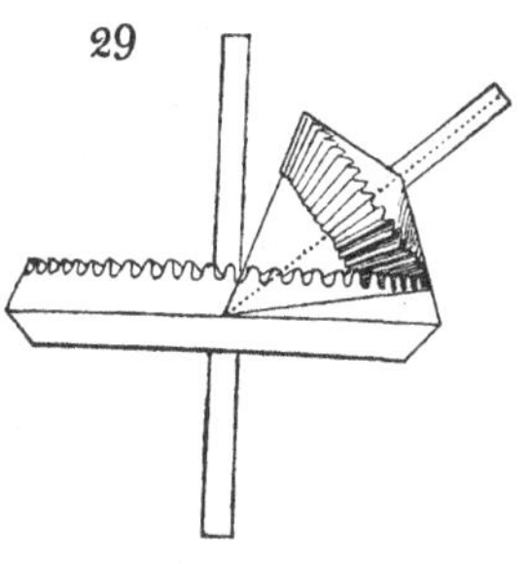

For every given position of the axes, however, we have a choice of two positions for the wheel which belongs to that shaft whose direction is carried past the other. In these last figures this wheel is placed below, but if it had been above, a different and smaller pair of cones would have been obtained for the given velocity ratio, in which these peculiarities of form would have been avoided.

67. When the axes are inclined to each other without meeting in direction, an intermediate double bevil-wheel may be employed, arranged as in Art. 48, or else frusta are employed, which are derived from the tangent cones of a pair of hyperboloids. (Art 47.)

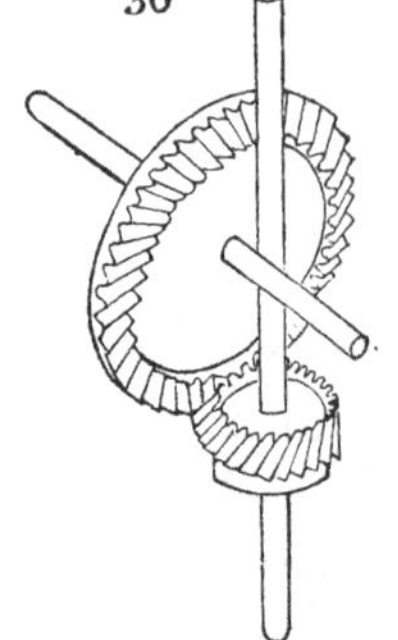
30

The direction of their teeth or flutes must be inclined to the base of the frustum, to enable them to come into contact; and the oblique position thus given to teeth has procured for wheels of this kind the name of *Skew Bevils*. If the teeth be cut in the direction of the generating line of each hyperboloid, they will obviously meet, since this line is the line of contact of the two surfaces.

To find this line upon a given frustum of the tangent cone, let fig. 31 be the plan of this frustum, l the center; set off lz equal to the shortest distance of the axes, (their common perpendicular) and divide it in k, so that lk is to kz as the mean radius of the frustum to the mean radius of that with which it is to work, draw km perpendicular to lz, and meeting the circumference of the conical surface at m. Perform a similar operation on the base of the frustum, by drawing a line parallel to km, and at the same distance lk from the center, meeting the circumference in p; join mp, which is plainly the line of direction of the teeth, (vide Art. 45).

31

We are also at liberty to employ the equally inclined line qn in the opposite direction, but care must be taken that in the two wheels that pair of directions be taken of which the inclinations correspond.

But this question may also be satisfied upon the principle of face-wheel geering, and was so disposed of by the older mechanists, the teeth being merely arranged on the principle already explained, so that they should pass at the instant of contact, in the same relative positions as if the axes had been parallel, or meeting in direction.

68. It has been already shewn that there is no rubbing friction when the point of contact of two edges is on the line of centers. Of this Dr Hooke was certainly aware, as appears from his remarkable contrivance to get rid of the friction of wheel-work. This, to use his own words, "I called the perfection of Wheel-work; an invention which I made and produced before the Royal Society in 1666."

"It is, in short, first, to make a piece of wheel-work so that both the wheel and pinion, though of never so small a size, shall have as great a number of teeth as shall be desired, and yet neither weaken the work, nor make the teeth so small as not to be practicable by any ordinary workman. Next, that the motion shall be so equally communicated from the wheel to the pinion, that the work being well made, there can be no inequality of force or motion communicated. Thirdly, that the point of touching and bearing shall be always in the line that joins the two centers together. Fourthly, that it shall have no manner of rubbing, nor be more difficult to be made than the common way of wheel-work, save only that workmen have not been accustomed to make it*."

This fourth condition of no rubbing is, however, as we have seen (Art. 35), necessarily included in the third.

* Vide Cutlerian Lectures, by R. Hooke, No. 2, entitled Animadversions on the first part of the Machina Cœlestis, 1674, p. 70.

First, then, if there be a certain large number of teeth required to be made in a small wheel, then must the wheel and pinion consist of several plates or wheels lying one beside the other, as in this figure *A*, where eight plates

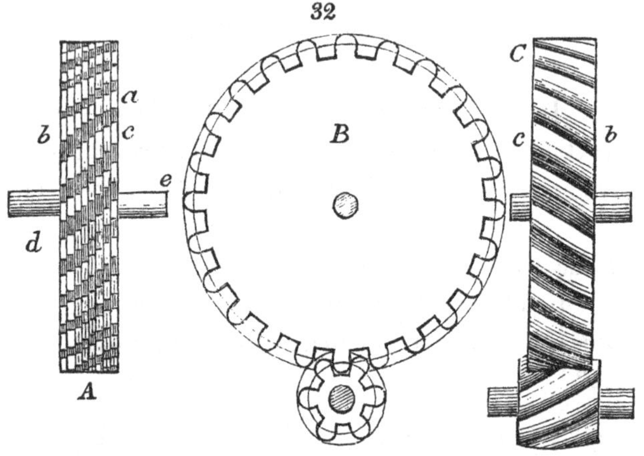

of equal thickness and size, are each cut into a wheel of twenty-five teeth, as shewn in front elevation at *B*; the wheels are fitted close together upon one arbor *de*, and fixed in such order that the teeth of the successive plates follow each other with such steps that the last tooth of each group may within one step answer to the first tooth of the next group. Thus, reckoning from *a* to *b*, the teeth follow each other in equidistant steps of such a magnitude that *b* is distant one such step from *c*, the first tooth of the next group.

The pinion being constructed upon a similar principle, and of the same number of plates, it is clear that the inequalities in the touching, bearing, or rubbing of such wheel-work, would be no more than what would be between the two next teeth of one of the sets, that is, about the same as in a wheel of 200 teeth, and yet the teeth are as large as those of a wheel of 25 teeth.

Secondly, if it be desired that the wheel and pinion should have infinite teeth, all the ends of the teeth must, by a diagonal slope, be filed off and reduced to a straight or rather a spiral edge, as in *C*, which may indeed be best made by one plate of a convenient thickness, which thickness must be more or less according to the bigness of the sloped tooth. And this is to be always observed in the cutting thereof, that the end of one slope tooth on the one side be full as forward as the beginning of the next tooth on the other; that is, that the end *b* of one tooth on the right side be full as low as *c*, the beginning of the next tooth on the left side.

Thus far I have employed nearly the words of Hooke, who has, however, said nothing respecting the *form* of the teeth, which must evidently, in the second system, be so shaped as to begin and end contact upon the very line of centers; the mode of effecting which will appear in the next chapter. The contact of the teeth will be at every instant at a single point, which point will, as the wheel revolves, travel from one side of the wheel to the other; a fresh contact always beginning on the first side, just before the last contact has quitted the other side. And as the point of contact is always on the line, or rather plane, of centers, it is strictly rolling, and there will be no sliding or friction between the teeth.

Hooke's system has been several times re-invented, for example, by Mr White, of Manchester, who patented it before 1808*; and endeavoured, in vain, to introduce it into the machinery of that place. The motion of such wheel-work is remarkably smooth and free from vibratory action, but it

* Vide White's Century of Inventions, 1822, Memoirs of Lit. and Phil. Soc. of Manchester, also Sheldrake, Theory of Inclined Plane Wheels, 1811. It has besides been reproduced as new in America, and lately in London, under the name of a Helix Lever.

has the defect of introducing an endlong pressure upon the axes, occasioned by the obliquity of the surfaces of contact to the planes of rotation. But there are many cases in which this property, when understood and provided for, would not be injurious. The first form of Hooke's geering, in which it appears as separate concentric wheels, as at *A*, has been employed successfully in cases where smooth action is necessary*; and is free from the oblique pressure, but loses the advantage of the perfect rolling action.

ON PITCH.

69. Let N and n be the numbers of teeth of the driver and follower respectively, then as the teeth are equally spaced upon the circumference of the two wheels, these numbers are proportional to the circumferences and radii of their respective wheels; hence

$$\frac{N}{n} = \frac{R}{r} = \frac{P}{p} = \frac{l}{L}. \quad \text{(Vide Art. 42.)}$$

70. The *pitch circle* of a toothed wheel is the circle whose diameter is equal to that of a cylinder, the rolling action of which would be equivalent to that of the toothed wheel (Art. 50); therefore in the above equation R and r are the radii of the pitch circles of the driver and follower respectively; these rolling cylinders being the limit to which the toothed wheels approach, as their teeth are indefinitely diminished in size and increased in number, the distance of the axes remaining the same.

This circle is variously termed the pitch circle of the wheel, the primitive circle, or the geometrical circle. I

* I have seen it in a planing engine by Mr. Collier, of Manchester.

prefer the term pitch, as less liable to ambiguity, and as, I believe, the one most usually employed. In conical wheels the pitch circle will be the base of the frustum.

71. Let the circumference of the pitch circle be divided into equal parts, in number the same as that of the teeth to be given to the wheel; the length of one of these parts is termed the *pitch* of the teeth, or of the wheel, and evidently contains within itself the exact distance occupied by one complete *tooth and space*. The word *space* is employed here in its technical meaning, as denoting the hollow or gap that separates each tooth from the neighbouring one.

Let C be the pitch, D the diameter of the pitch circle, both expressed in inches and parts; and let N be the number of teeth, then $NC = \pi D$*; from which expression if any two of the quantities C, D, N be given, the third may be found. The arithmetical rules which are immediately deducible from this equation are in constant requisition amongst millwrights.

72. In English practice it has been found convenient to employ only a given number of standard values for the pitch, instead of using an indefinite number. The values most commonly chosen are 1in., $1\frac{1}{8}$in., $1\frac{1}{4}$in., $1\frac{1}{2}$in., 2in., $2\frac{1}{2}$in., 3in. And it very rarely happens that any intermediate values are necessary. Below inch pitch the values $\frac{1}{4}$, $\frac{3}{8}$, $\frac{1}{2}$, $\frac{5}{8}$, and $\frac{3}{4}$, are perhaps sufficient.

These remarks apply to cast-iron wheels principally, as the great utility of this system of definite values for the pitch resides in its limiting the number of founders' patterns. Cast-iron teeth of less than $\frac{1}{4}$in. pitch are seldom employed; and, for machinery of a less size than this, the wheels would

* Where $\pi = 3.1415$. The millwrights commonly use $\frac{22}{7}$ for π.

be cut out of disks of metal in a cutting engine. Nevertheless the same system of sizes might be introduced with advantage into wheels of this latter kind.

73. Since the values of C are few and definite, the use of the expression $NC = \pi D$ may be facilitated by calculating beforehand the values of $\frac{C}{\pi}$ and $\frac{\pi}{C}$ that belong to these cases.

For $N = \frac{\pi}{C} . D$, and $D = \frac{C}{\pi} . N$; and the following Table furnishes the factor corresponding to each of the established values of the pitch, by the use of which the number of teeth may be readily found for any given diameter, or *vice versa.*

Pitch in inches.	$\frac{\pi}{C}$	$\frac{C}{\pi}$
3	1.0472	.9548
$2\frac{1}{2}$	1.2566	.7958
2	1.5708	.6366
$1\frac{1}{2}$	2.0944	.4774
$1\frac{1}{4}$	2.5132	.3978
$1\frac{1}{8}$	2.7924	.3580
1	3.1416	.3182
$\frac{3}{4}$	4.1888	.2386
$\frac{5}{8}$	5.0265	.1988
$\frac{1}{2}$	6.2832	.1590
$\frac{3}{8}$	8.3776	.1194
$\frac{1}{4}$	12.5664	.0796

EXAMPLES.

Given, a wheel of 42 teeth, 2 inch pitch, to find the diameter of the pitch circle. Here the factor corresponding to the pitch is .6366 which multiplied by 42 gives 26.7 inches for the diameter required.

Given, a wheel of four feet diameter, $2\frac{1}{2}$ pitch, to find the number of teeth; the factor is 1.257 which multiplied by 48, the diameter in inches, gives 60 for the number of teeth.

Given, a wheel of $30\frac{1}{2}$ inches diameter, and 96 teeth, to find the pitch. Here $\frac{D}{N} = \frac{30.5}{96} = .317 = \frac{C}{\pi}$; which value of $\frac{C}{\pi}$ corresponds in the Table to inch pitch.

Questions of this kind are continually occurring in the execution of machinery; and simple as the calculation may appear to a mathematician, they require more multiplication and division than is always at the command of a workman. By way of simplifying the expression of the relations between the size of the teeth, their number, and the diameter of the pitch circle, a different mode of sizing the teeth in small machinery has been adopted in Manchester, which may be thus explained.

74. Suppose the diameter of the pitch circle to be divided into as many equal parts as the wheel has teeth; and let one of these parts be taken for a modulus instead of the pitch hitherto employed; and accordingly, let the few necessary values be assigned to it in simple fractions of the inch. Call this new modulus the *diametral pitch* of a wheel, to distinguish it from the common pitch, which may be named the *circular pitch*, and let M be the diametral pitch;

$\therefore \dfrac{D}{N} = M$, and, as M is a simple fraction of the inch, let $M = \dfrac{1}{m}$; $\therefore mD = N$, in which N and m are always whole numbers.

The values of m, commonly employed, are 20, 16, 14, 12, 10, 9, 8, 7, 6, 5, 4, 3; and all wheels being made to correspond to one of the classes indicated by these numbers, the diameter or number of teeth of any required wheel is ascertained with much less calculation than in the common system of circular pitch.

This Table* shews the value of the circular pitch C, corresponding to the selected values of m already given.

m	C, in decimals of inch.	C, in inches to nearest $\frac{1}{16}$.
3	1.047	1
4	.785	$\frac{3}{4}$
5	.628	$\frac{5}{8}$
6	.524	$\frac{1}{2}$
7	.449	$\frac{7}{16}$
8	.393	$\frac{3}{8}$
9	.349	
10	.314	$\frac{5}{16}$
12	.262	$\frac{1}{4}$
14	.224	
16	.196	$\frac{3}{16}$
20	.157	$\frac{1}{8}$

* This table is founded on the practice of the well-known factory of Sharp, Roberts, and Co., at Manchester, and may therefore be relied on as exhibiting the present most perfect methods employed in the smaller class of mill-work, or cast-iron mechanism. In this system, a wheel in which $m = 10$ would be called a ten-pitch wheel, and so on.

Since $\frac{D}{N} = M$, we have $M = \frac{C}{\pi}$; therefore the diametral pitch is the quantity which has been calculated in the second column of the Table in page 58. In fact, it is easy to see that this scheme differs from the first, merely in expressing in small whole numbers the quantity $\frac{\pi}{C}$ instead of C.

In small machinery, of the kind that would be classed as clock or watch-work, and in which the wheels are cut out of plain disks by means of a cutting engine, the size of the teeth is often denoted by stating the number of them contained in an inch of the circumference, which may vary from about four to twenty-five. The word pitch is unknown to clockmakers, and their pitch circle is termed the geometrical circle; but, for the sake of uniformity, I shall apply the term pitch indifferently to all kinds of wheel-work. In cut wheels it is necessary to calculate the pitch for the purpose of obtaining the size of the cutter, which, as it operates by cutting out the spaces between the teeth, ought of course to be exactly of the same form and breadth as those spaces. When the number of teeth and geometrical diameter of a wheel are given, the pitch of these small teeth may be determined, in decimals of the inch, from the general expressions already given for the teeth of mill-work; and after the forms of the teeth have been described according to the methods contained in the next chapter, the shape and size of the cutter will be obtained.

CHAPTER III.

ELEMENTARY COMBINATIONS.

CLASS A. { DIRECTIONAL RELATION CONSTANT. VELOCITY RATIO CONSTANT. }

DIVISION B. COMMUNICATION OF MOTION BY SLIDING CONTACT.

---

75. THE axes being supposed parallel, it appears (Art. 33), that in sliding contact, the angular velocities are in the inverse ratio of the segments into which the normal of the curves, at the point of contact, divides the line of centers.

Any curve then being assumed for the edge of one revolving piece, if we can assign such a form to the edge of another revolving piece that the common normal of the two curves shall divide the line of centers in a fixed point, in all positions of contact, then will these curves preserve a constant angular velocity ratio, when one is made to move the other by sliding contact. Before, however, I proceed to develope this general principle, I shall, for the sake of simplicity, give the several ordinary solutions of the problem, and after that shew how they are included with others under this proposition.

76. *First solution.*—Let A, B be the centers of motion, AB the line of centers divided as usual, in T, in the inverse proportion of the angular velocities; describe through T the respective pitch circles, and let abc be a portion of an epicycloid whose base is the pitch circle aT, and whose describing circle has the same

diameter as the pitch circle Tb, and let b be a pin whose diameter is exceedingly small, so that it may be considered as a mathematical line. Then if the curve abc be cut out of a thin plate, and caused to turn round the center A, and the pin b carried by a piece capable of turning round the center B, the motion communicated from the edge to the pin will fulfil the required conditions. For at the beginning of the motion let

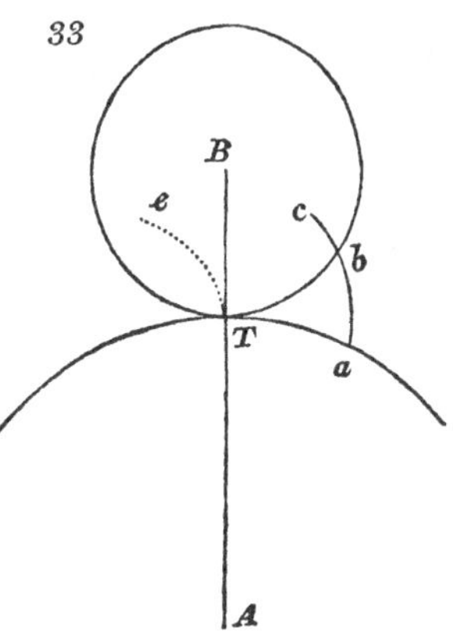

Te be the position of the curve; therefore, the pin b will coincide with T, and if the curve move into any other position abc driving the pin to b, the arc Ta will be equal to Tb; for Tb is an arc of the describing circle, and therefore, if it were made to roll on Ta, the point b would trace an epicycloidal arc coinciding with ba, and the point b would coincide with a. But the arcs Ta, Tb are also those described by the two pitch circles respectively, in moving from T to the second position; and since these equal arcs are described in the same time, the angular velocity ratio of the two pieces is constant, and the same as if the motion had been produced by the rolling contact of the pitch circle*.

Otherwise, by the known property of the epicycloid, the normal to any point b passes through the point of contingence T of its describing circle and its base circle. But these latter circles are the two pitch circles of the combination; and since the normal of the curve ab at the point of the contact

* For the properties of Cycloidal Curves, vide Peacock's Examples, p. 186. Young's Nat. Philosophy, Vol. II. p. 555. De la Hire sur les Epicycloides, &c.

is thus shewn to pass through a constant point T of the line of centers, the angular velocity ratio of the circles will be constant and equal to the inverse ratio of their radii, by Art. 75.

77. *Second solution.*—A, B being, as before, the centers of motion, T the point of contingence of the pitch circles. Let abc be an arc of an epicycloid whose describing circle is TbB, of half the diameter of the pitch circle FTd. From the center B draw a radial line through the describing point b, meeting the circle in d; then will this line touch the epicycloid in b. Let motion be communicated by contact from the curved edge abc, which revolves round A, to the radial line Bbd, which revolves round B; and let the beginning of the motion be reckoned from the position in which a coincides with T, and, therefore, d with a. In moving to any other position of contact abc, Bbd; Ta, Td, will be the arcs simultaneously described by the two pitch circles. Now TBb is an angle at the circumference of the circle TbB, and TBd an angle at the center of the circle TdF; therefore Tb measures an angle double of Td. Also the radius of Tb is half that of Td; therefore the arc $Tb = Td$. Again, TbB is the describing circle of the epicycloid abc, and Ta its base; $\therefore$ $Tb = Ta$; whence $Td = Ta$, that is, the arcs of the pitch circles described from the beginning of the motion are equal, and consequently the angular velocity ratio constant, and the

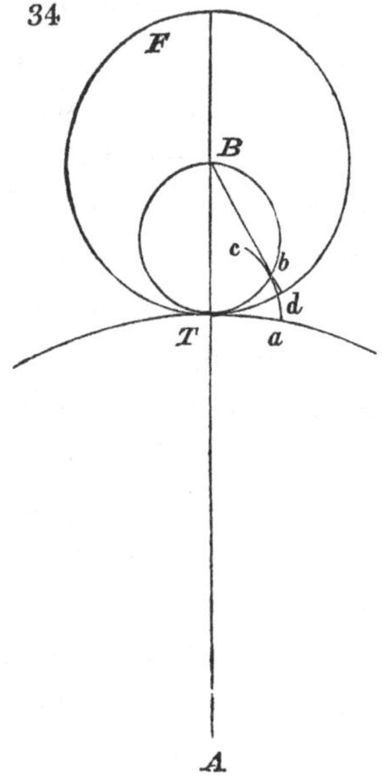

same as would be obtained by the rolling contact of the pitch circles.

Otherwise; as before, the normal of contact at *b* passes through the constant point *T* of the line of centers, and therefore divides it into a pair of constant segments; whence by Art. 75, the angular velocity ratio is constant.

Cor. The point of contact *b*, between the curve *ac* and the radial line *Bd*, is always situated in the circle *TbB*, described through *T*, with a diameter equal to the radius of the pitch circle of the radial line, and having its center upon the line of centers. This circle is therefore the *locus of contact.*

78. *Third solution.*—*A* and *B* being, as before, the centers of motion, *T* the point of contingence of the pitch circles. Let a describing circle *Tbk* be taken of any diameter, and with it describe an epicycloid *TC* by rolling on the outside of the pitch circle *Tm*, and an hypocycloid *TF* by rolling on the inside of the pitch circle *Tn*. Let these curves be cut out and made to revolve in contact, round their respective centers of motion *A* and *B*, until they come into a new position where *abc* is the epicycloid and *ebf* the hypocycloid. By the known properties of the curves they will have their common point *b* in the circumference of the describing circle *Tb*, when its center *O* is on the line of centers, and they will also have a common tangent there.

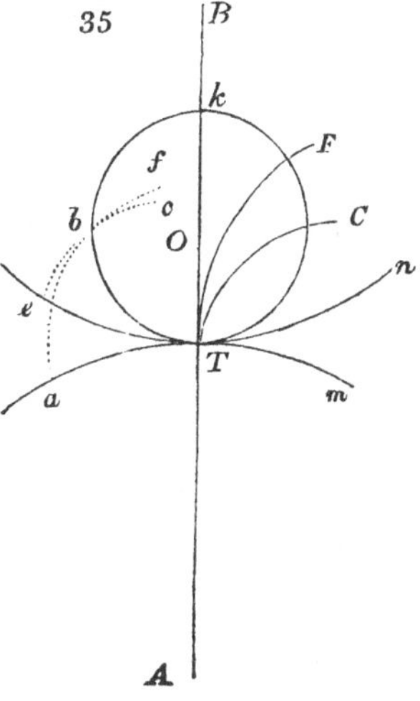

Also, if the describing circle *Tbk* were to roll upon *Te* from its present position, it would describe the curve *be* with the point *b*, and this point would come to *e*; therefore the arc *Tb* is equal to the arc *Te*, and similarly, the arc *Tb* is equal to the arc *Ta*; $\therefore$ $Te = Ta$. But these are the arcs respectively described by the two pitch circles in moving from the first position to the second; therefore, as before, the angular velocity ratio is constant and equal to that which would be obtained by the rolling contact of the pitch circles.

Otherwise; as before, the constancy of the angular velocity ratio may be shewn from the known property of the curves by which the normal from the point *b* passes through *T*.

This third solution includes the two former ones, for it is known that if the diameter of the describing circle of an hypocycloid be made equal to the radius of the base, the hypocycloid becomes a straight line coinciding with a diameter of the latter; and thus the second solution is obtained. Also, if the describing circle of the hypocycloid equal the circle of the base, the hypocycloid is reduced to a point in its circumference, and thus the first solution is obtained.

79. *Fourth solution*.—Let *A*, *B* be the centers of motion, *T* the point of contingence of the pitch circles. Through *T* draw *DTE* inclined at any angle to the line of centers, from *A* and *B* drop perpendiculars *AD*, *BE* upon *DTE*, and with radii *AD*, *BE* and centers *A* and *B* describe the circles to which *DE* will be a common tangent. Also we have $\frac{BE}{DA} = \frac{BT}{AT}$ by similar triangles *TAD*, *TBE*.

Through the point T describe an involute KTH of the circle DH, and an involute FTG of the circle FE. If these involutes be made to turn round the centers A and B respectively, and to remain in contact, the perimetral velocities of the pitch circles will be equal.

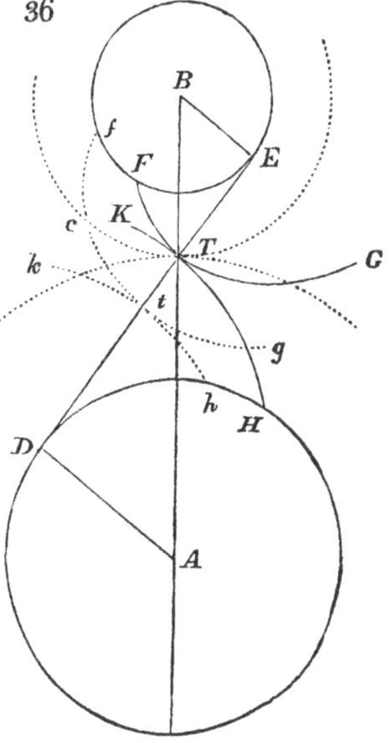

For, let kth, ftg be new positions of the involutes, the point of contact t will be always in the line DE, and Hh, Ff are the arcs respectively described by the base circles of the involutes. But

$$Hh = DH - Dh = DT - Dt = Et - ET = Ef - EF = Ff.$$

And since these arcs are equal, the perimetral velocities of the base circles are equal, and the angular velocity ratio constant.

But $AD : BE :: AT : BT$ by construction; that is, the radii of the bases are proportional to the radii of the pitch circles. Whence it follows that the perimetral velocities of the pitch circles are also equal, and the angular velocity ratio the same as that which would be obtained by making their circumferences act upon each other by rolling contact.

Otherwise; because the normal to any point of contact t of the involutes coincides with the common tangent of their bases, this normal is a fixed line, and passes through a fixed point T of the line of centers, which also shews, as before, the constancy of the angular velocity ratio.

80. If the distance of the centers A, B be altered, but so that the involutes may still remain in contact, then it can be shewn, in exactly the same manner, that the velocity of the circumferences of the bases will be equal; and, therefore, that the ratio of the angular motion of the two curves will remain unaltered. This is a property which distinguishes the involute from the other curves that have been given, and is of some practical importance; for when these curves are employed for the teeth of wheels, it is not only unnecessary to fix the centers of their wheels at a precise distance, but a derangement of the centers, from wearing or settlement in the frame-work, does not impair the action of the teeth. In every other pair of curves that have been assigned, a variation in the distance destroys the equal ratio of the motion, by destroying the principle of their connexion.

81. For every given pair of pitch circles an infinite number of pairs of involutes may be assigned, that will answer the conditions required; for the inclination of DTE to the line of centers is arbitrary, and every change of inclination produces a new pair of bases and of involutes.

82. *Fifth, or general solution.*—To return to the general principle (Art. 75). It appears that, from the properties of the curves in the cases already given, the normal to the point of contact passes through the constant point T of the line of centers, and that the problems already solved admit of demonstration upon that property alone. But if instead of employing a circle as a describing curve, other curves be employed, then a new set of forms applicable to our purpose will be obtained.

To shew this, we may employ the following Theorem* *It is always possible to find a curve, which by revolving*

* Airy on the Teeth of Wheels: Cam. Phil. Tr. Vol. II. p. 279.

upon a given curve, shall, by some describing point, in the manner of a trochoid, generate a second given curve, provided that the normals from all points of the second curve meet the first.

To prove this, let AB (fig. 37) be the first curve, AC the second, from the points C and E, which are very near, draw the normals CD, EF; if a describing point P be taken, and PQ, PR, be made respectively equal to CD, EF, and QR equal to DF, and this process be continued, a curve will be formed, which, by revolving upon BA, will, by the describing point P, generate the curve AC. For if Q coincide with D, then R will afterwards coincide with F; and so on for all succeeding points, since $QR = DF$. Also, $DC = QP$, &c. And the angles made by these with the tangents are equal, for the cosines of these angles, drawing DG, QS perpendicular to EF, PR are $\frac{FG}{FD}$ and $\frac{RS}{RQ}$, in which the numerators are the differences of equal lines, and the denominators are equal. Hence, P rolling on AB will describe AC. And the formation of the curve RQ is always possible, because RQ is greater than RS, for FD is necessarily greater than FG.

As an example of this, suppose it were required to find the curve, which, revolving on one straight line AB (fig. 38), would generate another straight line AP. Since the angles made by the line PQ with the tangent must be constant, it follows that the curve would be the logarithmic spiral, P being its pole.

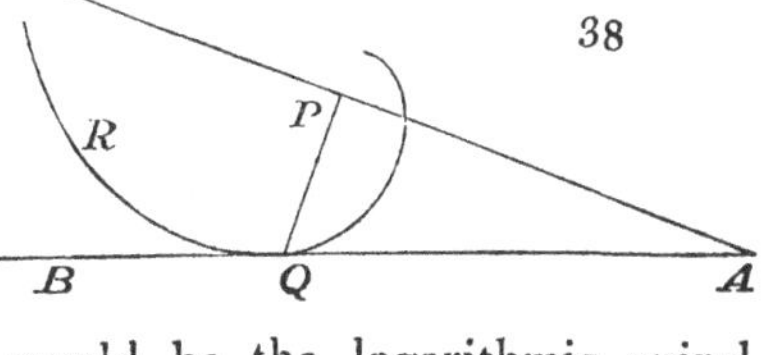

83. If the tooth HD (fig. 39) be generated by the revolution of any curve on the outside of the pitch circle HT, and if DK be generated by the revolution of the same curve in the same direction, in the inside of the pitch circle KT, then the normal at the point of contact of the teeth will pass through T. For, let the generating curve be brought into the position LT, so as to touch the circle HT at T, DT will be the normal of HD at D; and that the teeth may be in contact, the same generating curve in the other circle must touch KT at T, in which case it will coincide with this; D therefore will be in the surfaces of both of the teeth, and TD the normal of both at that point; therefore they will touch at D, and the line of action TD will pass through the fixed point T*; which being true in every position, the angular velocity ratio will be constant, and equal to that which would be obtained from the rolling contact of the pitch circles.

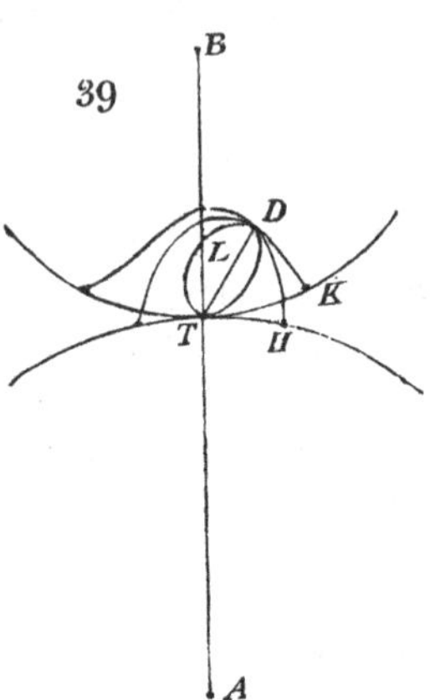

84. We are now able to solve the problem in its most general form. *Given, the form of the teeth of one wheel to find the form of those of another that they may work together correctly*†. Describe the pitch circles of the required wheels. Find the curve which, revolving upon the one, will describe the given tooth. Make the same curve

* Airy on the Teeth of Wheels: Cam. Phil. Tr. p. 280.

† The possibility of doing this was known to De la Hire, who gave an imperfect method for the purpose in the Traité sur les Epicycloides. The method in the text appears to have been first enunciated by Dr. Young, but without demonstration. (Nat. Ph. Vol. I. p. 176). The complete demonstration is due to Professor Airy, and I have given it nearly in his own words, in Articles 82, 83, and 84, and in the following note.

revolve within the other, and with the same describing point it will generate the tooth required.

That these forms may be applicable in practice, however, it is necessary that the curvature of the convexity of one tooth should be greater than that of the concavity of the other, or else that both should be convex*.

85. This problem† also admits of a simple practical solution, in the following manner.

Take a pair of boards *A* and *B*, whose edges are formed into arcs of the given pitch circles. Attach to one of them the shape of the proposed tooth *C*, and to the other a piece of drawing-paper *D*, the tooth being slightly raised above the surface of the board to allow the paper to pass under it. Keep the circular edges of the boards in contact, and make them roll together.

40

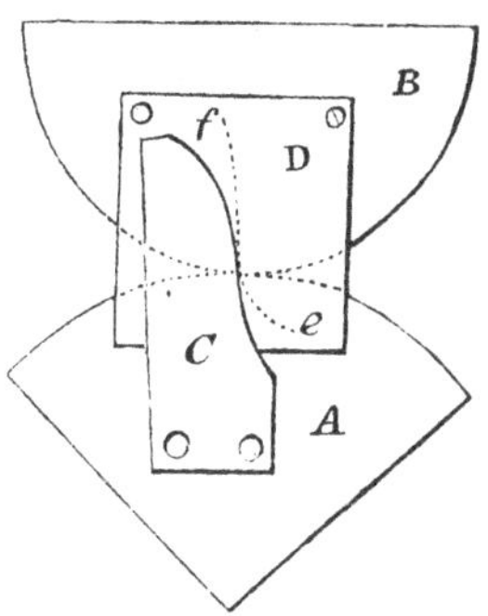

Draw upon *D*, in a sufficient number of successive positions, the outlines of the edge of *C*. A curve *ef*, which touches all these successive lines, will be the corresponding tooth required for *B*. For by the very mode in which it

* In the involutes, fig. 36, page 67, the separation of the circles of the bases would seem to exclude them from this general proposition. But, however, in the involute *ct* the normal *Et* is inclined at a constant angle to *BT*, and therefore to the tangent of the pitch circle at *T*, and the constructions just given shew that the involute *ct* may be generated by the revolution of a logarithmic spiral upon the pitch circle *cT*; the describing point being the pole of the spiral, and the angle between its radius and tangent the same as the angle made by *ET* with the tangent of the circle at *T*. In the same way, the revolution of this spiral within the second pitch circle *kT* will generate another involute *kt*, which will work correctly with the first.

The portions of the two involutes which lie respectively within and without the pitch circles, as *TG*, *TH*, being thus included in the general proposition, the remaining portions *TF*, *KT* can be in the same manner included in it.

† Transactions of Civil Engineers, Vol. II. p. 89.

has been obtained, it will, if cut out, touch *C* in every position; and therefore, the contact of these two curves *C* and *ef* will exactly replace the rolling action of the pitch circles.

Many forms of *C* tried in this manner, will prove untractable, for some of the successive portions of its edge may cover up and interfere with parts of the curve *ef* that have been previously drawn. In fact, although it be geometrically possible to assign a form *ef* to work with any given form *C*, it by no means follows that this is practically true, and indeed it does not appear that any new forms deduced from this general principle are likely to adapt themselves to practice, so well as those which are included under the four ordinary solutions. I shall therefore proceed to shew how these are to be adapted to the teeth of wheels.

ON THE TEETH OF WHEELS.

86. THE formation and arrangement of the teeth of wheels forms so important and interesting a branch of our subject, that I have thought it better to allot a separate Section of this Chapter to it. For the convenience of reference, it will be seen that I have distinguished, by number, the several solutions of the problem which requires curves to be found that will produce a constant velocity ratio when revolving together in sliding contact; and I shall now proceed to shew, in order, how these solutions are to be applied to the formation of the teeth of wheels.

To apply the first solution to the formation of the teeth of wheels.

87. This solution shews that an epicycloid traced on the pitch circle of the driver, by a describing circle equal to the pitch circle of the follower, will drive a pin in the circumference of the follower with the same motion as if the pitch circles rolled together. Let the pitch circles (fig. 41) be divided respectively into a number of equal parts, *ed*, *dg*, *gh*, &c.... *fa*, *ab*, *bc*, &c.... corresponding to the number of teeth proposed to be given to them; let fine pins be fixed into the follower at the points *e*, *d*, *g*, *h*, &c.... and let a series of epicycloidal arcs *fk*, *ka*, *al*, *lb*, &c.... be traced with a describing circle equal to the pitch circle of the follower, and through the points *f*, *a*, *b*, ... alternately to right and left, meeting at *k*, *l*.... If motion be given to the driver in the direction of the arrow, then the curved face *ak*, will press against the pin *d*, and move it in the same direc-

tion. But as the motion continues the pin *d* will slide up-

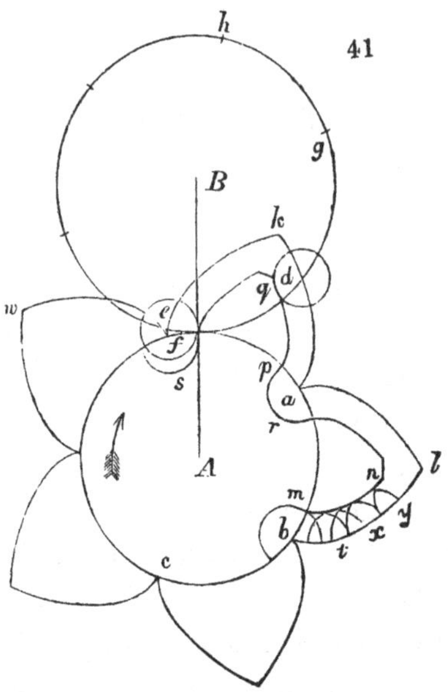

wards until it reaches *k*, when this tooth and pin will quit contact. Before this happens the pin *e* will have reached the point *f*, and the face *fw* of the next tooth will have commenced a similar action upon the pin *e*, which will in like manner be succeeded by the next pair; and so on continually.

88. But the demonstration supposes the pins to be mathematical points having no sensible diameter, which is practically impossible. Take therefore, a sufficient number of points *t*, *x*, *y*,... in the epicycloidal face of the tooth *bl*, and with a radius equal to that which the pin requires describe a series of small arcs, and draw a curve *mn* touching them all. Repeat this operation upon every tooth, so as to produce curves *sq*, *qp*, *rn*... respectively parallel to the original epicycloids. For example, let the curve *pq* be substituted for the epicycloid *ak*, and at the same time a pin of the

given radius be substituted for the point *d*. In every relative position of contact between this new pin and the curve *pq* the epicycloid *ak* will pass through its center *d*. For by the mode of its description the circle must touch the curve *pq*, when its center is in any point of the epicycloid. Therefore the tooth *w* derived from the epicycloid will drive a pin of any required diameter, exactly in the same manner as the original curve would have driven the mathematical point. A space *pr* must also be cut out within the pitch circle of the driver and between the bases of the teeth, to allow the pin to pass. But as the sides of this space never touch the pin, the form of it is immaterial, provided it be made sufficiently large to ensure that there shall be no accidental contact.

89. This solution is applicable to trundles or pin-wheels of all kinds (Art. 56). In the figure it appears, that while any given tooth *ka* is in contact with, and drives a pin *d*, the back *kf* of this tooth will be in contact with the succeeding pin *e*; and consequently, if the motion of the driver were reversed, the back of the tooth would begin to drive the tooth *e* without any shake taking place, and the wheels would work as well in one direction as the other. This perfection is unattainable in practice, as the smallest error in excess of the figure, or position of the tooth, or pin, would cause the teeth to wedge themselves fast between the two contiguous pins. It is necessary to allow a small space for play between the teeth and pins, and this play is termed *backlash*. The same principle and phrase applies to all forms of teeth which are capable of being so arranged as to work in both directions.

90. When the pin is reduced to a mathematical point, the contact of any tooth *ak* begins at the moment its base *a*

has reached the line of centers; and during the action of the tooth the point of contact gradually slides upwards, remaining always in the pitch circle of the pin-wheel, and at the same time it recedes from the line of centers until the contact is finally terminated at the point of the tooth k; the action being wholly confined to the recess from the line of centers. But if, on the other hand, the *pin-wheel* were made to *drive the teeth*, the reverse would happen; the contact would begin at the top of the teeth, and end at their base, and the action would be confined to the approach to the line of centers.

Now, in practice, the friction which takes place between surfaces whose points of contact are approaching the line of centers, is found to be of a much more vibratory and injurious character than that which happens while the points of contact are receding from it. It is therefore necessary to avoid the first kind of contact as much as possible, and for this reason the teeth are always given to the drivers, and the pins to the followers, in this kind of wheel-work. For the most part, the diameter of the pin is made equal to that of the tooth, with an allowance for play equal to one tenth of the pitch. The radius of the pin will be, therefore, rather less than a quarter of the pitch. When the stave has a sensible diameter, the first contact will take place, as before, when the center of the stave reaches the line of centers, and therefore at a distance before that line equal to the radius of the stave, or rather less than a quarter of the pitch.

But, plainly, one tooth must not quit contact before the succeeding tooth is engaged; therefore, when the point f has reached the line of centers, the tooth pq must not have quitted contact with the pin d; and the point q, when contact ceases, must therefore be at an angular distance from the line of centers, equal at least to half the distance *fa*, or half the pitch; so that in a pin-wheel the action that takes place before coming to the line of centers, is less than half that which must take place after passing it.

91. *A rack* may be considered as a wheel, the radius of whose pitch line is infinite (Art. 59); and on this hypothesis the form of its teeth may be derived from those of spur-wheels with finite radii, by very simple considerations.

The rack may drive or follow; in the first case the pins will be given to the wheel, and in the second case to the rack.

Now if the rack drive, the line *Ta*, fig. 33, (which is an arc of the pitch circle of the driver) will become a right line perpendicular to the line of centers, and *abc* will become a cycloid.

The teeth of the rack, fig. 42, must be derived from the

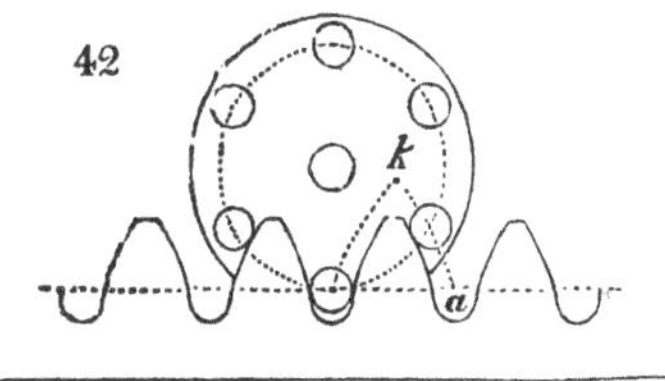

cycloid *ka*, by the method already explained, of tracing a parallel curve at a distance from it equal to the radius of the pin.

If, however, the rack be driven, as in fig. 43, then the arc *Tb*, fig. 33, will become a right line, and *abc* will become the involute of the pitch circle of the driver *Ta*. From which involute a parallel curve might be obtained, as before, for the teeth of the pinion; but this is unnecessary, inasmuch as this process would merely reproduce the same involute in a different position.

43

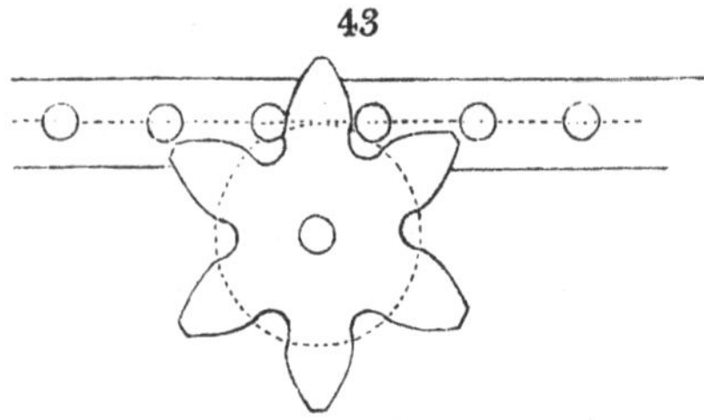

It follows, that to describe the teeth of a wheel which is to drive a pin rack, involutes of its pitch circle must be traced to right and left alternately, and at a distance from each other rather greater than the diameter of the pins.

92. In a similar way *an annular wheel* may either *drive or follow*.

If it *drive*, the pitch circle *Ta*, fig. 33, will become concave; and if the radius of the pins be small, the sides of the teeth will be hypocycloids, as at *pq*, fig. 44, traced by

44

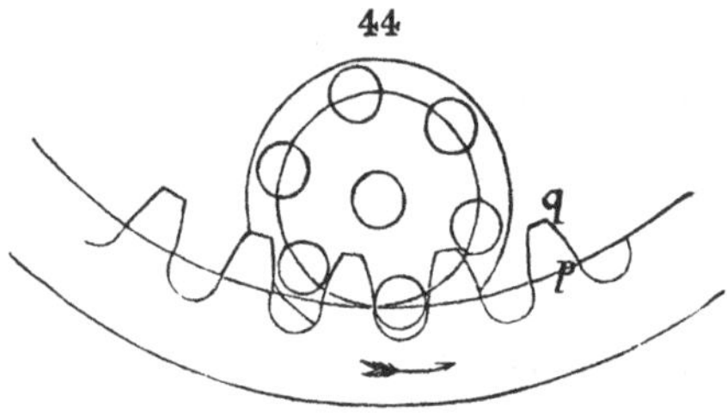

the rolling of the pitch circle of the follower within the pitch circle of the driver; or, as before, if the radius of the pins be considerable, then the sides of the teeth will be drawn parallel to the hypocycloids at a normal distance equal to the radius of the pins.

If the annular wheel *follow*, it will carry the pins, and the teeth of the driver will be traced by rolling the *inside* of the annular pitch circle upon the *outside* of that of the

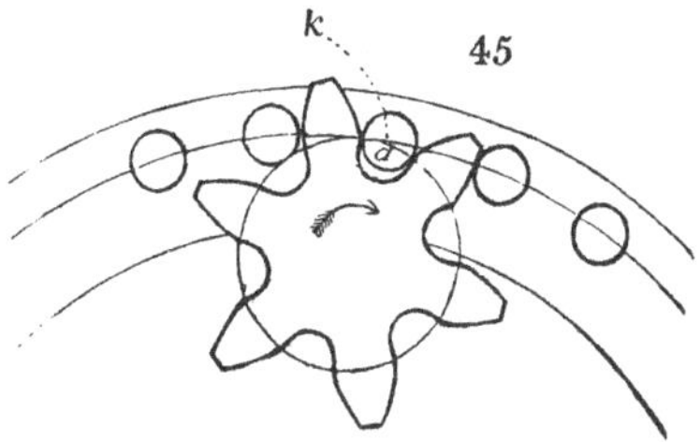

driver, making, as before, the true edge of the teeth equidistant from the epicycloid so obtained, ka, fig. 45, by a distance equal to the radius of the pin.

93. *To find the smallest number of teeth or pins that can be employed, when the pins have no sensible diameter.*

Let T, d, be two successive pins in a pin-wheel, Tda the tooth of the driver, and let the pin d coincide with the point of the tooth Tda, at the moment the next pin T arrives at the line of centers; then one tooth ceases its action at the moment the next tooth begins.

46

Let $AT = R$, $BT = r$, $BAd = \theta$,

$ABd = \phi$.

Now, from the nature of the curve ad, Ta which is equal to the pitch must be equal to $Td = r\phi$; and the angle BAd includes in the position of the figure half a tooth or half the pitch; $\therefore 2R\theta = r\phi$.

If the pin d had not quite reached the extremity of the tooth, when T arrived at the line of centers, TAd would have been less than half the pitch angle; but the action of the wheels would not be interrupted, but rather improved; whereas, on the contrary, were TAd greater than half the pitch angle, one tooth would quit its pin before the next could begin contact; therefore, we may have TAd equal to, or less than, half the pitch angle, but not greater;

$$\text{or } 2R\theta \left\{ \begin{matrix} = \\ < \end{matrix} \right\} r\phi.$$

$$\text{Now } \frac{Bd}{AB} = \frac{\sin BAd}{\sin AdB};$$

$$\text{that is, } \frac{r}{R+r} = \frac{\sin \theta}{\sin(\phi+\theta)} = \frac{\sin \theta}{\sin\left(1+\frac{2R}{r}\right)\theta};$$

in which equation, substituting different values of the ratio $\frac{R}{r}$, it will appear whether the value of θ is sufficiently small to answer the conditions; for example, let $R = r$;

$$\therefore \frac{1}{2} = \frac{\sin \theta}{\sin 3\theta}, \text{ or } 2 \sin \theta = \sin 3\theta = 3 \sin \theta - 4 \sin^3\theta;$$

$$\therefore \sin \theta = \frac{1}{2}, \text{ and } \theta = 30^0;$$

by which it appears that six teeth and six pins will exactly fulfil the conditions, and that the pin will exactly reach the extremity of its tooth when the next pin comes into action. Also any number greater than six may be employed, but with less than six the action will be interrupted.

If $r = 2R$, $\cos \theta = \frac{3}{4}$, and $\theta = 41^0.36$; $\therefore 2\theta = 83^0.12'$; which corresponds to four teeth and a fraction; the smallest whole numbers are five teeth to drive ten pins.

94. In this manner the following set of results were obtained.

A pinion of four pins may be driven by a wheel of any number of teeth greater than about sixteen, but a pinion of three pins cannot be driven even by a rack, that is, by a wheel of an infinite number of teeth.

Five pins may be driven by any number of teeth greater than about ten.

Six is the least number that admits of being employed in the case of the number of teeth and pins being equal.

Five teeth will drive a pin-wheel of any number from eight upwards, and four teeth require at least twelve pins; but three teeth will just drive a pin-rack, and consequently will not work with a wheel.

It must be recollected, that in this class of wheel-work the pins are always given to the follower.

95. In the last Article the pin was supposed to be a mathematical point; but as this is impracticable, let us examine the question, supposing the pin to have a sensible radius.

It has been shewn (Art. 87) that the form of tooth for such a stave is derived from the epicycloid *ak* (fig. 47), that

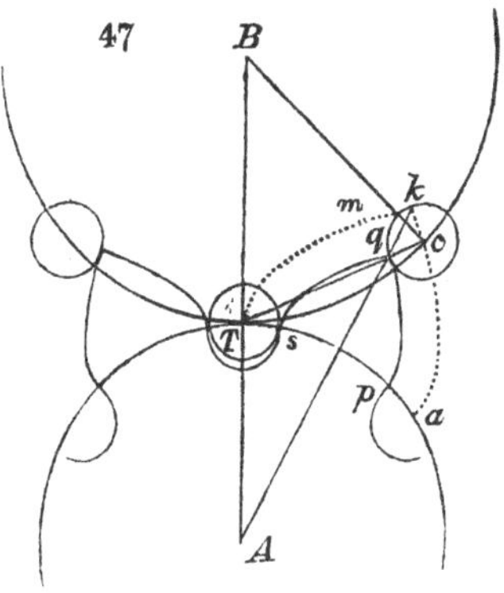

would serve when the stave is reduced to a point; by tracing a curve pq at a normal distance from it, equal to the radius cq of the stave. Let pqs be such a tooth, then, if it be quitting contact at the moment the next stave and tooth are coming into action, the center of this next stave T must coincide with the line of centers; and as the line Tc, which joins the center of the pin c with the tangent point T of the pitch circles, is the normal to the epicycloid ak, it necessarily passes through the point of contingence of the curve pq and the stave: this point q will also be the extremity of the tooth.

Let TBc (the pitch angle of the pin-wheel) $= \phi$,
and BAq (half the pitch angle of the toothed wheel) $= \theta$;
let $AT = R$, $BT = r$, and cq, the radius of the pin, $= \rho$;

$$\therefore r\phi = 2R\theta \quad (1).$$

* But $$cm = cq \cdot \frac{\sin cqm}{\sin cmq} = \rho \cdot \frac{\cos\left(\frac{\phi}{2} + \theta\right)}{\sin(\phi + \theta)}.$$

Also, $$\frac{\sin BAm}{\sin AmB} = \frac{Bm}{AB},$$

that is, $$\frac{\sin\theta}{\sin(\phi+\theta)} = \frac{r - (cm)}{R + r} = \frac{r - \rho \cdot \frac{\cos\left(\frac{\phi}{2} + \theta\right)}{\sin(\phi+\theta)}}{R + r}.$$

From this equation θ may be eliminated by (1.)

Let k be the ratio of the diameter of the pin to the pitch, which is the most convenient term in which to express the result;

$$\therefore k = \frac{2\rho}{2R\theta}; \quad \therefore \rho = kR\theta.$$

Substituting this value of ρ, and arranging the terms, we finally obtain

* In fig. 47, m should be at the intersection of Bc and Ak.

$$k = \frac{\sin\left(\frac{2R+r}{2R}\cdot\phi\right) - \frac{R+r}{r}\cdot\sin\left(\frac{r\phi}{2R}\right)}{\frac{\phi}{2}\cdot\cos\left(\frac{R+r}{2R}\cdot\phi\right)}.$$

From this equation, by substituting in each particular case the value of ϕ, and of $\frac{R}{r}$, the necessary diameter of k will be obtained; which will cause one tooth to quit contact at the instant the other begins. Should k come out negative, the case is thus shewn to be impossible; and if zero, then it corresponds to the arrangement in which the pin is a mathematical point. In practice it would not answer to arrange teeth so that one pair should quit contact at the instant the next pair begins it, because the least wearing or inaccuracy would cause an interruption in the action. It is necessary, therefore, to allow more teeth than our Tables will shew, or to make the stave of less diameter and the tooth of greater.

I have not thought it necessary to give the diagram for the case of annular wheels, but I have inserted the results which apply to them in the table. They may be obtained from the formulæ, by considering that R and r lie on the same side, instead of opposite, and therefore R and r must have opposite signs; also the angle AmB will, for the same reason, be taken equal to the difference, instead of the sum of θ and ϕ.

96. The following Tables shew that diameter of the stave or pin in parts of the pitch which allows one pair of teeth and pins to quit contact at the instant the next pair begin it.

The impossible cases are marked –, but when the character + is inserted, the necessary diameter of the stave is greater than half the pitch, and consequently all such cases may be employed in practice.

TABLE I.

Pinion drives, and Staves are given to the Wheel.								
	Value of $\frac{r}{R}$.	Diameter of Stave. Number of Teeth in the Pinion.						
		2	3	4	5	6	7	8
Annular Wheel.	3	.63	+	+	+	+	+	+
	4	.28	+	+	+	+	+	+
	8	–	.64	+	+	+	+	+
	Rack	–	.34	.73	+	+	+	+
Spur-Wheel.	8	–	–	.58	+	+	+	+
	6	–	–	.51	+	+	+	+
	5	–	–	.46	+	+	+	+
	4	–	–	.37	+	+	+	+
	3	–	–	.18	.59	+	+	+
	2	–	–	–	.37	.63	.75	+
	1	–	–	–	–	0	.38	.57

TABLE II.

Wheel drives, and Staves are given to the Pinion.								
	Value of $\frac{R}{r}$.	Diameter of Stave. Number of Staves in the Pinion.						
		2	3	4	5	6	7	8
Spur-Wheel.	1	–	–	–	–	0	.38	.57
	2	–	–	–	.20	.51	.66	+
	3	–	–	–	.39	+	+	+
	4	–	–	.01	.46	+	+	+
	5	–	–	.10	.50	+	+	+
	6	–	–	.16	+	+	+	+
	8	–	–	.22	+	+	+	+
	10	–	–	.26	+	+	+	+
	Rack	–	–	.38	+	+	+	+
Annular Wheel.	8	–	.01	.49	+	+	+	+
	6	–	.10	+	+	+	+	+
	4	–	.23	+	+	+	+	+

97. *Example.* A wheel is required to drive a pinion of one fourth of its diameter; to find the least number of teeth and pins that can be employed.

This example belongs to the second table; and in the line appropriated to $\frac{R}{r} = 4$ it appears that if four staves be given to the pinion, and consequently sixteen teeth to the wheel, the diameter of the stave is reduced to the hundredth part of the pitch; but that if the numbers 5 and 20 be employed, the pin may be made nearly half the pitch. In practice it would not be safe, therefore, to employ less numbers than 6, 24, or 7, 28.

To apply the second solution to the formation of the teeth of wheels.

98. The forms of teeth derived from this solution are the most generally employed at present, they having been found the best adapted for metal wheels, whereas those which have been derived from the first solution belong rather to the ancient practice of wooden mill-work, although they may still be occasionally employed in metal work, as pin-wheels.

Fig. 48 represents a pair of wheels whose teeth are derived from the second solution. A and B are their centers of motion, T the point of contingence of the pitch circles; and as the forms of the teeth in each wheel are obtained from the same principles, either wheel will act as driver or follower. The complete side of each tooth, as cTa, or hTg, is made up of two parts, one of which lies within the pitch circle, and the other without; the portion aT or Tg that lies without the pitch circle is technically termed the *face* of the tooth, and that which lies within as Th or Tc is termed its *flank*, which terms I shall employ.

With respect to the portions *Tc*, *Tg* of the pair of teeth *gTh*, *cTa*, *Tc* is a radial line to *A*, and *Tg* an arc

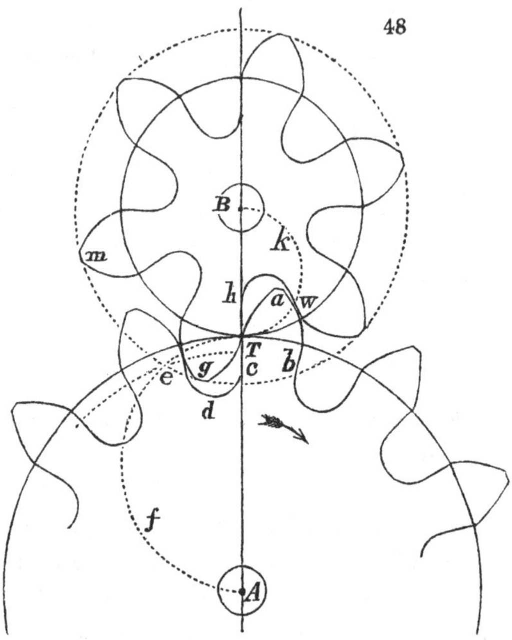

of an epicycloid whose describing circle is *Tefa*, equal in diameter to the radius *TA* of the lower pitch circle. On the other hand, *Th* is a radial line to *B*, and *Ta* an arc of an epicycloid whose describing circle is *TkB*, equal in diameter to the radius *TB* of the upper pitch circle; that is, the flanks or portions of teeth in both wheels that lie within their respective pitch circles are radial lines, and the faces, or those that lie without, are arcs of epicycloids traced in each wheel with a describing circle equal in diameter to the pitch radius of the other wheel. By the second solution, therefore, each flank and face will act in contact to produce a constant angular velocity ratio, but the action of each pair will be confined to its own side of the line of centers.

As the two sides of each tooth are precisely alike, and symmetrical to a line joining the centers of the wheel and point of the tooth, the wheels will turn each other in either direction at pleasure. The form of the curved line *cde* which connects each tooth with the next is indifferent, provided it afford sufficient room for the point of the opposite tooth; for it manifestly never comes into contact action, since that is entirely confined to the portions of the tooth before described. The curved part *cde* is termed the *clearing*.

99. To examine the action of the teeth, let the lower wheel of the figure be the driver, and let it revolve in the direction of the arrow; therefore the right sides of its teeth will press the left sides of the follower's teeth. Now, the locus of contact is the semicircle *feT* during the approach to the line of centers, and the semicircle *TkB* during the recess. The contact, therefore, of every pair of teeth begins at the root of the driver's tooth, that is, at that point of the flank which is nearest the center, and proceeds gradually outwards till it ceases at the point of the tooth. But in the follower the contrary action takes place. The contact begins at the point of its teeth, and ends at their root. This is evident, since the path of the point of contact is the sinuous line *eTk*.

Also, in every pair of teeth the extent of face that comes into contact action is much greater than the extent of flank with which it works. For, let *Tg* be a given length of the curve of a tooth in the upper wheel, then, to find the required length of flank in the lower wheel, describe with radius *Bg* an arc of a circle *gm*, intersecting the locus of contact *Tef* in *e*; therefore *e* will be the radial distance of the first point of contact of the flank with *g*, and

$AT - Ae$ the length of flank through which the action is continued; which is manifestly less than the face Tg.

100. *To find the smallest number of teeth that can be employed when the teeth of the driver are epicycloids whose describing circle is half the pitch circle of the follower, and the teeth of the follower radial lines having no sensible thickness.*

Radial teeth of this kind might be formed by inserting thin plates of metal edgewise into the surface of a block, in the same way that pins are when employed for teeth; and this arrangement falls under the second solution, as well as the last, although the form of the teeth appears different.

In fig. 49, B is the center of the follower, A of the driver, Tda one of the teeth of the latter, and Bdm* the radial tooth of the follower, with which the face ad has been in contact during its motion from T to a.

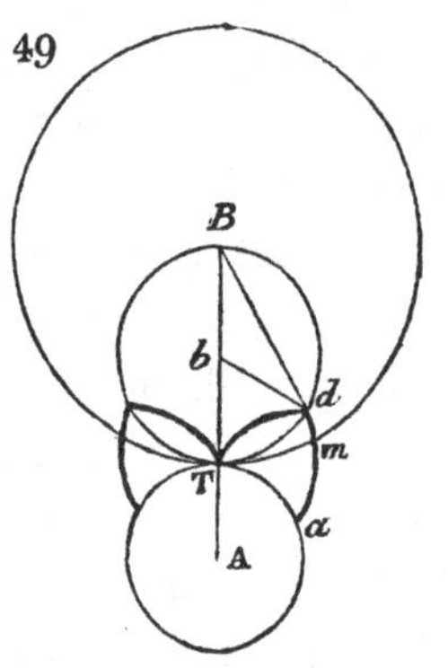

The semicircle TdB described upon the radius TB is the locus of contact; let the apex d of the tooth ad be quitting contact at the same moment that the succeeding tooth begins it; therefore d will lie in the semicircle Tdb, and the base of the succeeding tooth coincide with T.

Join bd, then comparing this figure with fig. 46, Art. 93, it will appear that in fig. 49, if b were the center of a pin-wheel, and d the pin acting with the tooth ad, Tbd would be the pitch angle that would cause the tooth ad to quit contact with the pin at the moment the next began it; but

* The line from d to m is obliterated in the wood-cut, but can easily be supplied, since it is the mere prolongation of Bd.

TBd is the similar pitch angle in the case of radial teeth, and $TBd = \frac{1}{2} Tbd$.

The least number of radii, therefore, that will work with a given number of epicycloidal teeth is equal to twice the least number of pins.

The results obtained upon this principle, from the formula of Art. 93, are as follows.

A pinion of 7 radii may be driven by a wheel of 56 teeth and upwards.
.... 8 radii .... 16 teeth.
.... 9 radii .... 12 teeth.
.... 10 is the least number when equal numbers of teeth and radii are employed.
.... 9 teeth will drive a wheel of 10 radii & upwards.
.... 8 teeth .... 11 ....
.... 7 teeth .... 12 ....
.... 6 teeth .... 12 ....
.... 5 teeth .... 16 ....
.... 4 teeth .... 24 ....
.... 3 teeth will drive a rack whose teeth are straight, and have no sensible thickness.

101. Although it appears from these tables that a pinion of three teeth will but just drive a rack, and that four is the least that can be employed to drive a wheel, supposing the radii to be very narrow, yet two teeth may be made to answer this purpose very practically by fixing them in two planes, as in fig. 50.

B represents a disk to which teeth $c, c, c, \dots$ $d, d, \dots$ are fixed alternately on one side and on the other, the sides or rather flanks of these teeth are straight, and radiate in direction from the center of B; and the extreme diameter of B measured from the opposite extremities of the teeth is equal to that of its pitch circle. The driver is formed of a pair of double epicycloids, of which *A* is in the plane of the upper teeth *c*, *c*, *c*, ... and *a* in the plane of the lower teeth *d*, *d*.... The describing circle of these epicycloids is of course equal to half the pitch circle of the follower. The action of this combination is very smooth.

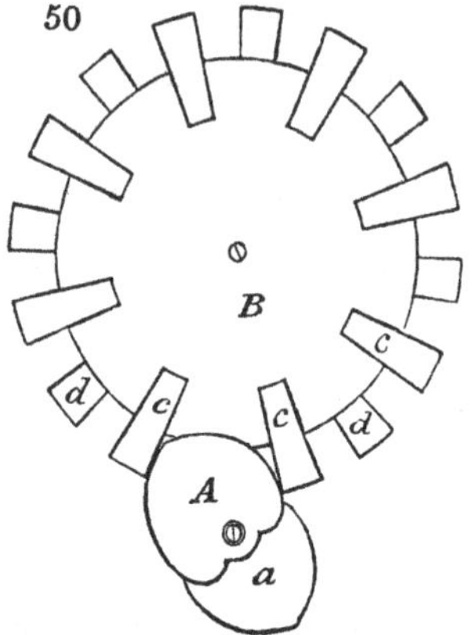

A pinion of one tooth communicating a constant angular velocity ratio between parallel axes appears absolutely impossible.

The endless screw is equivalent, however, as we shall see, to a single tooth.

102. *To shew the geometrical conditions that limit the employment of low-numbered pinions, when the teeth are formed in the usual manner, as in fig.* 48.

The usual general construction and letters being made, fig. 51. Let *TBd* be the angle through which it is desired that the contact of the tooth *ad* should continue after passing the line of centers. Therefore, as the contact is now ended, the point of contact will be at the extremity *d* of the tooth. Join *Td*, which will be perpendicular to the radius *Bdm*. Join *Ad*. Then, since *a* was in contact with *m* at the line of centers, the arc $Ta = Tm$, and is given, being that proportion of the pitch through which the contact of the teeth is required to continue. Also *af* is half the tooth, if the tooth be pointed, or else, if it be blunted by a certain quantity, then *af* is half the tooth diminished by that quantity; and in either case is given. Now *ka* is equal to the pitch, and must contain one tooth, and the space between; and since

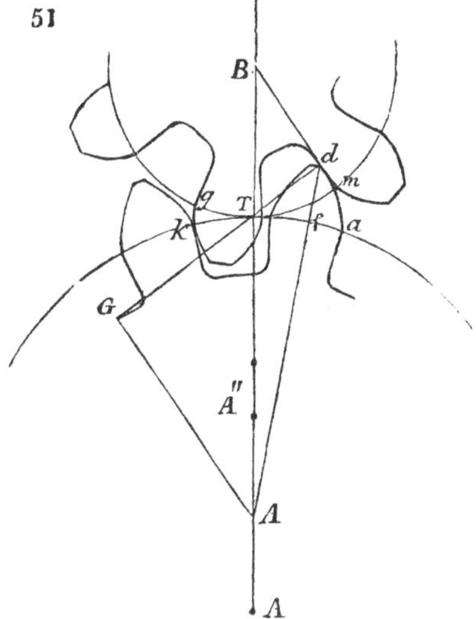

af cannot be greater than half a tooth, and may be less, therefore kf must contain at least half a tooth and a space, always supposing the tooth and space to be equal. Now for every given wheel BTm, and value of TBd, a value of TA may be assigned that will make kf exactly equal to a space and a half tooth, and in that case the tooth will be pointed.

If a greater value $TA_{,}$ be taken, the point f will fall nearer to a, and af will become less than half a tooth; so that the tooth may be blunted: but if a less value $TA_{,,}$ be taken, then the point f will fall nearer to T, and kf will become too small to contain the space and remaining half tooth. If the teeth of the wheel radius $TA_{,,}$ were set out, it would be found that the epicycloidal arcs on the two sides of df would intersect between d and f, and thus make the tooth too short to continue its action through the required arc Ta.

Let N and n be the numbers of teeth in a pair of wheels whose teeth are of the kind described, and whose action after passing the line of centers is given; it appears then that for every value of N a value of n may be assigned, a less number than which will make the action of the teeth impossible; and it is of some practical importance to determine these limiting values of n in every case, that we may avoid setting out impossible pairs of numbers in wheel-work.

103. A formula may be investigated thus: produce dT towards G, and from A draw AG perpendicular to and meeting it in G;

$$\text{then } \frac{\tan GAd}{\tan GAT} = \frac{Gd}{GT} = \frac{AB}{AT};$$

$$\text{or } \frac{\tan(TBd + TAd)}{\tan TBd} = \frac{AT + BT}{AT}$$

Now the angle TBd and the radius BT are given by the conditions, and also the arc Ta, which is the supposed arc of action; whence Tf is known;

$$\text{also } TAd = \frac{Tf}{AT}.$$

But if we attempt to extract the value of AT from the above expression, it will be found to be so involved as to make a direct solution of the equation impossible, although approximations may be obtained.

However, on account of the practical importance of the question, I have arranged in the following Tables the exact required results, which I derived organically from the diagram of fig. 51, by constructing it on a large scale with moveable rulers.

TABLE I. FOR SPUR-WHEELS.

TABLE of the *least* numbers of teeth that will work with given pinions. (Tooth = Space).

	Number of Teeth in given Pinion.	Least Number of Teeth in Wheel,	
		if Wheel drives.	if Pinion drives.
Arc of action, Ta = pitch.	5	impossible	impossible
	6	....	176
	7	....	52
	8	....	35
	9	....	27
	10	rack	23
	11	54	21
	12	30	19
	13	24	18
	14	20	17
	15	17	16
	16	15	...
Arc of action, $Ta = \frac{3}{4}$ pitch.	3	impossible	impossible
	4	....	35
	5	....	19
	6	....	14
	7	31	12
	8	16	10
	9	12	10
	10	10	10
Arc of action, $Ta = \frac{2}{3}$ pitch.	2	impossible	impossible
	3	....	36
	4	....	15
	5	....	13
	6	20	10
	7	11	9
	8	8	8

TABLE II. FOR ANNULAR WHEELS.

TABLE of the *greatest* numbers of teeth that will work with given pinions. (Tooth = Space).

	Number of Teeth in given Pinion.	Greatest Number of Teeth in Annular Wheel, if Wheel drives.	Greatest Number of Teeth in Annular Wheel, if Pinion drives.
Arc of action, Ta = pitch.	2	impossible	5
	3	....	12
	4	....	26
	5	....	85
	7	14	any number
	8	25	....
	9	60	....
Arc of action, $Ta = \frac{3}{4}$ pitch.	2	impossible	10
	3	....	77
	4	5	any number
	5	12	....
	6	77	....
Arc of action, $Ta = \frac{2}{3}$ pitch.	2	impossible	14
	4	8	any number
	5	64	....

N. B. The case of annular wheels differs from that of spur-wheels in this respect, that, with a given pinion a small-numbered wheel works with a greater angle of action than a large-numbered one, and therefore we have to assign the *greatest* number that will work with each given pinion. This will easily appear if a similar diagram to 51 be constructed for the case of annular wheels.

104. In these Tables I have supposed the tooth of the wheel to equal the space throughout, and have given the

whole of the limiting cases, and under three suppositions: first, that the arc of action Ta shall be equal to the pitch, in which case, if required, the teeth of the follower may be cut down to the pitch circle, and the contact of the teeth thus confined to their recess from the line of centers; for since the action of each pair of teeth continues through a space equal to the pitch, it is clear that at the moment one pair quits contact the next will begin. However, as some allowance must be made for errors of workmanship, it is better to allow the teeth to act a little before they come to the line of centers; or else, by selecting numbers removed from the limiting cases in the Table, to enable the teeth to continue in action through a greater space than one pitch. This principle will be examined more at length presently.

The limiting numbers under two other suppositions are inserted in the Tables, namely, that the arc of action Ta, shall equal $\frac{3}{4}$ and $\frac{2}{3}$ of the pitch, and when these are employed it is of course necessary that an arc of action, at least equal to $\frac{1}{4}$ and $\frac{1}{3}$ of the pitch respectively, shall take place between the teeth before they reach the line of centers.

It appears that a smaller pinion may be employed to drive than to follow. Thus, when the action begins at the line of centers the least wheel that can drive a pinion of eleven is 54, but the same pinion can drive a wheel of 21 and upwards; again, nothing less than a rack can drive a pinion of ten, but this pinion can drive a wheel of 23, and upwards. No pinion of less than ten leaves can be driven, but pinions as low as six may be employed to drive any number above those in the Table. And, lastly, the least pair of equal pinions that will work together is sixteen. These limits being geometrically exact, it is better in practice to allow more teeth than the Table assigns.

105. Other problems of the same nature as those already given might be suggested; as for example, to find the least numbers that can be employed when, without considering the relative action before and after the line of centers, the teeth are supposed to be drawn, as in fig. 48, with entire points both in the driver and follower, and the tooth equal the space; on which suppositions it would be found that the least possible number of teeth in a pair of equal wheels is *five*, that *four* will just work with *six*, and *three* with about *twelve*, and that *two* will not even work with a rack.

106. *To adapt the second solution to racks.*—If we suppose the lower pitch circle of fig. 48 to become a right line, we shall obtain a rack, and the epicycloidal faces *ab* of the rack teeth will become *cycloids*, because their describing circle *BkT* now rolls upon a right line, but the radial flanks *hT* of the pinion will remain unaltered. On the other hand, when the radius *TA* is thus increased to an infinite magnitude the describing circle *Tfa* coincides with the pitch circle whose center is *A*, and they unite in one straight line, tangent to the upper pitch circle at *T*; which line is, as already stated, the pitch line of the rack. But the curved faces *Tg*... of the upper pitch circle being thus described by the rolling of a tangent upon its circumference, are *involutes* of the circle, and the straight flanks *Tc* of the rack-teeth become parallel to each other and perpendicular to its pitch line.

Also, because *Tf* the locus of contact now coincides with the pitch line of the rack, therefore the action of the faces of the wheel-teeth is confined to that single point of each rack-tooth which lies upon the pitch line.

Fig. 52 represents a pinion and rack constructed upon the above principles, from which it appears, that, supposing the rack to be the driver, and to move in the direction of the arrow, the locus of contact will be the right line aT

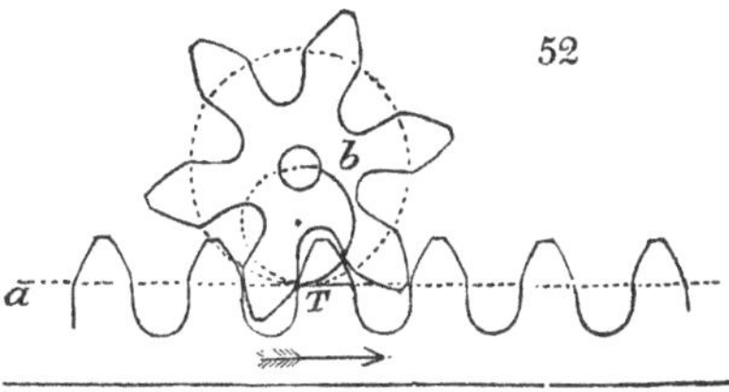

during the approach to the line of centers, and the semicircle Tb during the recess from that line. If the pinion drive, then the contact will take place in the semicircle on approaching the line of centers, and in the pitch line on receding from it. But as there is a great disadvantage in confining the action and consequent abrasion to a single point of the teeth, I am inclined to think that this method of forming rack-teeth, although most universally adopted, is bad, and that the forms derived from the succeeding solutions will be found to wear better. Nevertheless, this injurious action may be abridged or destroyed by cutting the teeth of the pinion shorter, or reducing it to the diameter of the pitch circle; but then if the pinion drive, as it generally does, we fall into the other difficulty of confining its action entirely to the approach to the line of centers.

To find the length of the teeth of wheels formed according to the second solution.

107. The length of the tooth will in all cases appear from the setting out, according to the rules already laid down; but it is more convenient to have some general principles for this purpose. It has been already stated, that the *true diameter* or radius of a wheel is that which is measured from the extremities of the teeth, in opposition to the *geometrical diameter*, or diameter of *the pitch circle*. Let R be the radius of the pitch circle, and E the projection of the tooth beyond it, and U the true radius; therefore $U = R + E$. Now this

addition E to the radius of the pitch circle is called by clockmakers the *addendum*, which term I shall, for convenience, employ. Let r, u, e be the geometrical radius, true radius, and addendum of a wheel, working with one of which the same quantities are respectively indicated by R, U, E;

$$\therefore \quad \frac{U}{u} = \frac{R+E}{r+e}.$$

As it is convenient to express the addendum in terms of the pitch $\left(= \frac{2\pi R}{N}\right)$,

$$\text{let } E = K \cdot \frac{2\pi R}{N}, \quad \text{and } e = k \cdot \frac{2\pi r}{n},$$

$$\text{and as } \frac{R}{r} = \frac{N}{n}, \text{ we obtain}$$

$$\frac{U}{u} = \frac{N + 2\pi K}{n + 2\pi k}.$$

The practice of millwrights is to employ a constant addendum of $\frac{3}{10} \times$ pitch, whether the wheel be a driver or follower; putting, therefore, $K = k = .3$, we have

$$\frac{U}{u} = \frac{N + 1.885}{n + 1.885} = \frac{N+2}{n+2}, \text{ nearly};$$

that is to say, to find the ratio of the true diameters of a pair of wheels of a given number of teeth, add two to each term of the ratio of the numbers. When the pitch is expressed according to the method described in Art. 74, where the pitch diameter of the wheel is laid down from a scale whose unit is a tooth, the true diameter is at once given by adding two teeth to the number.

Watchmakers assign a different value to the addendum, according as the wheel in question is a driver or follower.

Various proportions are assigned by different writers. Our latest and best English work* on the subject gives the rule

$$\frac{U}{u} = \frac{N + 2.25}{n + 1.5};$$

where U is the true radius of the driver, and u of the follower, and K, k are equal to .36 and .24, or $\frac{3}{8}$ and $\frac{1}{4}$ nearly of the pitch. I shall proceed to investigate a principle for these rules, but will first state the entire general proportions which are at present usually given to the teeth of mill-work, and which may be considered to have arisen almost entirely from practice.

108. In fig. 53 is represented a portion of the circumference of a pair of mill-wheels in geer, whose pitch lines are

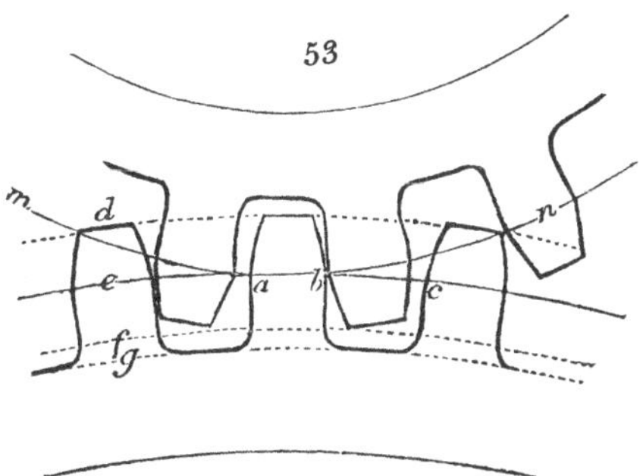

man, and *eac*; the forms of the teeth are those generally adopted in practice, and the rules for proportioning them are stated in fractions of the pitch, thus:

$$de = \text{Depth to pitch line} = \frac{3}{10} \text{ pitch.}$$

$$df = \text{Working Depth} = \frac{6}{10} \cdots\cdots$$

$$dg = \text{Whole Depth} = \frac{7}{10} \cdots\cdots$$

* Reid's Horology, p. 114.

$$ab = \text{Thickness of Tooth} = \frac{5}{11}\text{ pitch.}$$

$$bc = \text{Breadth of Space} = \frac{6}{11} \ldots\ldots$$

It thus appears that an allowance of $\frac{1}{11}$ pitch is made to prevent the sides of the teeth from getting jammed into the spaces, and an allowance of $\frac{1}{10}$ pitch to prevent the tops of the teeth from striking the bottoms of the spaces. These proportions differ slightly with different workmen and different localities.

109. The necessary length of the teeth may be assigned with sufficient precision as follows. Vide fig. 51, page 91.

$$Ad^2 = TA^2 + Td^2 - 2\,TA\,.\,Td\,.\cos ATd.$$

Let $AT = R$, $BT = r$, and the addendum $fd = E$;

$$\therefore\ Ad = R + E;$$

and let the angle $TBd = \theta$. This is the angle through which the contact will be continued after passing the line of centers, and may be termed the angle of *receding action*. Substituting these values in the above expressions, and arranging the terms, we obtain

$$\frac{R+E}{R} = \left\{1 + \frac{2Rr + r^2}{R^2} \times \sin^2\theta\right\}^{\frac{1}{2}}.$$

Expanding this expression by the binomial theorem, and putting for $\sin\theta$ the series $\theta - \frac{\theta^3}{6} + \&c\ldots$ we may reject terms including the fourth power of θ, and higher powers, for θ is a small angle in all practical cases; we thus obtain

$$\frac{E}{R} = \frac{2Rr + r^2}{2R^2} \times \theta^2.$$

It is convenient to express both the addendum and the arc of action in relation to the pitch.

$$\text{Let } C \text{ be the pitch} = \frac{2\pi R}{N} = \frac{2\pi r}{n};$$

$$\therefore \frac{E}{C} = \frac{E}{R} \times \frac{N}{2\pi}.$$

Let F be the ratio of the arc of action Tm $(= r\theta)$ to the pitch;

$$\therefore \theta = \frac{F}{r} \times \frac{2\pi r}{n} = \frac{2\pi F}{n}.$$

Substituting these values, we have

$$\frac{E}{C} (= K) = \pi F^2 \left(\frac{2}{n} + \frac{1}{N}\right) . \quad (1).$$

This is the addendum to the driver.

The addendum of the follower is obtained in the same manner, by reversing the diagram, and considering the driver and follower to change places; in which case, the arc of action Tm will be that which takes place before reaching the line of centers. Let e be the addendum to the follower, f the ratio of the arc of action before reaching the line of centers to the pitch, which arc may be termed that of *approaching action;* substitute these letters for the corresponding ones in (1), and counterchange N for n, and we have

$$\frac{e}{C} (= k) = \pi f^2 \left(\frac{2}{N} + \frac{1}{n}\right);$$

$$\therefore \frac{E}{e} = \frac{F^2}{f^2} \times \frac{2N + n}{2n + N}. \quad (2).$$

110. From these expressions rules may be obtained, by which the addendum can be assigned in every case, by help of a few preliminary principles.

In the first place (fig. 53), the addendum de is the projection of the tooth beyond the pitch circle, and there must be an extent of tooth or flank ef within the pitch circle sufficient to receive the corresponding projection of the tooth with which the wheel is acting, as well as a small additional space fg to prevent the teeth of one wheel from striking the bottom of the spaces of the other; the entire depth or rather length of a tooth is made up, therefore, of the sum of the addenda of the driver and follower, added to this allowance for clearing, which in practice is made $\frac{1}{10}$ of the pitch and termed *freedom*;

$$\therefore \text{ whole length of tooth} = E + e + \frac{C}{10}.$$

It is essentially necessary that each pair of teeth should continue in action until the next pair have come into contact, therefore the sum of the arcs of approaching and receding action, must be at least equal to the pitch, that is, $F + f = 1$. But it is better that they should continue in action longer than this, in order to divide the working pressure between more teeth, as well as to prevent the chance of one tooth escaping before the next begins. It is therefore unnecessary to proportion the addendum so accurately as to give the entire arc of action a constant length. It is merely required to find a value that will be sufficient in all cases to prevent the teeth from escaping too soon. Now the expression (1) shews at once that the greatest addendum is required for the smallest numbers of teeth when the arc of action is given; and hence a rule assigned for the small numbers will serve for all cases.

If equal wheels of 15 work together with an arc of receding action of $\frac{2}{3} \times$ pitch, the expression (1) will give $K = .28$ for the necessary addendum; therefore the millwrights' value ($K = .3$) is sufficient for all cases of higher numbers than 15. But for smaller numbers the addendum will be greater and must be calculated. For example, the limiting cases in the Table, (page 93) will all be found to require a much greater addendum, varying from about 63 to .5, in the different examples.

111. The arc through which the action of the teeth is continued is governed by the magnitude of the addendum; and as the arc of approach depends on the addendum of the follower, and the arc of recess on the addendum of the driver, we are at liberty to give these arcs any required proportion by properly adjusting these addenda.

Now considering merely that the friction which takes place before the line of centers is of a different and more injurious character than that which happens after passing that line, it would seem that the best method would be to exclude altogether any action between the teeth until the line of centers is passed, by giving no addendum to the follower whatever; thereby making its true diameter equal to its geometrical diameter. On the other hand, it has been shewn, (Art. 34), that the quantity of friction in both cases increases rapidly with the distance of the point of contact from the line of centers. If the action be entirely confined to one side of the line of centers, it must be continued to a proportionably greater distance from that line, and so the teeth at the extremity of their action may incur greater abrasion and friction than they have lost by avoiding contact before the line of centers.

The best method, then, is to adjust the addenda so that there shall be less action before coming to the line of centers than after it; but the exact proportion between these arcs of action cannot be assigned for want of proper data; for although the fact is certain, no experiments have been hitherto made to compare these two kinds of friction.

112. To examine the effect of a constant addendum upon the ratio of the arcs of approach and recess, put $E = e$ in (2);

$$\therefore \frac{F^2}{f^2} = \frac{2n + N}{2N + n} = \frac{2 + \frac{N}{n}}{1 + \frac{2N}{n}}.$$

When equal wheels work together, or $N = n$, then $f = F$, or the arcs of action before and after the line of centers are equal. When a wheel drives a pinion, N is greater than n, and f greater than F; but if a pinion drive a wheel, then n is greater than N, and F than f. In the first case, there is more action before the line of centers than after it, and in the second, the reverse. It appears, then, that the constant addendum of the millwrights produces an effect exactly contrary to the principles just laid down, in every case except that of a pinion driving a wheel; and this is one reason why the action in this case is so much smoother than when a wheel drives a pinion. In fact, any rule that fixes the proportion of the addenda will make the ratio of the two arcs of action vary exceedingly. However, it appears from the expression

$$\frac{E}{e} = \frac{F^2}{f^2} \times \frac{2N + n}{2n + N},$$

that the ratio of the addenda is constant when the ratio of the arcs of action and also of the number of teeth is constant; if, therefore, the ratio of the arcs of action is determined, a small table will give the ratio of the addenda corresponding to the principal ratios of numbers of teeth.

The following Table of values of $\frac{E}{e}$, is calculated for three different ratios of the two arcs of action; namely, supposing them to be equal, double, or in the proportion of about 2 to 3.

	Value of $\frac{N}{n}$.	Values of $\frac{E}{e}$.		
		$F = 2f$.	$F = \sqrt{2}f$.	$F = f$.
Rack follows.	Zero.	2	1	.5
Pinion drives.	$\frac{1}{10}$	2.3	1.1	.5
	$\frac{1}{7}$	2.4	1.2	.6
	$\frac{1}{5}$	2.5	1.3	—
	$\frac{1}{3}$	2.8	1.4	.7
	$\frac{1}{2}$	3.2	1.6	.8
Wheel drives.	1	4	2	1
	2	5	2.5	1.2
	4	6	3	1.5
	6	6.5	3.2	1.6
	10	7	3.3	—
Rack drives.	Infinite.	8	4	2

Example. In clocks and watches the wheels always drive the pinions, and the ratio of their numbers varies from 8 to 10. In Mr Reid's rule (Art. 107) the ratio of the addenda is $\frac{225}{150} = 1.5$; but from the third column of the Table it appears that this is scarcely enough even to give an equal action before and after the line of centers, and that it would be better to take a ratio of three, which would give the simpler rule,

$$\frac{U}{u} = \frac{N+3}{n+1}.$$

This rule gives an addendum of about $\frac{1}{2}$ the pitch to the driver, and $\frac{1}{6}$ to the follower; and may safely be adopted when the wheels drive, or if the wheels be equal; but when the pinion drives, then

$$\frac{U}{u} = \frac{N+2.5}{n+1.5}, \quad \text{or} \quad \frac{U}{u} = \frac{N+2}{n+2}, \text{ will be better}$$

To apply the third solution (Art. 78) *to the formation of the teeth of wheels.*

113. Teeth whose forms are derived from the previous solutions, and especially the latter, are the most commonly adopted in practice; but they are subject to this inconvenience: a wheel of a given pitch and number of teeth, for example 40, if it be made to work correctly with a wheel of 50 teeth, will not suit a wheel of any other number, as 100. This is obvious, for the diameter of the describing circle by which the epicycloid is traced must be made equal to the radius of the pitch circle of the wheel with which the teeth are to work, and will therefore be, in this example, twice as large in the second case as in the first, producing different epicycloids.

In the modern practice of making cast-iron wheels this objection is a very serious one, as it compels the founder to make a new pattern of a wheel of a given pitch with 40 teeth, for every combination that it may be required to make of such a wheel with others; and so on for wheels of every other number.

Besides, it often happens in machinery that one wheel is required to drive at the same time two or more wheels whose numbers of teeth are different, and in this case the teeth cannot be correctly formed at all on the principles hitherto explained.

In cast wheels, then, it is especially essential that the teeth should be shaped so as to allow a given wheel to work correctly with any other wheel of the same pitch; and this may be done by employing the following corollary from the third solution*.

114. If for a set of wheels of the same pitch a *constant describing circle* be taken and employed to trace those por-

* Transactions of Civil Engineers, Vol. II. p. 91.

tions of the teeth which project beyond each pitch line by rolling on the exterior circumference, and those which lie within it by rolling on its interior circumference, then any two wheels of this set will work correctly together.

115. Fig. 54 represents a pair of wheels of such a set.

Here *A*, *B* are the centers of motion as usual. *TdD* or *TgG* the constant describing circle. This is employed to trace the faces or portions of the teeth that lie beyond the pitch circle *FTf* of the driver, as *qr*, by rolling upon it,

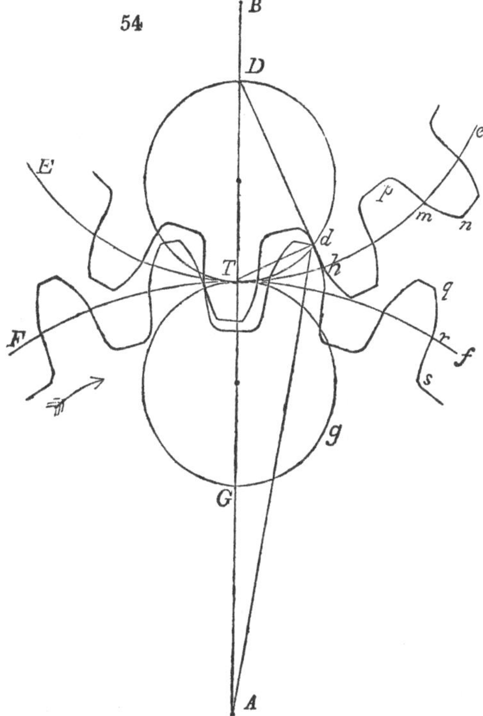

and the flanks or portions that lie within the pitch circle *ETe* of the follower, as *pm*, by rolling within it; consequently, by the third solution, these curves will work together with a constant velocity ratio, and the describing circle TdD will be the locus of contact; which beginning upon the line of centers between the point r of the driving tooth, and the point m of the following tooth, will gradually recede from the driver's center A, and approach the follower's center B; the teeth finally quitting contact at the point q of the driver, and the root p of the follower, their action being confined to their recess from the line of centers.

In the same manner, the same constant describing circle at TgG is employed to trace the flanks rs which lie within the pitch circle FTf of the driver, and the faces mn which lie without the pitch circle ETe of the follower; TGg will be the locus of contact which begins between the root s of the driver and the point n of the follower, and is confined to the approach of the teeth to the line of centers.

But as a constant describing circle is used for the whole set, it is clear that this demonstration will apply to any pair of the wheels that may be placed in action together; for whether the point of contact lie on one side or other of the line of centers, we have an epicycloid working with an hypocycloid, and both have been drawn by the same describing circle; that is, by the constant circle of the set. Also any wheel may be taken either for a driver, or a follower.

116. Nevertheless, the diameter of the describing circle must not be made *greater* than the radius of the pitch circle of any of the wheels, as the effect of this would be to produce a tooth much smaller at the root than at the pitch circle; a fault which is partly incurred in the common form where the describing circle is equal in diameter to the radius of the pitch circle, as in fig. 48; for as the flanks of the teeth are radial, they are nearer together at the root of the tooth than on the pitch circle.

On the contrary, when the describing circle is *less* in diameter than the radius of the pitch circle, the root of the tooth spreads, as in fig. 54, and it acquires a very strong form. Nevertheless, if this be carried to excess by making the describing circle too small, the curvature of the epicycloidal faces will be injuriously increased, and the teeth become too short. The best rule appears to be, that the diameter of the constant describing circle in a given set of wheels shall be made equal to the least radius of the set.

117. With respect to the length of the teeth, that may in every case be determined by construction, thus:

Since TdD is the locus of contact, take Th equal to the arc of the pitch circle, through which it is required that the teeth shall remain in contact after passing the line of centers, that is, to the arc of receding action. Describe the hypocycloidal arc hd, then will d be the last point of contact; consequently, Ad the true radius of the driver, and dh the necessary length of the flank of the follower. A similar construction on the other side of the line of centers will give the length of the follower's teeth and the flanks of the driver.

118. Otherwise the necessary length may be computed in a similar manner to that of Art. 109; for comparing fig. 54 with fig. 51, it will appear that the diameter TD of the describing circle in fig. 54 is equivalent to the diameter TB of the follower in fig. 51; and since Th, the arc of action in fig. 54, is equal to the arc Td, that is, to $TD \times$ angle TDd, we shall obtain for the driver, exactly as in Art. 109, the formula

$$\frac{E}{C} = \pi F^2 \left(\frac{2}{N_1} + \frac{1}{N}\right);$$

where N_1 is the number of teeth which belongs to a wheel

whose radius is the diameter of the constant describing circle; and for the follower

$$\frac{e}{C} = \pi f^2 \left(\frac{2}{N_1} + \frac{1}{n}\right).$$

119. But as the wheels in question constitute a set, any pair of which are expected to work together, there can be no different proportions for driver and follower, since any wheel may be called upon to perform either function. Recollecting, therefore, that if the addendum of a wheel be too small, the teeth will quit hold of each other too soon, but that too large an addendum introduces no other inconvenience than an unnecessary length of tooth, we may find the necessary addendum for the set thus.

$$\frac{E}{C} = \pi F^2 \left(\frac{2}{N_1} + \frac{1}{N}\right),$$

is the general formula for the addendum to every wheel in the set, in which as N decreases, E increases; but the smallest value of N, by Art. 116, is N_1;

$$\therefore\ E = \frac{3\pi C f^2}{N_1},$$

is the greatest necessary value of E. Let the smallest wheel of the set have 16 teeth, and let the arc of action equal $\frac{3}{4}$ pitch. Then it will be found that the usual constant addendum of $\frac{3}{10}$ of the pitch may be safely used for wheels of 19 and upwards, but that a greater addendum must be given to the wheels 16, 17, and 18; the first requiring about $\frac{3}{8}$ of the pitch.

120. But it was also shewn in Art. 112, that the practice of employing a constant addendum under the second solution had the mischievous effect of making the arc of action before the line of centers greater than the receding arc. To examine the effect of the constant addendum in the present system:—

Let F, f be the arcs of action of two wheels, N, n their numbers of teeth;

$$\therefore \frac{E}{C} = \pi F^2 \left(\frac{2N + N_1}{N N_1}\right) = \pi f^2 \left(\frac{2n + N_1}{n N_1}\right),$$

$$\therefore \frac{F^2}{f^2} = \frac{\frac{2nN}{N_1} + N}{\frac{2nN}{N_1} + n}:$$

which shews that the arc of action that belongs to the greater number of teeth is the greater of the two; so that when a constant addendum is employed, if the wheel drives the pinion, the arc of action after the line of centers is greater than that before that line, and *vice versa;* which is the reverse of what happens in the second solution, and removes the objection to the constant addendum in the first case, but introduces it in the second.

Of course, the most complete system would be to make two sets of wheels, one for each case, with the addenda separately calculated for each; but the increase of expence occasioned by the making of two patterns for each wheel is sufficient to prevent the practical use of such a system, unless in very particular instances.

121. The smallest numbers of teeth that this system admits of may be derived from the same Table that has been given for the radial teeth. For fig. 51 applies also to this case, in the manner explained in Art 118, if BT be the diameter of the describing circle. To apply the Table, page 93, the numbers that indicate Followers, must be interpreted as denoting the number of teeth that would correspond to a wheel whose radius equals the diameter of the describing circle.

Example.—The arc of receding action is equal to the pitch, and the describing circle corresponds to a wheel of twelve teeth. Thirty teeth is the least wheel that will drive, and of course a wheel of any number greater than this may be employed. But if the arc of action equal $\frac{2}{3}$ of the pitch, then the same describing circle being employed, any number of teeth greater than twelve may be used, and so on.

122. *To apply the third solution to racks.* When rack-teeth are formed, as in the usual manner, according to the second solution, by making their flanks straight and the teeth of the pinions involutes, we have seen that the action on one side of the line of centers is confined to a constant point in each rack-tooth, because the pitch line of the rack is the locus of contact. This may be avoided by taking any describing circle Tkm, and employing it to describe cy-

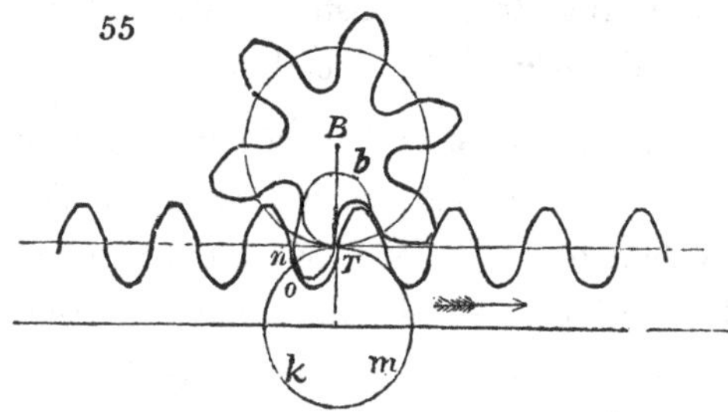

cloidal flanks, as no for the rack-teeth, by rolling on its pitch line nT, and then by describing the faces of the teeth of the wheel with the same describing circle, in which case the contact will no longer be confined to the pitch line of the rack, but will be found in To; and will consequently be distributed over the distance on, which may be made as small as we please by increasing the diameter of the describing circle. If the circle Tmk be the constant describing circle of a set of wheels, then any one of them will work with the rack.

To apply the fourth solution (Art. 79) *to the formation of the teeth of wheels**.

123. Involute teeth differ from the epicycloidal teeth derived from the second and third solution, in having the entire side of the tooth, both face and flank, formed of a continuous curve; whereas, as we have seen, the side of an epicycloidal tooth is made up of two different curves joined at the pitch circle.

Fig. 56 represents a pair of wheels with involute teeth. *A*, *B* the centers of motion, *T* the point of contingence of the

56

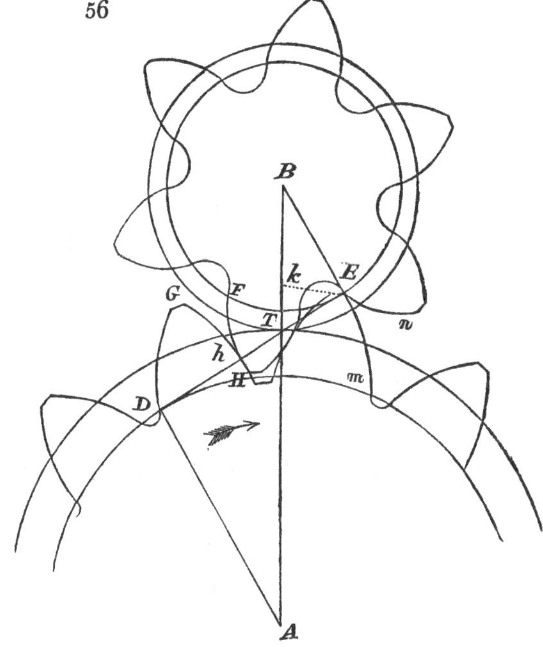

pitch circles; *BE*, *AD* the radii of the bases of the involutes, *ED* their common tangent, and therefore, the locus of contact of the teeth. As in the teeth already described, the contact lies within the pitch circle of the driver during the

* The involute was first suggested for this purpose by Euler, in his second paper on the Teeth of Wheels. N. C. Petr. xi. 209.

approach to the line of centers, and within that of the follower during the recess from that line.

Referring to fig. 36, page 67, it appears, that as the action of the curves begins at D, and T is the point of contact at the line of centers of the teeth TH and TG; therefore TH must have moved through an arc DH in its approach to that line. But $DT =$ arc DH, since TH is an involute of DH; $\therefore$ angle of action before the line of centers, or

$$\frac{DH}{DA} = \frac{DT}{DA},$$

and the arc of action upon the pitch circle

$$= \frac{AT \times DT}{DA}.$$

In like manner, as the tooth TK recedes from the line of centers until it finally quits contact at E, it can be shewn that this receding arc of action upon the pitch circle

$$= \frac{BT \times ET}{BE};$$

$$\therefore \frac{\text{approaching arc}}{\text{receding arc}} = \frac{AT \times DT \times BE}{BT \times ET \times DA} = \frac{AT}{BT} = \frac{DA}{BE}.$$

The arcs of action in a pair of involute teeth before and after the line of centers, are therefore, in the direct proportion of the radii of the bases of the driver and follower respectively. This of course supposes that the teeth are each made sufficiently long to extend to the base of the opposite tooth, as at mE, fig. 56.

124. However, by reducing the length of the teeth the quantity of action may be altered at pleasure. For example, in the tooth FH, fig. 56. With radius BH and center B, describe an arc of a circle cutting DE in h; then, supposing as before, that the lower wheel is the driver, h will be the

first contact, and it can be shewn, as in the last Art., that the actions before and after the line of centers are as hT to TE.

125. Although the contact action of the teeth is confined to the outside of the bases, yet it is necessary, as in epicycloidal teeth, to form clearing curves (Art. 98) within the bases; for example, the nearest point of *contact* of the tooth mE to the center B, is E; but if we describe with radius AE and center A an arc Ek meeting the line of centers in k, then k will be the nearest *approach* of the point of the tooth E to the center B, and a clearing hollow must be formed within the base circle, whose depth is at least equal to k, as shewn in the figures.

126. The two pitch circles being given, (fig. 56,) and the required angle of action TBE, the radii of the bases are easily found; for $BE = BT \times \cos TBE$.

Comparing the diagram $ATBE$ of fig. 56 with $ATBd$ in fig. 51, it will appear that they are identical in their relations to the teeth, and that the same formulæ (Art. 109)

$$\frac{E}{C} = \pi F^2 \left(\frac{2}{n} + \frac{1}{N}\right), \text{ and } \frac{e}{C} = \pi f^2 \left(\frac{2}{n} + \frac{1}{N}\right),$$

will apply to the involutes, but only at the points E or D, when the contact coincides with the bases. They will therefore give the addendum required to enable the teeth to continue their action to the base of the opposite wheel, but will not apply to all other positions of contact as they do for epicycloids.

127. The plan of this work excludes the examination of the relations of pressure; but in this case, it is necessary to remark, that a great objection to involute teeth is founded

upon the obliquity of their action, by which a much more considerable divergent pressure is thrown upon the axes than in the other forms of teeth. The action of epicycloidal teeth is, in fact, perpendicular to the line of centers at the instant of crossing it; but that of involute teeth is constantly in the direction of the common tangent of their bases, and is therefore oblique to the line of centers*. This injurious property is balanced by the advantages that a variation of the distances of their centers does not destroy the action of the teeth, and that any two wheels of the same pitch will work together; but this last property, I have shewn (Art. 114) to be possessed also by some arrangements of the epicycloidal teeth. In smaller machinery, constructed rather for the modification of motion than for the transmission of force, this oblique action ceases to be objectionable, and the other advantages of involute teeth will then recommend them in preference to all others.

Such teeth manifestly possess greater strength of form than epicycloidal teeth, at least than those with radial flanks, and I shall proceed to shew that they admit of a greater reduction of back-lash than any other kind.

128. For in fig. 56, suppose the teeth to be so described that no back-lash exists, that is to say, that both sides of the acting teeth are in contact at once, which is theoretically possible in all forms of teeth when they are symmetrical to a radius, but which, as already stated (Art. 89), is not possible in practice, because a slight error *in excess*, in the form of any tooth, would cause it to wedge itself fast into its corresponding space.

Now if the distance of the centers of these wheels be increased, this double contact will be destroyed, although the

* In fig. 56 the arc of action and obliquity are made, for the sake of distinctness, greater than would be necessary in practice.

action of the teeth in effecting a constant velocity ratio will not be impaired. A back-lash will therefore be introduced, which will be the greater the more the wheels are withdrawn from each other. In any given pair of involute wheels, therefore, we can, by properly adjusting by trial the distance of their centers, reduce the back-lash to the least quantity that will allow the teeth to act without jamming. This advantage is possessed by no other form, and particularly recommends these teeth for dial-work, or any such kinds of mechanism, in which the back-lash is mischievous.

129. *To apply involutes to rack-teeth.*

Describe a pitch circle, (fig. 57,) radius BT, and draw AC a tangent at T for a pitch line to the rack; let the

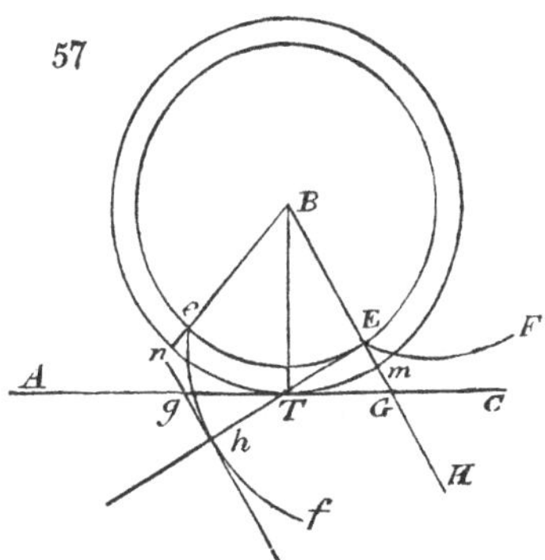

circle whose radius is BE be the base of an involute EF, and let the tooth of the rack be bounded by a straight line EGH, making an angle EGA with the pitch line equal to BTE. If the involute be moved to ef, it will drive the sloped tooth to gh, always touching it in the line ETh; and the velocity of the circumference of the pitch circle will always equal that of the pitch line: for

$$Gg = \frac{Eh}{\sin EGT},$$

also $Eh = \text{arc } Ee$, by the property of the involute

$$= \text{arc } mn \times \frac{BE}{BT}$$

$$= \text{arc } mn \times \sin BTE;$$

$$= \text{arc } mn \times \sin EGT; \quad \therefore \; Gg = \text{arc } mn.$$

This may be shewn from fig. 56, page 113. For let the radius of the wheel AT become infinite, then will the pitch line be a straight line passing through T, and touching the pitch

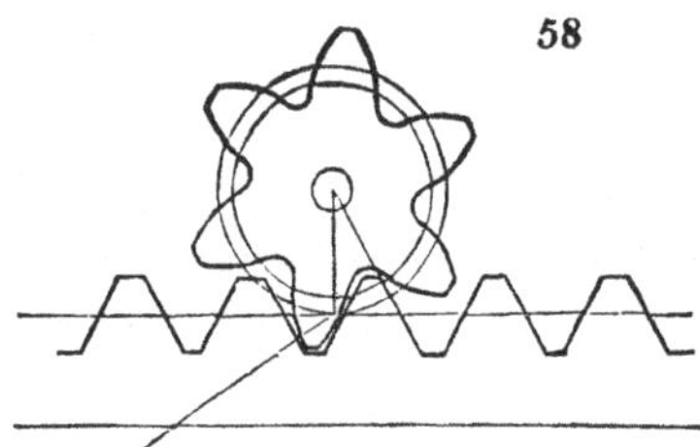

circle of the wheel whose center is B, and the involutes GH, Em will become right lines perpendicular to the line ETD. Thus is obtained a rack with straight-sided sloping teeth, as in fig. 58.

Hence a wheel with involute teeth will work with a rack whose teeth are straight-sided and inclined to the pitch line at an angle θ, provided

$$\frac{\text{radius of base}}{\text{radius of pitch circle}} = \sin \theta.$$

In such a rack, the locus of contact being the tangent line ETh, the contact will not be confined to a single point of the tooth, as it is in the common involute rack teeth, (Art. 106) which are derived from that particular case of this figure, in which the radius of the base coinciding with that of the pitch circle, the line ETh coincides with the pitch line of the rack. But a rack with sloping teeth will be pressed downwards by a resolved portion of the working pressure, and this appears to me to be in many cases advantageous, and destructive of vibration.

To approximate to the true form of a tooth by arcs of circles.

130. The portion of curve employed in a tooth is so short, that a circular arc might be substituted for it with sufficient accuracy for all practical purposes, if its center and radius were determined upon correct principles.

In fact, practically the edges of teeth are always made arcs of circles, but unfortunately, these arcs are often struck from the merest empirical rules, such as setting the point of the compasses in the pitch line on one side of the tooth, in order to strike the other, and vice versa, or similar absurdities*. Teeth have even been set out by forming their edges into semicircles struck alternately without and within the pitch circle; these are technically known by the name of hollows and rounds.

Some millwrights, with equal neglect of principle, give their teeth plane faces passing through the axis of the wheel, expecting them to wear themselves in a short time into proper forms. But the best workmen endeavour to give to their wheels teeth of the epicycloidal form, according to the rules laid down in Camus†, or in Buchanan's Treatise on Mill-work‡, which are immediately derived from Camus. In truth, the question is one of great practical importance; I do not mean to say, that it is necessary, or even practicable, to shape the teeth of small wheels into exact epicycloids or involutes, such as those which have been described in the preceding pages; but I do assert, that unless the rules for shaping them be derived from such considerations, so as to approximate their form to the true ones, as nearly as

* Vide Imison's School of Arts, or Gray's Experienced Millwright.

† Camus on the Teeth of Wheels, 1806 and 1837.

‡ 1808, 1823 and 1841.

possible, that the action of the machines will be irregular and noisy, producing those vibrations which must be familiar to all who have been in the habit of examining machinery, and which are above all things conducive to the wearing out and disintegration of every part of the mechanism. The investigation of the proper curves for the teeth of wheels is, therefore, by no means one of mere curiosity, although this has been sometimes hastily asserted. One proof of the necessity of attending to the exact theoretical forms, is the acknowledged impossibility of making one wheel to work with two others whose numbers of teeth are different, by means of the usual rules.

131. The method employed by the best workmen for shaping the teeth of a proposed wheel, or of a pattern from which to cast one, is as follows :

The shape of a single tooth adapted for this wheel is traced in the true epicycloidal form, by means of *templets*, that is, of a pair of boards whose edges are cut to the curvature of the pitch circle, and describing circle respectively, and which may be termed the pitch templet and the describing templet. The latter carries a describing point in its circumference, and by rolling its edge upon that of the pitch templet, the arc required for the face of the tooth is traced upon the drawing board*.

This done, the workman finds with his compasses, by trial, a center and small radius, by which an arc of a circle can be described, that will coincide as nearly as he can manage to make it with the templet-traced epicycloid.

* If the method I have recommended under the third solution (Art. 114) be adopted, then one describing templet will serve for the entire set; but since this templet is required to trace hypocycloids for the flanks, as well as epicycloids for the faces, every pitch templet must have a convex and a concave edge, both shaped into an arc of the pitch circle of the wheel in question. The concave edge is not required upon the common system (Art. 98), because the flanks are radial lines.

Then, having struck upon the fronts of the rough cogs a circle which is concentric with the pitch circle, and whose distance from it is equal to that of the center of his small arc, he adjusts his compasses to the small radius, and always keeping one point in the circle just described, he steps with the other to each cog in succession, they having been previously divided into equal parts corresponding to the given pitch and breadth of the teeth; upon each cog he describes two arcs, one to the right and one to the left, which serve him as guides in shaping and finishing the acting faces.

132. The practical convenience of this method is very great, and appears to require only a more commodious and certain method of determining the center and radius of the approximate arc.

The first method that suggests itself, is to find the center and radius of the circle of curvature at some intermediate point between the extremities of the curve selected for the teeth, and to substitute an arc of this circle in lieu of the actual curve. But the determination of the required circles may be effected upon general principles, without taking individual curves into the considerations. In fact, Euler, in his elaborate paper on the Teeth of Wheels*, undertook to investigate a general expression for curves that possess the property of revolving in contact with a constant velocity ratio, which he effected by determining the relation between their radii of curvature; and suggested that in practice small arcs of the circles of curvature thus obtained would probably suffice for the sides of teeth. He accordingly gave some geometrical constructions for this purpose, but the hint thus supplied was neglected by every subsequent writer, partly, perhaps, by reason of the abstruse manner in which he treated the subject.

* N. C. Pet. xi. 209.

Besides, in the mechanical practice of that day, it is probable that the millwrights would have regarded any method founded upon geometrical considerations, as a useless refinement, while the theoretical mechanician would have considered the substitution of a circle for the exact curve, however accurately determined, to be too coarse an approximation.

At present, the necessity for precise forms on the one hand, and the practical limits to such precision on the other, are beginning to be better understood; and by following out the views suggested by Euler's paper, I have succeeded in simplifying the investigation of the required circles, and in adapting, and even introducing the method into modern practice.

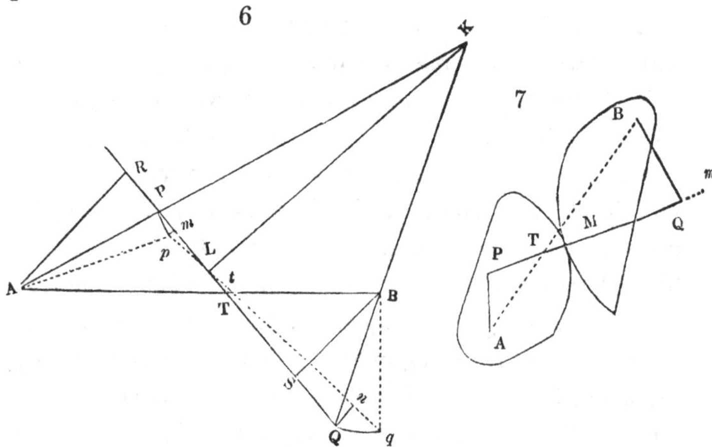

133. A simple construction is sufficient to give the centers and radii of the arcs in any required case. For it has been shewn (Art. 33,) that the action of a pair of curves by contact is equivalent at every moment to that of a pair of radii *AP*, *BQ* (fig. 7,) connected by a link *PQ*, *P* and *Q* being the respective centers of curvature of the curves at the point of contact. Now (fig. 6) the angular velocity ratio

between the radii AP, BQ is that of the segments $BT : AT$, into which the link divides the line of centers (Art. 32); and if the rods be moved into a new position, this ratio becomes $Bt : At$, which is greater or less than the former, according as the point t moves to one side or other of the point T.,

But if the point L, which is the intersection of two successive positions of the link, happen to coincide with T, the ratio of the segments will be the same in both positions, and the angular velocity ratio constant at that instant.

If then the rods and links of fig. 7 be placed in such a relative position that L and T may unite, and the curves in contact be replaced by arcs of circles described from centers P and Q through *any point* M of the line PQ, the angular velocity ratio of these curved pieces will be perfectly constant at the moment of their reaching the position that makes M the point of contact, and the ratio will not vary essentially during a small angular motion on each side of this position.

134. As this constancy of the velocity ratio depends only upon the centers of the arcs, they may be struck through any common point of the line of action PQ, as at m, beyond both the centers. Only that if this point lie between the centers P, Q, as at M, the arcs and edges will be convex, but if the point lie beyond the centers, as at m, the edge corresponding to the most distant center P, will be concave.

135. It follows, that to find a pair of centers that possess the property of communicating motion in a constant velocity ratio, it is only necessary to construct the diagram, (fig. 6) in such a manner, that the point L shall fall on the line of centers. But (by Art. 32. Cor. 2,) L is that point

of PQ which is met by a perpendicular from K, the intersection of the directions of the radius rods AP, BQ. Whence the following construction.

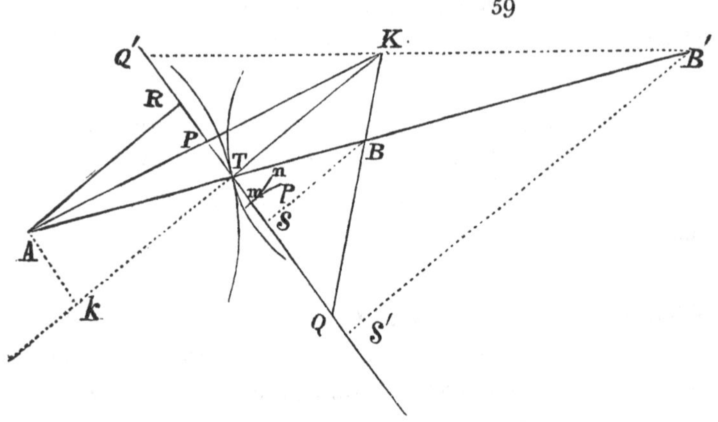

Let A, B be the centers of motion of the wheels, T the point of contingence of the pitch circles; through T draw PTQ making any angle with the line of centers, and upon it assume P as a center, from which the circular side is to be described for a tooth of a wheel whose center of motion is A. To find the corresponding center for the wheel which turns upon B, draw TK perpendicular to PTQ, produce AP to meet it in K, join KB and produce it to meet PTQ in Q; then will Q be the required center.

And a small arc mn, struck from P as a tooth for the wheel whose center of motion is A, will work correctly with an arc mp, struck from Q through m, and employed as a tooth to the wheel whose center of motion is B.

If B be so placed that the angle KBT is acute, as for example at B', then will Q fall at Q' on the same side of T as P, but beyond it; the effect of this is to make the tooth mp concave instead of convex.

But if the angle $KBT = PTA$, KB will become parallel to PT, and the point Q being thus removed to an infinite distance, the arc mp or tooth of the wheel whose center of motion is B, will be a right line perpendicular to PT.

136. The distance of the centers from T may be calculated as follows.

Draw AR perpendicular to PT.

Let $KT = C$, $AT = R$, $PT = D$, $ATP = \theta$, then by similar triangles, ARP, PTK,

$$KT = \frac{PT \times AR}{PR} = \frac{PT \times AR}{TR - PT},$$

$$\text{or } C = \frac{DR \,.\, \sin\theta}{R \,.\, \cos\theta - D}; \quad \therefore \; D = \frac{RC \cos\theta}{C + R\sin\theta},$$

and similarly, drawing BS perpendicular to TQ, and putting

$$BT = r, \; QT = d,$$

we have for the corresponding arc mp,

$$d = \frac{rC\cos\theta}{C + r\sin\theta}.$$

But if a concave tooth be employed, draw $B'S'$ perpendicular to PTQ, then

$$KT = \frac{Q'T \times B'S'}{Q'T + TS'}, \quad \text{whence } d = \frac{Cr\cos\theta}{r\sin\theta - C}.$$

137. If the side of the tooth be made to consist of *a single arc*, a very simple rule may be obtained; for suppose KT to be infinite, then will AP and BQ become perpendicular to the line PTQ, and the points P, Q will come to R, S respectively. Let the arcs of the teeth be struck through T, let θ be the angle ATP, which the line PTQ makes with the line of centers, and let R be the radius AT of the wheel, and $D = TR$ be the required distance of the center of the tooth from the point T;

$$\therefore D = R \cos \theta$$

is independent of the wheel with which it is to work, as well as of the pitch and number of teeth of its own wheel.

If therefore θ be made constant in a set of wheels, any two of them will work together, and their teeth are easily described as follows. Assume $\theta = 75^0\ 30'$, which is a very convenient value;

$$\therefore D = \frac{R}{4};$$

for $\cos 75^0\ 30' = .25038 = \frac{1}{4}$ very nearly.

138. Let A be the center, AT the radius of the pitch circle of a proposed wheel. Draw TP making an angle ATP

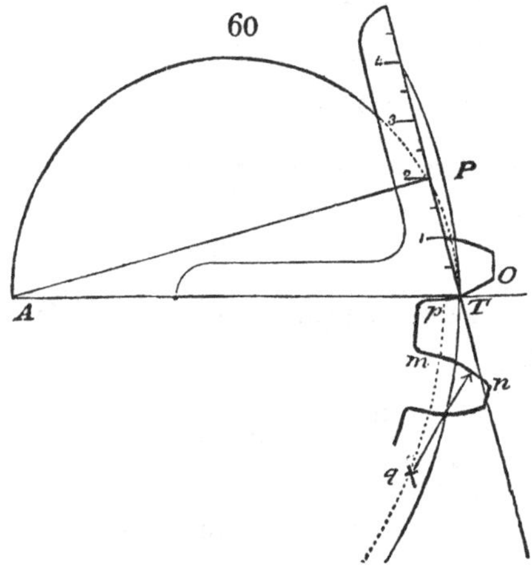

of $75^0\ 30'$ with the radius, and drop a perpendicular AP upon TP, $\left(\text{or describe a semicircle upon } AT \text{ and set off } TP = \frac{AT}{4}\right)$, then will P be the center from which an arc op, described through T, will be the side of the tooth required.

Or more conveniently, let a bevil of $75^0\ 30'$ be made of brass or card-paper, as in the figure, of which the side TP is graduated into a scale of quarter-inches and tenths. If this bevil be laid upon the radius AT, so that its point T coincides with the pitch circle, the center point P will be found at once, by reading off the radius of the wheel in inches upon the reduced scale. Thus the radius AT in the figure, is two inches long, and the point P is found at 2 upon the scale.

To describe the other teeth, draw with center A and radius AP, a circle within the pitch circle, dotted in the figure, this will be the locus of the centers of the teeth; then having previously divided the pitch circle, take the constant radius PT in the compasses, and keeping one point in the dotted circle, step from tooth to tooth and describe the arcs, first to the right and then to the left, as for example, mn is described from q and pO from P.

If Op were an arc of an involute having the circle Ppq for its base, PT would be its radius of curvature at T. These teeth, therefore, approximate to involute teeth, and they possess in common with them the oblique action, the power of acting with wheels of any number of teeth, and the adjustment of back-lash; but, as the sides of the teeth consist each of a single arc, there is but one position of action in which the angular velocity ratio is strictly constant, namely, when the point of contact is on the line of centers.

139. By making the side of each tooth consist of *two arcs* joined at the pitch circle, and struck in such wise that the exact point of action of the one shall lie a little before the line of centers, say at the distance of half the pitch, and the exact point of the other at the same distance beyond that line, an abundant degree of exactitude will be obtained for all practical purposes.

To describe the teeth of such a pair of wheels, let A (fig. 6) be the center of motion of a proposed wheel, B the center of motion of the wheel with which it is to work, T the point of contingence of the pitch circles. Draw QTq making an

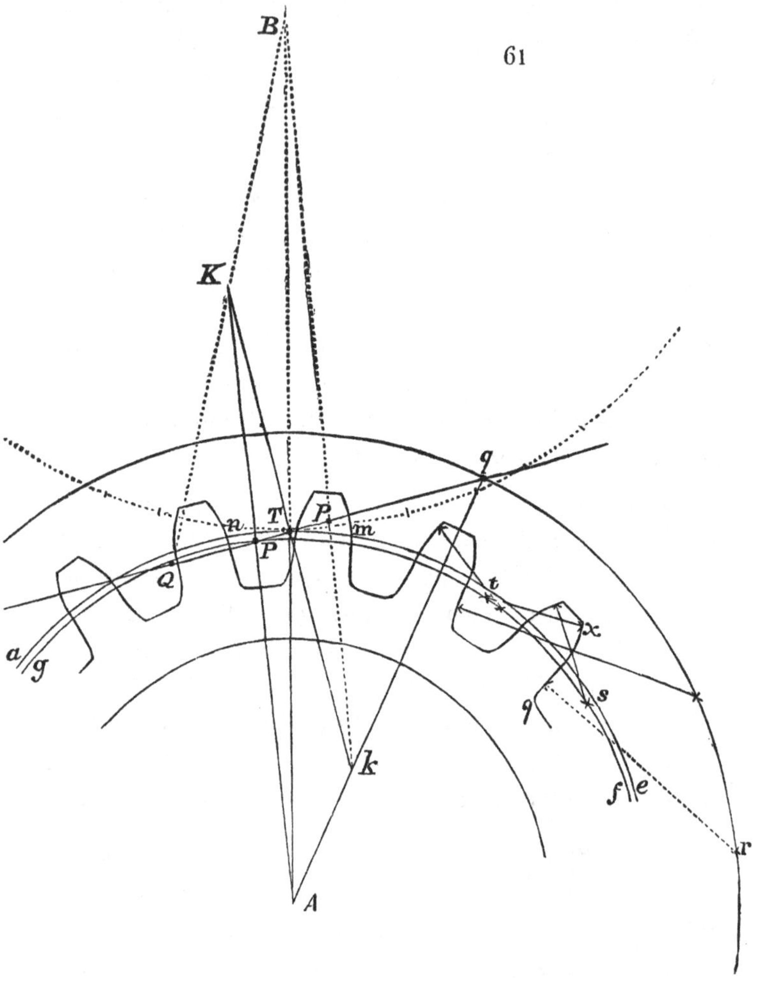

angle of 75° with the line of centers. (This angle is in fact arbitrary, but by various trials I find 75° to give the best form to the teeth.)

Draw kTK perpendicular to QTq, and set off TK and Tk equal to each other, and less than either AT or TB. Join AK and BK, producing the latter to Q, then P and Q

are a pair of tooth-centers. Take a point m on the pitch circle aTe, at the distance of half the pitch from T, and on the opposite side to the tooth-centers. A convex arc struck from P through m on the outside of this pitch circle will work correctly with a concave arc struck from Q through the same point, and within the other pitch circle.

To describe the faces of the teeth of the lower wheel we may proceed as in the last example, thus: draw with center A a circle through P, which will be the locus of the centers of the small arcs; and having previously divided the pitch circle for the reception of the teeth, take the constant radius Pm in the compasses, and keeping one point in the circle Pf, describe the faces of the teeth to the right and left outside the pitch circle, as shewn in the figure at t and s.

A similar proceeding will give the flanks of the teeth of the upper wheel.

To obtain the flanks of the lower wheel and faces of the upper wheel, join Bk and Ak, producing the latter to q, then will p and q be another pair of centers, from which let arcs be struck through a point n, at the distance of half the pitch beyond T, but within the pitch circle of A and without that of B. The action of these arcs will be exact at the distance of half the pitch from T.

To complete the teeth of the lower wheel already begun, describe from A with radius Aq, a circle for the locus of the centers of the flanks of these teeth, and with the constant radius equal to qn step from tooth to tooth, describing the flanks in the manner shewn in the figure, as at r and q.

140. From the construction it appears that these teeth of the lower wheel would work correctly with a wheel of any radius, provided the points K and k remain constant; for a change in the position of B, on the line of

centers, only affects the points Q, p, which belong to its own teeth, but does not disturb the points P, q, from which the teeth of the lower wheel have been described.

In short, if any number of wheels be in the above manner described, in which the lines Qq, Kk, preserve the same angular position with respect to the line of centers and the same distances KT, kT, then any two of these wheels will work together. The distance KT may be determined for a set of wheels by considering that if A approach T, Aq becomes parallel to Tq, and q is at that moment at an infinite distance; the flank of the tooth becoming a right line perpendicular to Tq. If A approach still nearer, q appears on the opposite side of T, and the flank becomes convex, giving a very awkward form to the tooth.

The greatest value therefore that can be given to KT, must be one which when employed with the smallest radius of the set, will make Aq parallel to Tq; therefore, if $R_{,}$ be this smallest radius, we have

$$KT = R_{,} \times \sin QTA, \text{ or } C = R_{,} \times \sin \theta;$$

which substituted in the formula (Art. 136), gives

$$PT = D = \frac{RR_{,} \cos \theta}{R_{,} + R}, \text{ and } qT = d = \frac{RR_{,} \cos \theta}{R - R_{,}}.$$

141. By assuming constant values for $R_{,}$ and θ in a set of wheels, the values of D and d which correspond to different numbers and pitches, may be calculated and arranged in tables for use, so as to supersede the necessity of making the construction in every case. Thus the tables which follow in fig. 62 were obtained by assuming twelve teeth as the least number to be given to a wheel, and $\theta = 75^{\circ}$.

The unit of length in which the values of D and d are expressed is one twentieth of an inch, that being sufficiently small to avoid errors of a practical magnitude.

THE ODONTOGRAPH.

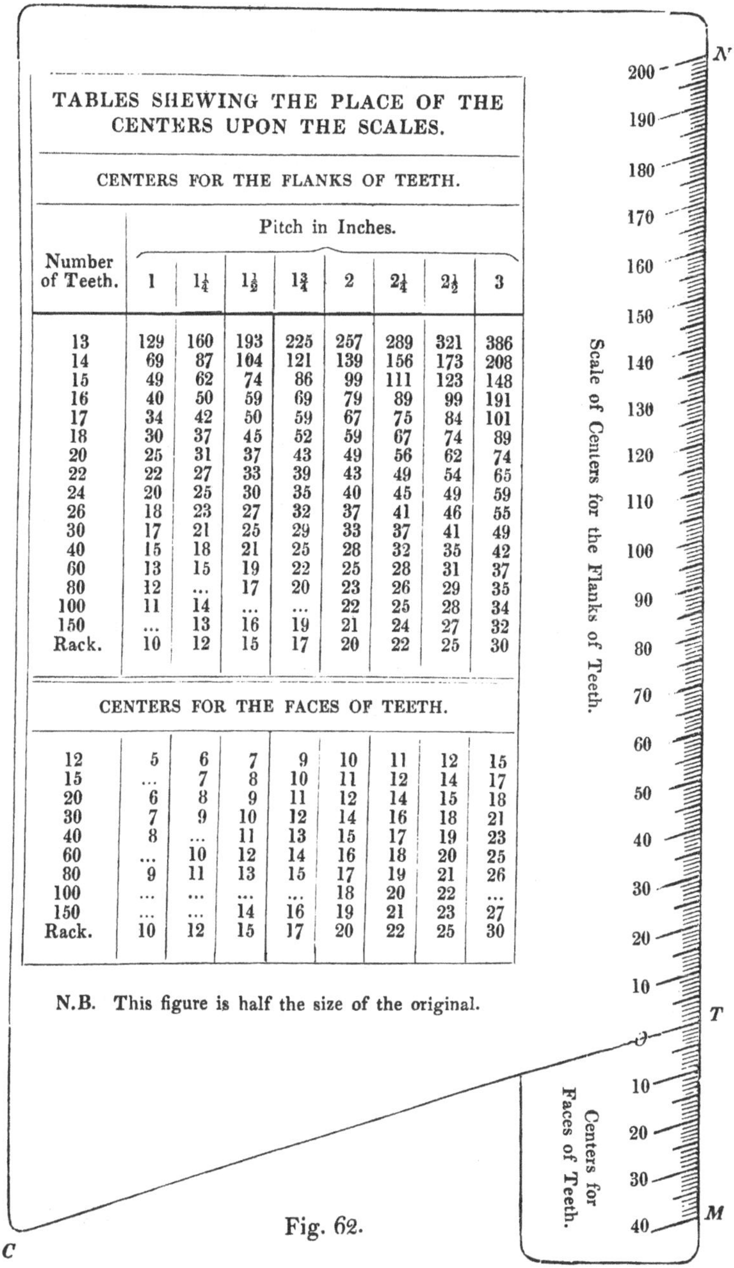

TABLES SHEWING THE PLACE OF THE CENTERS UPON THE SCALES.

CENTERS FOR THE FLANKS OF TEETH.

Number of Teeth.	Pitch in Inches.							
	1	$1\frac{1}{4}$	$1\frac{1}{2}$	$1\frac{3}{4}$	2	$2\frac{1}{4}$	$2\frac{1}{2}$	3
13	129	160	193	225	257	289	321	386
14	69	87	104	121	139	156	173	208
15	49	62	74	86	99	111	123	148
16	40	50	59	69	79	89	99	191
17	34	42	50	59	67	75	84	101
18	30	37	45	52	59	67	74	89
20	25	31	37	43	49	56	62	74
22	22	27	33	39	43	49	54	65
24	20	25	30	35	40	45	49	59
26	18	23	27	32	37	41	46	55
30	17	21	25	29	33	37	41	49
40	15	18	21	25	28	32	35	42
60	13	15	19	22	25	28	31	37
80	12	...	17	20	23	26	29	35
100	11	14	...	...	22	25	28	34
150	...	13	16	19	21	24	27	32
Rack.	10	12	15	17	20	22	25	30

CENTERS FOR THE FACES OF TEETH.

Number of Teeth.	1	$1\frac{1}{4}$	$1\frac{1}{2}$	$1\frac{3}{4}$	2	$2\frac{1}{4}$	$2\frac{1}{2}$	3
12	5	6	7	9	10	11	12	15
15	...	7	8	10	11	12	14	17
20	6	8	9	11	12	14	15	18
30	7	9	10	12	14	16	18	21
40	8	...	11	13	15	17	19	23
60	...	10	12	14	16	18	20	25
80	9	11	13	15	17	19	21	26
100	...	...	...	...	18	20	22	...
150	...	...	14	16	19	21	23	27
Rack.	10	12	15	17	20	22	25	30

N.B. This figure is half the size of the original.

Fig. 62.

The reduction of this system to a divided scale is necessarily more complex than when a single arc only is employed. I have endeavoured, however, to put it into a form which shall be sufficiently easy in practice, and have ventured to name the instrument an *Odontograph*. It is at present employed in some of the best factories, and, as I am informed, with complete success.

Fig. 62 represents the Odontograph exactly half the size of the original; but as it is merely formed out of a sheet of card-paper, this figure will enable any one to make it for use. The side *NTM* which corresponds to the line *QTq* in fig. 61, is straight, and the line *TC* makes an angle of exactly 75° with it, and corresponds to the radius *AT* of the wheel. This side *NTM* is graduated into a scale of half inches, each half inch being divided into ten parts, and the half inch divisions are numbered both ways from *T*.

142. One example will shew the mode of using this instrument. Let it be required to describe the form of a

63

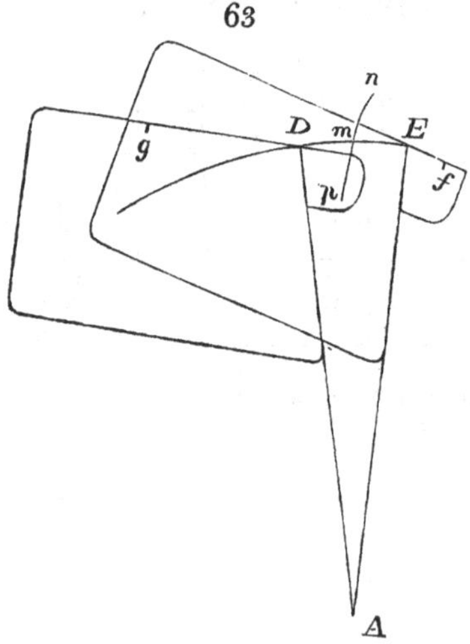

tooth for a wheel of 29 teeth, of 3 inches pitch. Describe

from a centre *A*, fig. 63, an arc of the given pitch circle, and upon it set off *DE*, equal to the pitch, and bisected in *m*. Draw radial lines *DA*, *EA*. For the arc within the pitch circle apply the slant edge of the instrument to the radial line *AD*, placing its extremity *D* on the pitch circle, as in the figure. In the Table headed, *Centers for the Flanks of Teeth*, look down the column of 3 inch pitch, and opposite to 30 teeth, which is the nearest number to that required, will be found the number 49. The point *g* indicated on the drawing-board by the position of this number on the scale of equal parts, marked *Scale of Centers for the Flanks of Teeth*, is the center required, from which the arc *mp* must be drawn with the radius *gm*.

The center for the arc *mn*, or face, which lies outside the pitch circle is formed in a manner precisely similar, by applying the slant edge of the instrument to the radial line *EA*. The number 21 obtained from the lower table, will indicate the position *f* of the required center upon the lower scale. In using the instrument, it is only necessary to recollect, that the scale employed and the point *m* always lie on the two opposite sides of the radial line to which the instrument is applied.

The curve *nmp* is also true for an annular wheel of the same radius and number of teeth, *n* becoming the root and *p* the point of the teeth. For a rack, the pitch line *DE* becomes a right line, and *DA*, *EA*, perpendiculars to it, at a distance equal to the pitch.

143. Numbers for pitches not inserted in the tables may be obtained by direct proportion from the column of some other pitch: thus for 4-inch pitch, by doubling those of 2-inch, and for half-inch pitch by halving those of inch-pitch. Also, no tabular numbers are given for twelve teeth in the upper table, because within the pitch circle their teeth are radial lines*.

144. But if it be not required that wheels shall work in a set, the construction of fig. 58 may be readily adapted to particular cases: thus, if a pin-wheel be required, the pin is evidently already a tooth, whose acting edge is an arc of a circle. And supposing K to remove to an infinite distance, AP and BQ will become perpendicular to PTQ, and the points P and Q coincide respectively with R and S.

* In fact, in the actual instrument I have inserted columns for $\frac{1}{4}$, $\frac{3}{8}$, $\frac{1}{2}$, $\frac{5}{8}$, $\frac{3}{4}$, and $3\frac{1}{2}$ pitch, which are omitted in fig. 62 for want of room, and are indeed scarcely necessary, as the numbers are so easily obtained from the columns given.

It is unnecessary to have numbers corresponding to every wheel, for the error produced by taking those which belong to the nearest as directed, is so small as to be unappreciable in practice. I have calculated the amount and nature of these errors by way of obtaining a principle for the number and arrangement of the wheels selected. It is unnecessary to go at length into these calculations, which result from very simple considerations, but I will briefly state the results.

The difference of form between the tooth of one wheel and of another is due to two causes, (1) the difference of curvature, which is provided for in the Odontograph by placing the compasses at the different points of the scale of equal parts, (2) the variation of the angle DAE (fig. 63), which is met by placing the instrument upon the two radii in succession.

The first cause is the only one with which these calculations are concerned. Now in three inch pitch the greatest difference of form produced by mere curvature in the portion of tooth which lies beyond the pitch circle, is only ·04 inch between the extreme cases of a pinion of twelve and a rack, and in the acting part of the arc within the pitch circle is ·1 inch, so that as all the other forms lie between these, it is clear that if we select only four or five examples for the outer side of the tooth and ten or twelve for the inner side, that we can never incur an error of more than the $\frac{1}{200}$th of an inch in three inch pitch by always taking the nearest number in the manner directed, and a proportionably smaller error in smaller pitches. But to ensure this, the selected numbers should be so taken, that their respective forms shall lie between the extremes at equal distances. Now it appears that the variation of form is much greater among the teeth of small numbers than among the larger ones, and that in fact the numbers in the two following series are so arranged that the curves corresponding to them possess this required property.

For the outer side of the tooth, 12, 14, 17, 21, 26, 34, 47, 73, 148, Rack.

For the inner side, 12, 13, 14, 15, 16, 17, 19, 22, 26, 33, 46, 87, Rack.

Now these numbers, although strictly correct, would be very inconvenient and uncouth in practice if employed for a table like that in question, where convenience manifestly requires that the numbers, if not consecutive, should always proceed either by twos or fives, or by whole tens, and so on. They are only given as guides in the selection, and by comparing them with the actual table, their use in the formation of the first column will be evident.

If S therefore be the center of the pin, R will be that of the tooth which is to drive it, and the point m should be assumed somewhere between T and S, and Tm may be about half the pitch, Sm being manifestly the radius of the pin.

Again, if the side of the tooth (of the left-hand wheel, for example) is required to be a radial line, in imitation of the second solution (Art. 98), this, as already explained (Art. 135), will remove its tooth-center to an infinite distance, and the point k will be found by drawing Ak perpendicular to kTK. Join Bk, and the intersection of this line with PTQ will give the center of the tooth which is to work with the radial tooth; also AR, the perpendicular from A upon PTQ, is the radial tooth, and R is the point through which the arc must be struck, and the angle RTA must be of such a magnitude as will make TR equal to about half the pitch, since R is the point at which the exact action takes place.

145. The Odontograph is also applicable to the obtaining a correct form for the *cutters* used in forming metal wheels out of plain disks; for since the form of the cutter is that of the space between two contiguous teeth, we have only to describe a pair of teeth in any given case, in order to obtain the form of the cutter. In making, however, a set of cutters, especially for small pitches, it is by no means necessary to make one for every wheel, as the forms for numbers of teeth that lie together are so nearly alike, that the errors of workmanship would entirely destroy the difference.

The variation of form, however, is much less among high numbers than in low ones. For example, the difference of form between a cutter for 150 teeth, and one for 300, is not greater than that between cutters for 16 and 17 teeth.

This being the case, it appeared worth while to investigate some rule by which the necessary cutters could be determined for a set of wheels, so as to incur the least possible chance of error. To this effect I have calculated, by a method sufficiently accurate for the purpose, the following series of what may be termed equidistant values of cutters; that is, a table of cutters so arranged, that the same difference of form exists between any two consecutive numbers.

TABLE OF EQUIDISTANT VALUES FOR CUTTERS.

	1	2	3	4	5	6	7	8	9	10	11	12	13	14	15	16	17	18	19	20	21	22	23	24	25
No. of Teeth.	Rack.	300	150	100	76	60	50	43	38	34	30	27	25	23	21	20	19	17	16	×	15	14	13	×	12

This will be a guide in the selection of the wheel to which each cutter shall be accurately adapted after it has been determined how many are necessary in a set. For example, if a single cutter were thought sufficient for very small wheels, it had better be accurately adapted to teeth of 25, for that value is intermediate between the two extremes. If three cutters are to suffice for the whole set, then 76, 25, and 15 must be selected, of which the cutter 76 may be used for all teeth from a rack to 38, the cutter 25 from 38 to 19, and the cutter 15 from 19 to 12, and so on. I find that in the shapes of cutters, the greatest difference of form is at the apex of the tooth, (that is, at the base of the cutter,) and amounts to ·25 inch in 2-inch pitch, when the teeth have the usual addendum; from this the difference may be ascertained for any smaller pitch, and as many cutters interposed as the workman's notion of his own powers of accuracy may induce him to think necessary.

Thus if the hundredth of an inch be his limit of accuracy in forming cutters, and he is making a set for half-inch pitch, where the difference of form is $\frac{1}{4} \times$ ·25 or ·06 nearly, then half a dozen cutters will be sufficient, and these must be made as nearly as possible to suit the wheels of 150, 50, 30, 21, 16, 13.

146. In the epicycloid *abc* (fig. 35, p. 65) join Tb, and let $TOb = \phi$, $AT = R$, and $Tk = 2r$, then radius of curvature at $b = \frac{4r(R+r)}{R+2r} \cdot \sin\frac{\phi}{2}$ (*Peacock's Examples*, p. 195), and this radius passes through T, for Tb is a normal to *abc* at b. Now $Tb = 2r \cdot \sin\frac{\phi}{2}$, and it makes an angle with the line of centers $= \frac{\pi - \phi}{2} = \theta$, suppose; therefore $\sin\frac{\phi}{2} = \cos\theta$. Hence the distance of the center of curvature at b from T

$$= \left\{\frac{4r \cdot (R+r)}{R+2r} - 2r\right\} \cdot \sin\frac{\phi}{2} = \frac{2Rr}{R+2r} \cdot \cos\theta,$$

which expression becomes identical with the value of D in Art. 136, if we put $2r = \frac{C}{\sin\theta}$.

It appears therefore that if, in fig. 59, *mn* were an arc of an epicycloid whose base were the pitch circle, and diameter of the describing circle $= \frac{KT}{\sin\theta}$, then would Pm be its radius of curvature at m; and in like manner $Q'm$ can be shewn to be the radius of curvature of the corresponding hypocycloid *mp*.

Consequently teeth described by this method approximate to epicycloidal teeth, and when described in sets by the Odontograph, approximate to those of the third solution (Art. 113). Hence the rules that have been given for the least numbers, and the length or addenda of all such teeth, may also be applied to these.

147. In all the figures of teeth hitherto given the teeth are symmetrical, so that they will act whether the wheels be turned one way or the other. If a machine be of such a nature that the wheels are only required to turn in one direction, the strength of the teeth may be doubled by an alteration of form, exhibited in fig. 64.

This represents a portion of the circumference of a pair

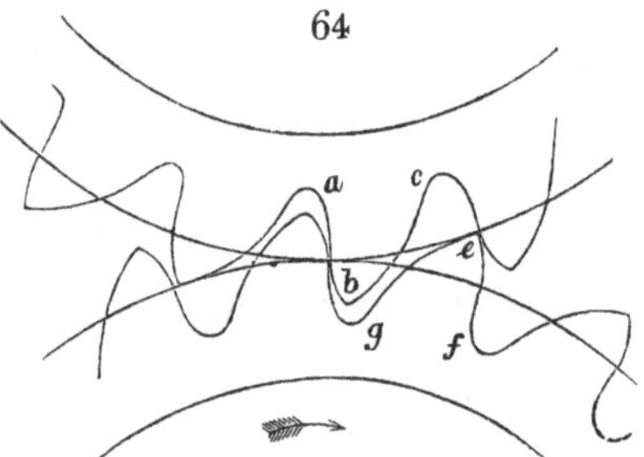

of wheels of which the lowest is the driver, and always moves in the direction of the arrow, consequently the right side of its teeth and the left side of the follower's teeth are the only portions that are ever called into action; and these sides are formed exactly as usual. But the back of each tooth, both in the driver and follower, is proposed to be bounded by an arc of an involute, as *eg* or *cb*.

The bases of these involutes being proportional to the pitch circles, they will during the motion be sure to clear each other, because, geometrically speaking, they would, if the wheels moved the reverse way, work together correctly; but the inclination of their common normal to the line of centers is too great for the transmission of pressure. The effect of this shape is to produce a very strong root, by taking away matter from the extremity of the tooth where the ordinary form has more than is required for strength, and adding it to the root.

148. In Hooke's system, under its second form (Art. 68), it has been shewn that the point of contact travels during the motion of the wheels from one side to the other; a fresh contact always beginning on the first side just before the last contact has quitted the other side. To ensure this, the teeth of the wheels in each section *B* (fig. 32) must be so formed that when the angular velocity ratio is constant the teeth may begin and end contact on the line of centers; otherwise, if the teeth were formed upon the principles of the previous Articles of this Chapter, it is evident that the sliding contact of the teeth before and after the line of centers would still remain. The simplest mode of effecting this object is to make the flanks of the teeth radial, as in the second solution, and their faces any arc of a circle that will lie *within* the epicycloidal face required by that solution. If, for example, the portion of tooth that lies beyond the pitch line be a complete semicircle whose center is upon that line, this condition will be complied with. I have described the teeth of *B*, fig. 32, in this manner. The figures *A* and *C* are nearly the same as Hooke's, but he has given no front view of his wheels, and has said nothing respecting the forms of the teeth.

To describe the teeth of wheels when their axes are not parallel.

149. *To describe the teeth of bevil-wheels*, let *ACT*, *ATD*, fig. 66, be the *pitch cones* of a pair of bevil-wheels described as in Arts. 43, 44; *AT* their line of contact. Let *AET* be any other cone also lying in contact with *ATD* along *AT*, and having its apex at *A*; therefore the axes of the three cones will be in the same plane *ABF*. Also the circumferences of their bases being at the same distance *AT* from *A*, will lie on the surface of a sphere whose center and radius are *A* and *AT*.

Let the three cones revolve round their axes with the

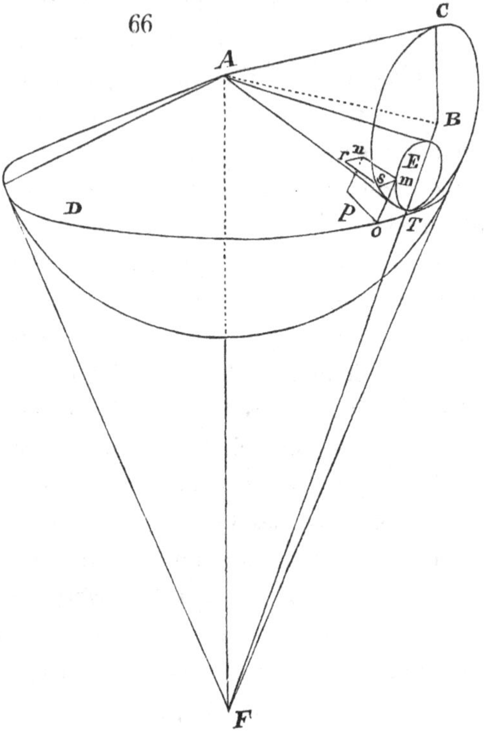

same relative velocity as would be produced by the rolling contact of their surfaces, then the line of contact will always be AT, and (calling the intermediate cone AET the *describing cone*) a line nm upon the surface of the describing cone directed to the common apex will generate one surface $ompn$ on the *outside* of the cone ATD, and another surface $smrn$ on the inside of the cone ACT.

Also, these surfaces will touch along the describing line nm, for since $ponm$ is generated by the rolling of the describing cone upon the surface of the cone ADT, the motion of nm is at every instant perpendicular to the line of contact AT; and therefore, the normal plane at nm to the surface generated by nm will pass through AT. And in like manner, the normal plane to the surface $rsnm$ will pass through AT'; therefore the surfaces touch along nm.

If these surfaces be employed as teeth, and the rotation of the cone ATD be communicated to the cone ACT by their contact action, the angular velocity ratio will, from the mode of their generation, be precisely the same as that produced by the rolling contact of the conical surfaces; for at the beginning of the motion op and rs coincide with AT, and in the position of the figure the arcs To, Ts respectively described by the bases of the two cones are each equal to Tm, and therefore themselves equal.

150. The arc om is an arc of a spherical epicycloid* whose base is the cone ADT, and describing cone the cone AET; and in like manner sm is an arc of a spherical hypocycloid whose base is the cone ATC, and describing cone AET. But in practice, the portion of spherical surface occupied by these arcs, when employed for teeth, is a narrow belt extending to a small distance only from ToD

* DEFINITION. If a cone ABC be made to roll upon another fixed cone $ADCE$ in such a manner that their summits A always coincide; then a tracing

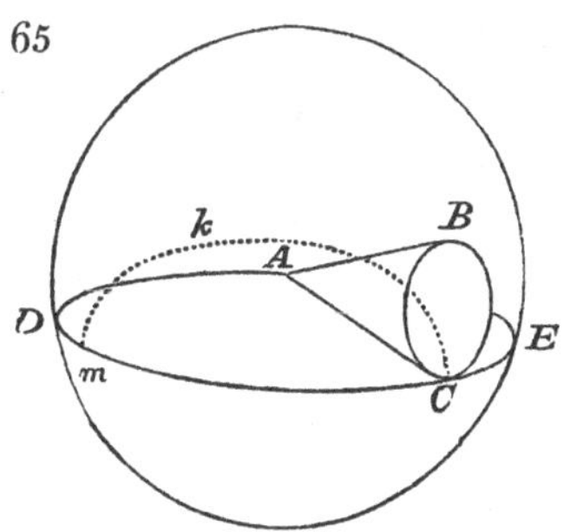

point C in the circumference of the base of the rolling cone will trace a kind of epicycloid Ckm, which will plainly lie on the surface of a sphere whose center is A and radius AC, whence this curve is termed a *spherical epicycloid*. If the cone roll on the concave surface of the base, the curve becomes a *spherical hypocycloid*.

and TsC. The surface, therefore, of cones tangent to the sphere along TD and TC may be substituted for that of the sphere itself, as follows: draw BTF perpendicular to AT, and intersecting the axes of the two cones in F and B; then BF revolving round AF will generate a conical surface tangent to the sphere along the base TD of the cone ADT, and the same line BF revolving round AB will generate a conical surface tangent to the sphere along the base TsC of the cone ACT.

And since the arc mo, which really lies in the spherical surface, is very short in practice, it may be supposed, without sensible error, to lie in the surface of the tangent cone FTD, and to be described with a circle whose diameter is equal to that of the base of the describing cone. And in like manner, the arc sm may be supposed to be described with a similar circle upon the surface of the tangent cone BTC.

151. Now by developing these conical tangent surfaces into planes we obtain a ready practical mode of describing the teeth, which was first suggested by Tredgold*.

Let AB, AC, fig. 67, be the axes meeting in A, AT the line of contact, l, k the rolling frusta described by Art. 44. Draw BTC perpendicular to AT, and meeting the axes in B and C; with center B and radius BT describe a circle Tf, and with center C and radius CT describe a circle Te. Also describe the frusta n and m which will be frusta of the tangent cones to the spherical surface at the bases of the rolling frusta l and k, as above explained. The circle Tf will be the developement of the face of n, and Te that of the face of m; and it follows from the demonstration in Art. 150, that if the circumferences of these circles be treated as the pitch lines of a pair of ordinary spur-wheels, and teeth

* Buchanan's Essays on Mill-work, by Tredgold, 1823, p. 103; or new edition, 1841, p. 59.

described upon them according to any of the rules laid down

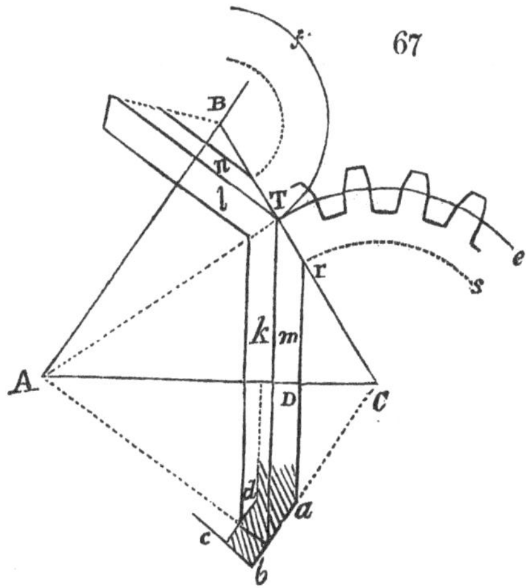

for such wheels, that these teeth when transferred to the conical surfaces will communicate the desired constant velocity ratio. The following practical mode of completing the bevil-wheel is easily deduced from the above.

152. Prepare a solid of revolution whose axis is AC, and the section of whose edge is represented at $abcd$, as bounded by two parallel conical surfaces ab, cd, and by a third cb, whose generating line is directed to A.

This surface is to be cut into teeth, and therefore the portion cb projects beyond the surface of the pitch cone, by a sufficient quantity to contain the projections of the teeth beyond that surface, as shewn at Te. For the surface ab is plainly the same as that which has been developed at $Ters$. The teeth there figured must be cut out of thin metal and wrapped round this conical surface, so as to allow their outlines to be traced upon it. They may then be cut out, observing that a line passing through A must lie in complete contact with every point of the side of the tooth contained between ba and cd, or in other words, that the acting surfaces of the teeth are generated by the motion of a line one of whose extremities always passes through A, and the other is made to follow the outline traced out upon the surface ab.

The usual method for large wheels is to develope also the interior surface cd, making a new construction for it precisely similar to that employed for the exterior surface ab.

If separate wooden cogs are employed, they are first fitted and fixed into their mortises, then the conical surfaces ab, bc, cd formed upon them in the lathe, and the outlines of the two ends ab, cd traced by patterns derived from the two constructions. They are then taken out separately, and easily shaped by careful planing in straight lines from one outline to the other. The same method is employed for the large wooden patterns that are used in casting wheels, and in which the teeth are made in separate pieces, to allow of this method of shaping.

153. Let the radius TD of the base of the frustum $k = R$, and the radius TC of the developed pitch circle $= r$. Also the semiangle TAD of the rolling cone $= K$; therefore $r = \frac{R}{\cos K}$. Whence the action of the teeth in any bevil-wheel is equivalent to that of a spur-wheel of the same pitch whose radius is $\frac{R}{\cos K}$; also if N be the number of teeth in the bevil-wheel, $\frac{N}{\cos K}$ will be those of the spur-wheel.

This is a reason for the superior action of bevil-wheels over spur-wheels of the same number of teeth, for spur-wheels always act the better the more teeth they have, and it appears that a bevil-wheel is always equivalent in its action to a spur-wheel of a greater number of teeth.

When a pair of bevils have equal numbers of teeth, and their axes are at right angles, they are termed *mitre-wheels;* in this case

$$\theta = 45^0, \text{ and } \frac{1}{\cos\theta} = 1.4;$$

therefore the action of a mitre-wheel is nearly equivalent to that of a spur-wheel with half as many more teeth.

154. *Face-wheel geering* (Art. 62) is almost driven out of practice by the employment of bevil-wheels; but it may be sometimes used with advantage, and its principles are worth investigating.

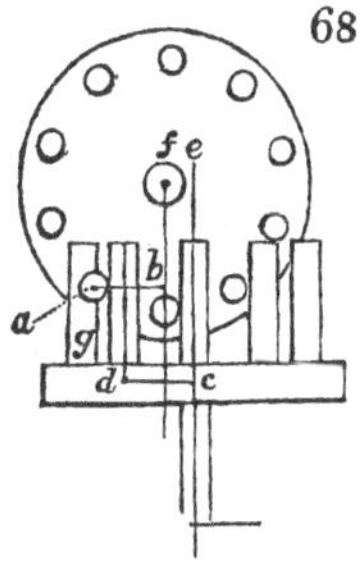

Let two face-wheels with cylindrical pins exactly alike in every respect be placed in geer, as in fig. 68, with their axes at right angles; not meeting in a point, but having their common perpendicular fe equal to the diameter of the pins. Then will these wheels revolve together with a constant angular velocity ratio.

68

For let the pin whose center is a in the upper wheel, be in contact with the pin whose axis is at d in the lower wheel. Draw fb parallel to the axis of the lower wheel, and ab perpendicular upon fb. Also through c the center of the lower wheel draw a line parallel to the axis of the upper wheel, and therefore perpendicular to the plane of the paper, and let dc be a perpendicular upon this line from the axis of the pin d, therefore ab is the sine of the angular distance of a from fb, which is parallel to the axis of the lower wheel, and dc is the sine of the angular distance of d from a line drawn through c parallel to the axis of the upper wheel. But a is removed to the left of d by a horizontal distance equal to the diameter of the pins, and b is removed to the left of c by a horizontal distance equal to *fe*, which is also by hypothesis equal to the diameter of the pins; therefore $ab = dc$, and the angular motion is equal.

The pin *a* appears in the figure to cut the pin *g*, but a little consideration will shew that the circular motion of the lower wheel removes this pin to a sufficient distance from the plane of the upper wheel to clear the ends of the pins of the latter.

155. If, however, which is generally the case, the diameter of the wheels be different and their axes meet, then supposing one of them, as in this figure, to have cylindrical pins or staves, the other must have cogs whose acting surfaces are those of solids of revolution. The axes of these solids may, or may not, coincide with the centers of the cogs. If they do, the cogs are easily formed in the lathe. The generating curve of these solids may be found as follows.

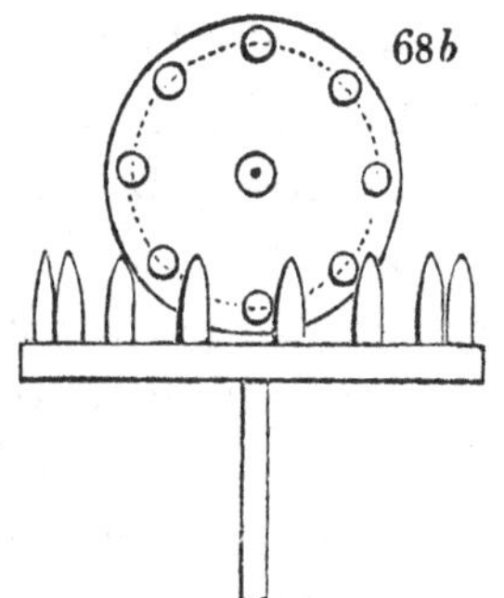

In fig. 69, *C* is the center of the pin-wheel, the pins of which are supposed to have no sensible diameter, the axis of the pin-wheel is perpendicular to the plane of the paper, and that of the cog-wheel is parallel to it, and meets the first axis in a point whose projection is *C*. PAP' is the pitch line of the pin-wheel, and $pAm_{,}$ the projection of the pitch line of the cog-wheel. *A* their point of contact.

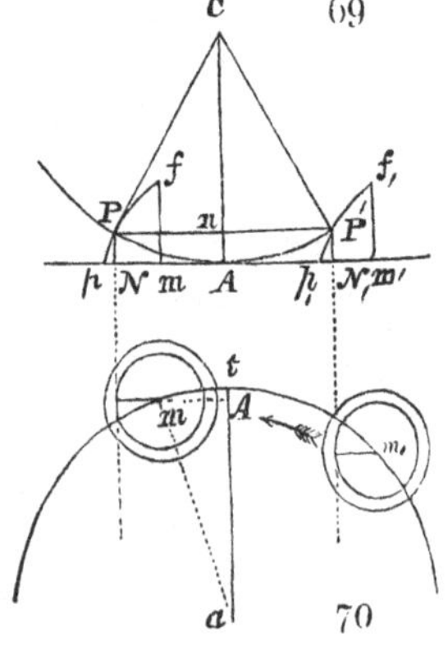

Let P be one of the pins, fm the axis of the solid of revolution, or cog, which is to work with it, pPf the generating curve of the solid. Fig. 70 is a plan of the cog-wheel, t the point of contact, m the seat of mf, and the concentric circles the plans of the cog; the large one at the level of mp, and the small one at the level of Pn.

Let the radius $AC = r$, $at = R$, and the angular distance of mf from the plane of centers, or $maA = \phi$;

$$\therefore \; Am = R \,.\, \sin \phi.$$

Let $mN = x$, $NP \; (= An) = y$, $ACP = \theta$, $mp = \rho$;

then we have (1) $y = r \,.\, \text{versin } \theta.$

$$x = r.\sin \theta - R \sin \phi, \text{ for } Nm = Pn - Am.$$

Also, since the velocities of the pitch circles are equal by supposition, and p and P coincide at A, therefore the arc AP in fig. 69, must be equal to the arc tm in fig. 70, + the radius mp of the base of the solid very nearly.

$$\therefore \; \phi = \frac{r\theta - \rho}{R}, \text{ and } x = r \sin \theta - R \sin \left\{ \frac{r\theta - \rho}{R} \right\} \; (2).$$

From (1) and (2) the curve pPf may be constructed by points, and a curve for a pin of any required diameter derived from it, by tracing it at a normal distance from pPf equal to the radius of the pin, as in the case of common trundles (Art. 87.).

156. The cog pPf, supposing it to drive, is necessarily moving in the direction of the arrow, and receding from the plane of centers; but if we consider the relative positions of the approaching pin $P_{\prime}$ and cog $p_{\prime}P_{\prime}f_{\prime}$ on the other side of the plane of centers, at an equal angular distance θ, and therefore with the same value of y, we get the corresponding value of x, or $x_{\prime} = R \sin \phi_{\prime} - r \cdot \sin \theta \; (\phi_{\prime} = m_{\prime}at)$,

$$\text{and } \phi_{,} = \frac{r\theta + \rho}{R};$$

$$\therefore\ R\phi_{,} - r\theta = \rho = r\theta - R\phi;$$

whence it follows that

$$R\,.\sin\phi_{,} - r\,.\sin\theta < r\,.\sin\theta - R\,.\sin\phi;$$

that is, $x_{,} < x$.

This curve $p_{,}P_{,}f_{,}$, therefore, is not the same as pPf, but will lie within it. But if the cogs are turned in the lathe, the axis of the solid of revolution will coincide with the center, and the smallest curve of the two must necessarily be used; and therefore the action will only be maintained while the cylindrical pin lies between the cog and the plane of centers; and as receding action is preferable to approaching action, it follows that the cylindrical pin must be given to the driver and the cogs to the follower, if the cogs be turned in a lathe. But fig. 70 shews that the point of contact of the cogs on one side of the line of centers, as $m_{,}$, is very nearly confined to the half of each cog which lies within the pitch circle, and that on the other side as m to the portion which lies without. By making the outer portion of each cog of the form $p_{,}P_{,}f_{,}$ and the inner portion of the form pPf, we may have action on both sides the plane of centers at pleasure.

157. This shews the *possibility* of forming the cogs of face-wheels so as to communicate motion with a constant velocity ratio. In practice, the form of the cogs may be obtained by finding two or three points for the curve pPf, which may be done on the drawing board by constructing a diagram similar to figs. 69 and 70, but in which the cogs and pins shall be placed in two or three different successive distances from the plane of centers. In small wooden mill-

work, the cogs used to be turned in the lathe and with round shanks, and consequently made complete solids of revolution, as in fig. 68 *b*; but in the larger wheels each cog had its acting face shaped into segments of solids of revolution of considerably greater diameter than the cog itself.

In this kind of geering, however, the surfaces of the teeth touch only in a single point*; while in bevil-geer, as we have seen (Art. 65), the contact is along a line directed to the point of intersection of the axes. The abrasion is therefore less in the latter, but the convenience of forming the cogs in the lathe sometimes occasions the face-geer to be used even now in light machinery or models.

In face-geering, a derangement in the relative position of the wheel and trundle, if it take place in a line parallel to the axis of the latter, will not interfere with the action of the geering.

158. The surfaces adapted for teeth in the case of rolling hyperboloids, Art. 45, might be obtained in a similar manner to those of rolling cones, by taking an intermediate describing hyperboloid; but it does not appear possible to derive from this any rules sufficiently simple for application. This kind of wheels is only employed to enable the two axes to pass each other, which is impossible in conical wheels; and, on account of the imperfection of their rolling action, explained in Art. 46, the axes should be brought as close together as possible, by which the solids will approximate nearly to a pair of rolling cones. The teeth should be small and numerous, and therefore the frusta should be

* To use the words of a practical American millwright, in speaking of wooden face geers, "the disadvantage of face geers is the smallness of the bearing, so that they wear out very fast, for if the bearing of cogs be small, and the stress so great that they cut one another, they will wear exceedingly fast; but if it be so large and the stress so light that they only polish one another, they will wear very long." —Oliver Evans, *Young Millwright's Guide*, Art. 80.

placed as far as convenient from the common perpendicular of the axes. When the frusta have been described by Art. 47, the forms of the teeth may be obtained with sufficient approximation by treating these frusta similarly to those in fig. 67, that is, draw a line perpendicular to tP at P (fig. 15, Art. 47), this will intersect the axis at some point beyond K; take this point for the apex of a cone whose base shall coincide with that of the rolling frustum KP, develope its surface, and describe the teeth as in Art. 51.

An interior surface, corresponding to cd in fig. 67, must also be developed and the teeth traced upon it; the relative position of these interior forms to those already traced upon the exterior surface, will be determined by drawing an inclined line at the pitch surface, according to the method of Art. 67.

The principal machine in which these skew bevils are employed is that which is known by the name of the Bobbin and Fly Frame, in the cotton manufacture.

159. *To communicate motion by means of involutes between two axes inclined without meeting**.

Fig. 56 represents, as already explained, a pair of wheels whose teeth are formed of arcs of involutes, the point of contact of which is always situated in the common tangent DE of the bases.

In this figure the wheels are in the same plane, and their axes consequently parallel. Suppose now that the plane of one wheel be inclined to the other by turning on the line DE, in the manner of a hinge, so that this line shall be the intersection of the two planes, but that the position of each wheel in its own plane with respect to this line shall not be altered.

* This property of involutes is due to M. Ollivier. Vide *Bulletin de la Soc. d'Encouragement.* tom. XXVIII. p. 430.

The inclination of the axes will be however changed, but they will not meet, and their common perpendicular will be equal to *DE*. Since *DE* is the locus of contact, it is clear that this motion will not disturb the angular position of either wheel in its own plane; and hence the angular velocity ratio of the wheels will remain constant and unaltered by the change of position. Involute wheels, therefore, may be employed to communicate a constant velocity ratio between axes that are inclined at any angle to each other, but which do not meet.

But the demonstration supposes the wheels to be very thin, since they coincide with the planes that meet in the line *DE*, and the invariable points of contact are situated in this line. The edge of one of the wheels must be in practice rounded so that it may touch the other teeth in a point only.

ON CAMS AND SCREWS.

---

160. Having disposed of the teeth of wheels, we may now return to the remaining combinations in which sliding contact is employed to communicate a constant velocity ratio between two pieces.

If the motion of these pieces be limited to a not very considerable angle, or if one of them moves in a short rectilinear path in the manner of a rack, any of the pairs of curves in the first part of this chapter (in Arts. 75 to 85) may be employed in the single forms there shewn, instead of being reduced to short arcs, and placed in successive order as teeth. To avoid unnecessary details, I shall confine myself to the examination of the cases in which one of these curves is reduced to a pin, as in the First Solution: for this method is generally preferred, and it has this advantage, that whereas greater friction is introduced when a long curved plate is substituted for a series of teeth*, the pin can be made into a roller, and thus the abrasion which would tend to destroy the form of the curved edge is transferred to the axis of the roller, which can be easily repaired when worn out.

161. In fig. 71, A is the centre of motion of a revolving plate in which a slit ab is pierced, having parallel sides so as to embrace and nearly fit a pin m, which is carried by a bar CD fitted between guides so as to be capable of sliding in the direction of its length.

* By carrying the point of contact farther from the line of centers (Art. 34.)

If the plate revolve in the direction of the arrow the inner side of the slit presses against the pin and moves it further from the center A, but when the plate revolves in the opposite direction the outer edge of the slit acts against the pin and moves it in the opposite direction.

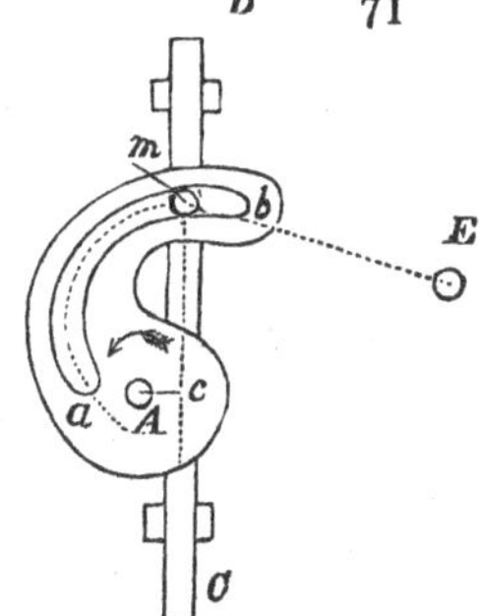

If the curved edges of the slit be involutes of the circle whose radius is Ac where Ac is a perpendicular upon the path mc of the bar, it appears from Art. 91 that the velocity ratio of plate and bar will be constant, and the linear velocity of the bar equal to that of the point c of the plate. But if any other velocity ratio be required, let Pc (fig. 72) be the path of the sliding bar, P the pin, A the center of the curve, aP the curve.

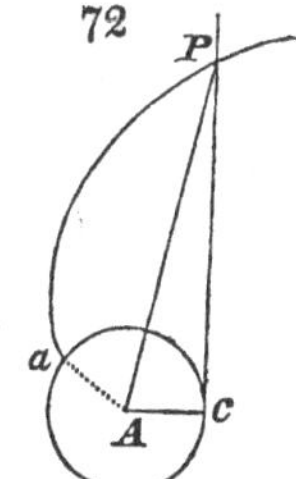

Let $cAP = \phi$, $PAa = \theta$, $Ac = a$, $AP = r$,

then while a has moved from c to a, let P have moved from c to P; so that $ca = m \times cP$; preserving a constant velocity ratio during the motion;

$$\therefore \theta + \phi = m \times \tan \phi.$$

$$\text{But } \tan \phi = \frac{\sqrt{r^2 - a^2}}{a}, \text{ and } \phi = \cos^{-1} \frac{a}{r};$$

$$\therefore \theta + \cos^{-1} \frac{a}{r} = \frac{m}{a} \sqrt{r^2 - a^2} \text{ is equation to curve.}$$

If the velocity of the circumference of the circle (radius Ac) equals the linear velocity of the bar,

$$ca = cP, \text{ and } \therefore m = 1;$$

$$\therefore\ \theta + \cos^{-1}\frac{a}{r} = \frac{\sqrt{r - a^2}}{a};$$

which is the equation to the involute of the circle as it ought to be*.

If, however, the line Pc of the follower's path pass through the center A, then since equal angles described to the curve are to produce equal differences of radial distance in the pin, the curve becomes evidently the spiral of Archimedes; a curve which, although, as we see, capable of communicating velocity in a constant ratio between a circular and rectilinear path, cannot be employed for the teeth of racks, because the pitch line passes through the center of the wheel.

162. Sometimes the pin, instead of being mounted on a slide, is carried by an arm revolving round a center E, as mE, and therefore describes an arc of a circle. The curve is then derived from the first solution (Art. 87), the line of centers AE having been previously divided, in the ratio of the required angular velocities.

The angular motion of the curved plate which is the driver is of course limited to the length of the slit ab, but this may be carried through several convolutions, as in fig. 73, where it is shewn in the form of a spiral groove, excavated in the face of a revolving plate, and communicating rectilinear motion to the bar Dm by means of the pin at its extremity m, which lies always in the groove.

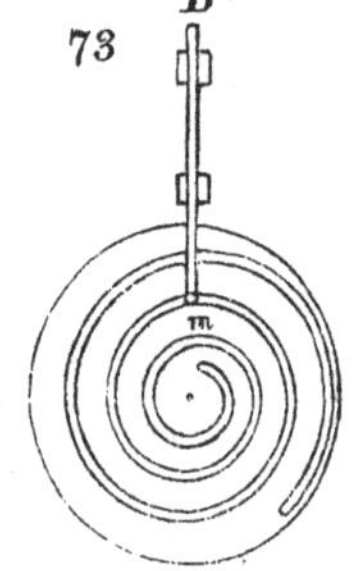

This may be termed a *flat screw* or *plane screw*.

163. Combinations of this kind assume a great many different forms, the complete exhibition of which belongs

* Peacock's Examples, p. 177.

rather to descriptive mechanism than to the plan of the present work. Thus, instead of employing the slit or groove, shewn in these figures, the object of which is to produce action in both directions, a single curved edge may be employed, and the returning action produced by a weight or spring, which may be applied to the bar so as to keep the pin constantly in contact with it.

Curved plates of this kind are termed Cams, or, when small, Tappets, and they are more used to produce varying velocity ratios than constant ones. For which reason I shall refer to Chap. VIII. for so meother forms in which they appear.

164. If the path both of driver and follower be rectilinear, the slit will become straight.

Let a plane rectangle CD move in its own plane, in a

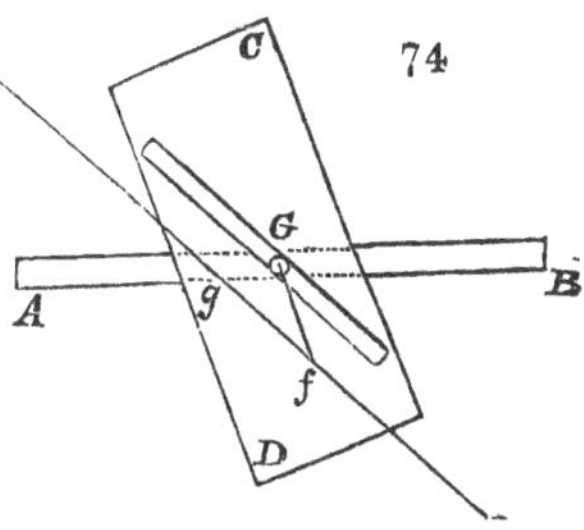

path parallel to its longest side, and have a straight slit cut in it making an angle θ with that side, and let a bar AB moving in the direction of its own length below this plane be provided with a projecting pin G which enters the slit, the slit making an angle ϕ with the path of this bar. Therefore the paths of the plane and bar make an angle $\overline{\theta + \phi}$ with each other.

If the plane move through a space $= Gf$, draw gf parallel to the first position of the slit, then g will be the new position of the pin, and Gg the space described by the pin or bar;

$$\therefore \frac{\text{velocity of plane}}{\text{velocity of bar}} = \frac{Gf}{Gg} = \frac{\sin Ggf}{\sin Gfg} = \frac{\sin \phi}{\sin \theta},$$

a constant ratio.

If the bar move perpendicularly to the plane, $\theta + \phi = \frac{\pi}{2}$,

$$\text{and } \frac{\text{velocity of plane}}{\text{velocity of bar}} = \tan \phi \text{ or } \frac{1}{\tan \theta}.$$

165. To return to the revolving plate and bar; if the path of the bar be not parallel to the plane of rotation of the plate, the latter must be formed into the cone or hyperboloid that would be generated by the rotation round its axis of the line which is the path of the pin, or other point of contact of the bar. Thus, in fig. 75, AB is the axis, CD the sliding bar, e its pin, the path cd of whose acting extremity is in this case supposed to meet the axis. If this line cd generate a cone D by revolving round AB, the pin will always lie at the same depth in any groove excavated in the conical surface. Also, if this surface be developed, the groove ef will be the spiral of Archimedes. It is unnecessary to follow into detail all the forms, curves, and combinations, that arise in this manner. One case only requires more particular attention.

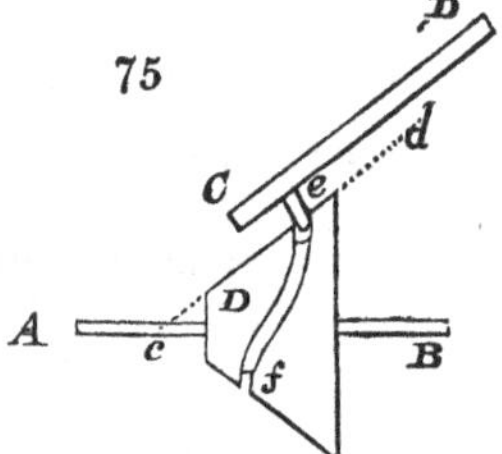

166. If the path of the bar CD be parallel to the axis of rotation AB, the conical surface upon which the groove is traced, will become a cylinder; and to produce a constant velocity ratio the spiral groove must be at every point

equally inclined to a line drawn upon the surface parallel to the axis.

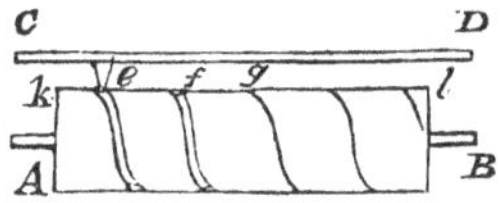

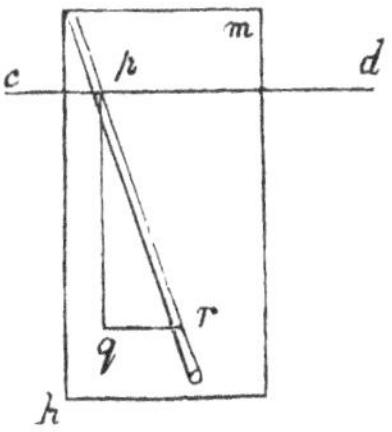

For it has been shewn that a plane surface mh, fig. 76, moving perpendicularly to a sliding bar cd, will communicate motion to it in a constant ratio, by means of a straight slit pr in which lies a pin fixed to the bar, and that

$$\frac{\text{velocity of plane}}{\text{velocity of bar}} = \tan\phi;$$

76

where ϕ is the angle rpd made by the slit with the path of the bar.

If this plane be wrapped round the cylinder, keeping its axis parallel to the path of the bar, the groove will become a spiral, inclined at the angle ϕ to a line drawn parallel to this axis. But the motion given to the bar by this spiral when the cylinder revolves will be exactly the same as if the plane had passed under it through the line kl and perpendicularly to the plane of the paper.

The velocity of the plane is now the velocity of rotation of the cylindrical surface, and therefore we have, if r be the radius of the cylinder, A its angular velocity, V the velocity of the bar,

$$\frac{rA}{V} = \tan\phi.$$

If the length of the plane be greater than the circumference of the cylinder, the spiral groove will encompass its surface through more than one revolution, and may, in this way, proceed in many convolutions from one extremity of the cylinder to the other, its inclination to the axis of the cylinder remaining constant and equal to ϕ; such a recurring spiral is termed *a screw*.

Draw pq, qr respectively perpendicular and parallel to the path of the bar; if pq is equal to the circumference of the cylinder, qr will be the distance between two successive convolutions of the screw, and $qr = \frac{2\pi r}{\tan\phi}$. This is termed the *pitch* of the screw, from its analogy to the pitch of a rack or toothed wheel. Every revolution of the screw carries the bar through a space equal to the pitch.

167. The screw is sometimes made in this elementary form, consisting of a simple spiral groove, with distant convolutions, which gives motion to a slide, by means of a pin fixed to the latter, and lying in the groove; for example, the screw by which the wick of the common Argand lamp is adjusted in height is always made in this form. But, generally, screws receive a more complex arrangement, in the following manner.

Firstly, the inclination of the spiral to the axis is made small, and the convolutions of the groove brought close together. The ridge which separates two contiguous grooves is a spiral precisely resembling that of the groove in inclination, and in the number and pitch of its convolutions. This ridge is termed the thread of the screw, and according to the form of its section, the screw is said to have a square thread as at *A*, an angular thread as at *B*, or a round thread as at *C*.

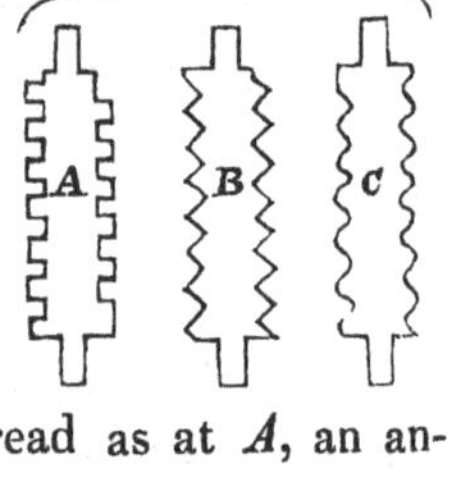

Secondly, instead of a single pin e let other pins f and g be also fixed to the bar opposite to the other convolutions; then, since each pin will receive an equal velocity from the revolving cylinder, the motion of the bar will be effected as before, with the advantage of an increased number of points of contact. But this series of pins is generally thrown into the shape of a short comb, the outline of which exactly fits that of the threads of the screw, as at *C*, fig. 78*. This is the most ancient form in which the screw was employed. It appears to be that which is described by Pappus†.

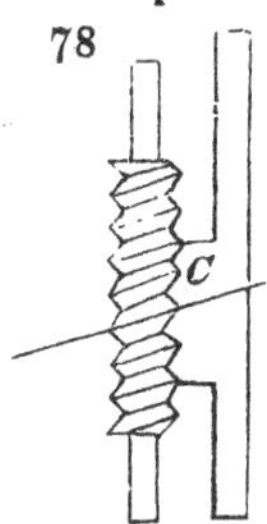

168. Most commonly, however, the piece which receives the action of the screw is provided with a cavity embracing the screw, and fitting its thread completely, as shewn in section in fig. 79, being in fact a hollow screw, corresponding in every respect to the solid screw. Such a piece is termed *a nut*, and the hollow screw, a *female screw*.

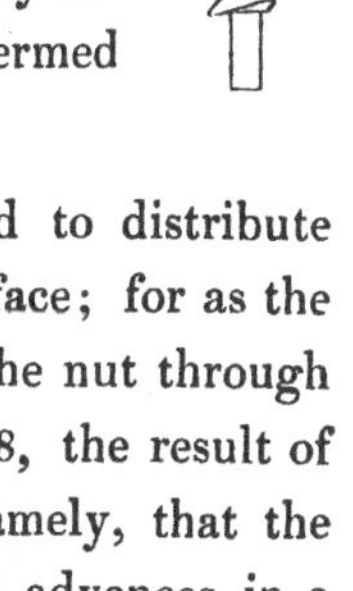

These modifications are only introduced to distribute the pressure of the screw upon a greater surface; for as the action of the thread upon every section of the nut through its axis is exactly the same as that of fig. 78, the result of all these conspiring actions is the same: namely, that the piece to which the comb or nut is attached advances in a direction parallel to the axis of the screw, and describes a space equal to as many pitches as the screw has performed revolutions.

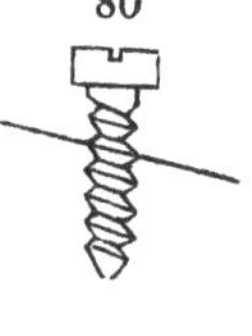

169. A screw may be *right handed* or *left handed*, that is, looking at the screw in a vertical position, the thread may incline upwards to the *right*, as in fig. 78, or to the *left*, as in fig. 80.

* The same expedient may be resorted to in the flat spiral of fig. 73, which is, in fact, a flat screw; and on the same principle a screw may be formed on a conical or hyperboloidal surface.

† Pappi Math. Col. Commandini. lib. VIII. p. 332.

170. When the comb or rack form (fig. 78) is used instead of a nut, this farther modification is sometimes employed, that the screw is made short and the rack lengthened, as in fig. 81. In both these cases, the length of the path that may be described by the bar, without allowing any portion of the screw or rack to quit contact at the extremities of the motion, will be the difference between the lengths of the screw and rack.

81

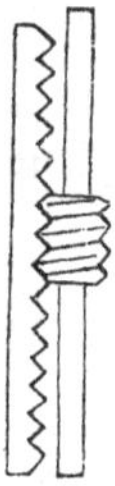

From this latter modification, we easily pass to the so-called *endless screw**. In this contrivance, the screw C is employed to communicate rotation to a revolving follower or wheel B. An axis Aa is mounted in a frame, so as to prevent its endlong motion, and provided with a short screw C. The wheel B has its edge notched into equidistant teeth of the same pitch as the thread of the screw with which they are in contact. If the screw axis be turned round, every revolution will cause one tooth of the wheel to pass the line of centers BC; and as this action puts no limit, from the nature of the contrivance, to the number of revolutions in the same direction, a screw fitted up in this mode is termed an endless screw, in opposition to the ordinary screw, which when turned round a certain number of times either way, terminates its own action by bringing the nut to the end of its thread; the term endless applying in this case not to the form but to the action of the screw.

82

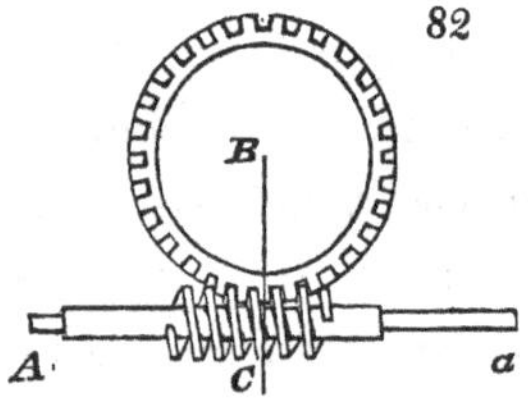

171. To determine the form which should be given to the thread and teeth in this contrivance, it may be re-

* Also described by Pappus in the Article already referred to; also lib. VIII, Prop. 24.

marked, that from the nature of a screw the section of its thread made by a plane passing through its axis is everywhere the same; and that if a series of such sections of the entire screw be made by planes at equal angular distances round the circle, a set of similar figures resembling a double rack (as in fig. 77,) will be obtained alike in the number and form of their teeth, but in which the teeth will gradually approach nearer and nearer to the extremity of the screw. The action of the screw upon the wheel-teeth, in revolving without end play, brings these successive sections into action upon the teeth, and produces exactly the same effect as if the screw were pushed endlong without rotation, in the manner of a rack*. But this latter supposition enables us to obtain the figure of the thread and teeth, upon the principles already given for the teeth of racks.

Fig. 83 is a transverse section of a wheel and endless

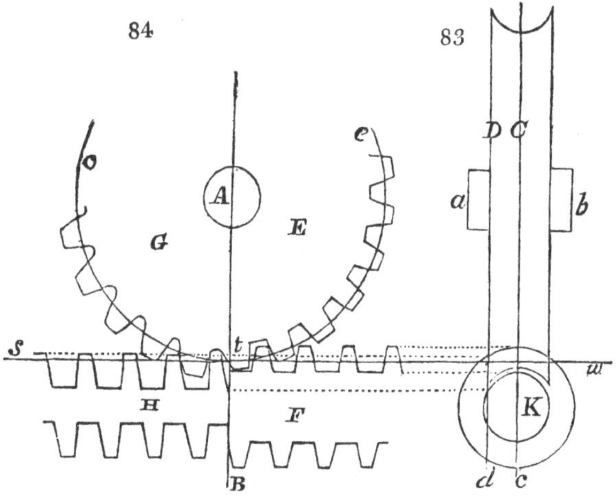

screw, made through the line of centers; *ab* the axis of the wheel, *K* that of the screw; fig. 84 represents the correspond-

* Thus if a screw be held to the light and turned round, the outline of its threads will appear to travel from one end of the screw to the other continually, in the manner of the teeth of a sliding rack.

ing sections, in which *AB* being the line of centers, the section to the right of this line is made by a plane passing through the axis of the screw, and through the line *Cc*, fig. 83; and the section to the left of the line of centers in fig. 84 is made by a plane passing through the line *Dd*, fig. 83 on one side of the axis of the screw, and parallel to the first. The effect of this is, that *F* is a direct section of the screw, while *H* is an oblique section: also, *cte* is the pitch circle of the wheel, and *stw* the pitch line of the screw, supposing it to act as a rack.

Nevertheless, according to the supposition already made, it appears that in these two sections, and in any other parallel to them within the wheel, the screw is required to act as a rack upon the teeth of the wheel. But whatever figure be given to the screw-thread, it is seen that the forms of these racks will necessarily be different in each section; for although the form of the thread is the same in all, it is cut at a different angle in each section, by which the teeth of *H* remote from the axis will be more prolonged and twisted in their form than those of *F* in the central section; and besides this, the successive racks will retire further from the center *A* of the wheel, as their section recedes from the axis of the screw; as shewn in the figure in which the rack-teeth *H* are lower than in *F*.

Now it has been already shewn (Art. 84), that any form of tooth being assumed, the corresponding tooth *may be* assigned.

The forms of the teeth in the central plane *E* may therefore be made to suit those of *F*, and the forms of the teeth in *G* may also suit those of *H*; and so on for every intermediate section. It is therefore *possible* to make an endless screw whose thread shall be in contact with the entire side of the tooth, provided the figure of the wheel-teeth be different in every section. Also, since in every section two or three pairs of teeth may be in simultaneous contact, the screw may be in contact along the entire side of all these teeth.

172. The practical difficulty of making the teeth of a wheel of which the form in every parallel section shall be different, is very simply overcome by making the screw cut the teeth, thus:

An endless screw is formed of steel, exactly the same as the proposed one, and this is notched regularly across its threads so as to convert it into a cutting instrument or tap, and then properly hardened. The wheel having had its teeth roughly cut in the proposed number, is mounted in its frame, together with the cutting screw, and the latter is turned in contact with it, and pressed gradually nearer and nearer, cutting out the teeth as it proceeds, till it has formed them to correspond exactly with its thread; it is then taken out and replaced by the smooth threaded screw.

173. The endless screw falls under the case of two revolving pieces whose axes are not parallel and never meet. It communicates motion very smoothly, and is equivalent to a wheel of a single tooth, because one revolution passes one tooth of the wheel across the plane of centers; but, generally speaking, can only be employed as a driver, on account of the great obliquity of its action.

174. In a cutting engine by Hindley of York, an endless screw of a different form was introduced, which is thus described by Smeaton:—" The endless screw was applied to a wheel of about thirteen inches diameter, very stout and strong, and cut into 360 teeth. The threads of this screw were not formed upon a cylindrical surface, but upon a

solid whose sides were terminated by arches of circles. The whole length contained fifteen threads, and as every thread (on the side next the wheel) pointed towards the center thereof, the whole fifteen were in contact together, and had been so ground with the wheel, that, to my great astonishment, I found the screw would turn round with the utmost freedom, interlocked with the teeth of the wheel, and would draw the wheel round without any shake or sticking, or the least sensation of inequality*."

"The screw was cut by the rotation of the point of a tool, carried by the wheel itself, the wheel being driven by an ordinary cylindrical endless screw."

Fig. 85 shews this form of endless screw, and fig. 86 is

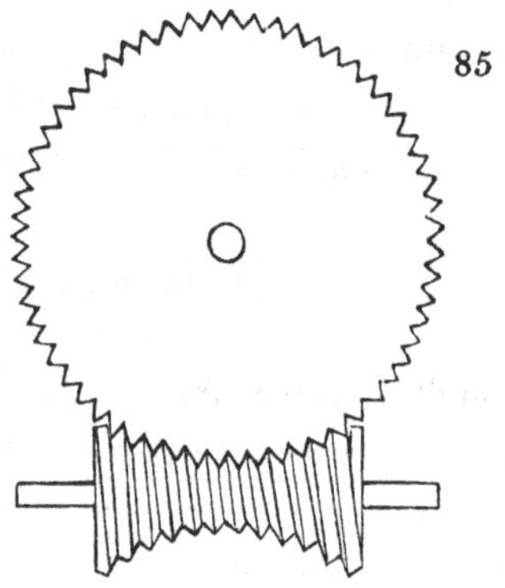

an arrangement to shew the manner of cutting the spiral thread upon the solid, in which *A* is a wheel driven by an endless screw *B*, of the common form; *C* a toothed wheel fixed to the axis of the endless screw and geering with another equal toothed wheel *D*, upon whose axis is mounted the smooth surfaced solid *E*, which it is desired to cut into Hindley's endless

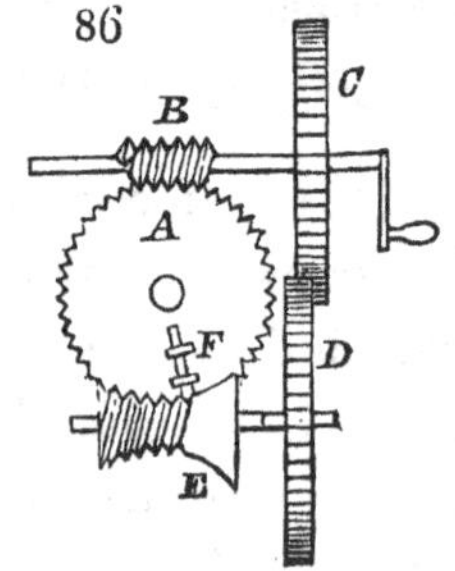

* Smeaton, p. 183, Miscellaneous Papers.

screw. For this purpose a cutting tooth *F* is clamped to the face of the wheel *A*. When the handle attached to the axis *BC* is turned round, the wheel *A* and solid *E* will revolve with the same relative velocity as *A* and *B*, and the tooth *F* will trace upon the surface of the solid a thread which will correspond to the conditions. For from the very mode of its formation the section of every thread through the axis will point to the center of the wheel. The axis of *E* lies considerably higher than that of *B*, to enable the solid *E* to clear the wheel *A*.

The edges of the section of the solid through its center, exactly fit the segment of the toothed wheel, but if a section be made by a plane parallel to this, the teeth will no longer be equally divided, as they are in the common screw; and therefore this kind of screw can only be in contact with each tooth along a line corresponding to its middle section. So that the advantage of this form over the common one is not so great as appears at first sight.

175. If the inclination of the thread of a screw to the axis be very great, one or more intermediate threads may be added, as in fig. 87. In which case the screw is said to be double, or triple, according to the number of separate spiral threads that are so placed on its surface. As every one of these threads will pass its own wheel-tooth across the line of centers, in each revolution of the screw, it follows, that as many teeth of the wheel will pass that line during one revolution of the screw as there are threads to the screw.

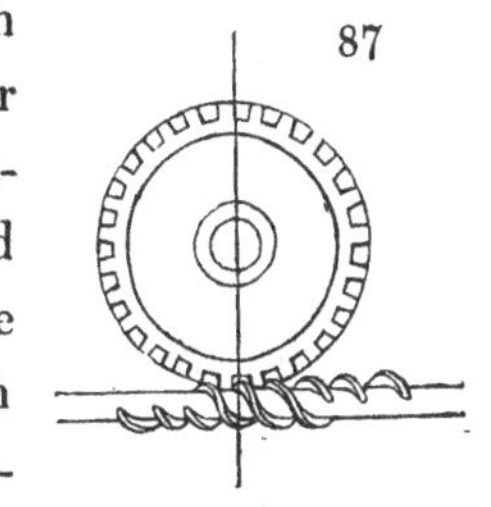

If we suppose the number of these threads to be considerable, for example, equal to those of the wheel-teeth, then the screw and wheel may be made exactly alike, as in fig. 88; which may serve as an example of the disguised forms which some common arrangements may assume.

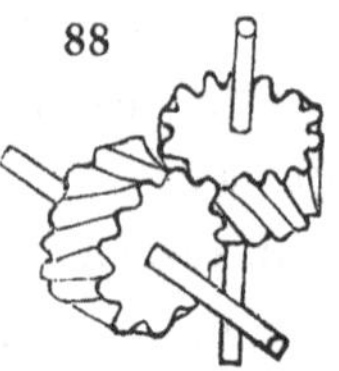

The old Piemont silk-mill is an example of disguised endless screws*.

176. In fig 89 is represented a method of communicating equal rotation by sliding contact between two axes whose directions if produced are parallel. *Aa Bb* are the axes, parallel in direction.

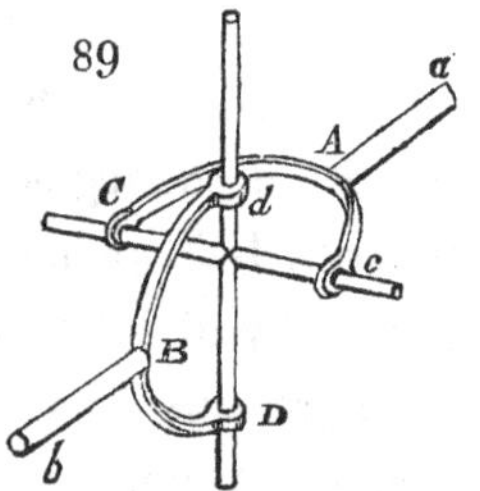

The axis *Aa* is furnished with a semicircular piece *CAc*, forming two equal branches, and terminated by sockets bored in a direction to intersect the axis at right angles. The axis *bB* is provided with a similar pair of branches *dbD*, and the whole is so adjusted that their four sockets lie in one plane perpendicular to the axes. A cross with straight polished arms is fitted into the sockets in the manner shewn in the figure; and its arms are of a diameter that allows them to slide freely each in its own socket. If one of the axes be made to revolve, it will communicate to the other by means of this cross a rotation precisely the same as its own.

For let fig. 90 be a section through the cross transverse to the axis, and let *AB* be the axes, and the circles be those described by their sockets respectively.

Then if *D* be a socket of *A*, the arm of the cross which passes through it must meet the center *A*; and in

* Described in Encyc. Methodique, Manufactures and Arts, tom. II. p. 31; and in Borgnis, Machines pour confectionner les étoffes, p. 160.

like manner if C be a socket of B, the arm CB must pass through B. Also, if D move to d, the new (or dotted) position of the cross will be formed by drawing dA through A, and Bc perpendicular to it through B the other axis; therefore C will be carried to c; and it is easy to see that the angle $DAd = CBc$. Therefore the angular motion of the axes is the same.

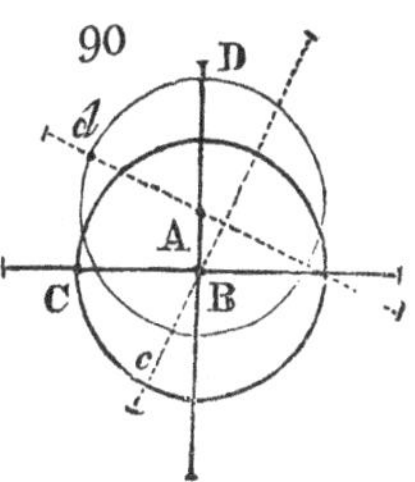

Also, every arm of the cross will slide through its socket and back again during each revolution, through a space equal to twice the perpendicular distance of the axes (AB).

* This arrangement is essentially the same as that of a coupling invented by the late Mr Oldham, and introduced by him into the machinery of the Banks of England and Ireland. His form of it is more solid, but not so well adapted for geometrical illustration as that which I have given. His axes are each terminated by a disk in which a transverse groove is planed, and the cross consisting of two square bars in different planes has each bar completely buried in the groove of its neighbouring disk.

CHAPTER IV.

ELEMENTARY COMBINATIONS.

CLASS A. { DIRECTIONAL RELATION CONSTANT.
VELOCITY RATIO CONSTANT.

DIVISION C. COMMUNICATION OF MOTION BY WRAPPING CONNECTORS.

---

177. ANY two curves revolving in the same plane whose wrapping connector (Art. 37) cuts the line of centers in a constant point, will preserve a constant angular velocity ratio. In practice, however, circles or rather cylinders only are employed, which revolve round their centers, and manifestly possess the required property. To enable the rotation to proceed in the same direction indefinitely, the band which serves as a wrapping connector has its two ends joined so as to form an *endless band*, which embraces a portion of the circumference of each circle or *pully*, and is stretched sufficiently tight to enable it to adhere to and communicate its motion to the edge.

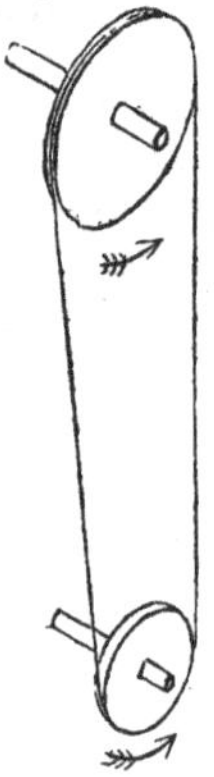

The band may be *direct*, that is, with parallel sides, as in fig. 91. or it may be *crossed*, as in fig. 92. In the first case the axes or pullies will both revolve in the same direction, in the latter case in opposite directions.

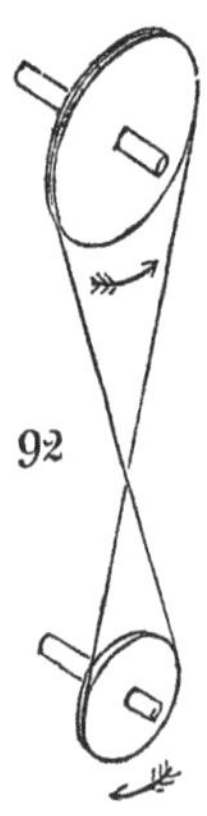

178. Motion communicated in this manner is remarkably smooth, and free from noise and vibration, and on this account, as well as from the extreme

simplicity of the method, it is always preferred to every other, unless the motion require to be conveyed in an exact ratio.

As the communication of motion between the wheels and band is entirely maintained by the frictional adhesion between them, it may happen that this may occasionally fail, and the band will slip over the pully. This, if not excessive, is an advantageous property of the contrivance, because it enables the machinery to give way when unusual obstructions or resistances are opposed to it, and so prevents breakage and accident. For example, if the pully to which the motion is communicated were to be suddenly stopped, the driving pully, instead of receiving the shock and transmitting it to the whole of the machinery in connexion with it, would slip round until the friction of the band upon the two pullies had gradually destroyed its motion.

But if motion is to be transmitted in an exact ratio, such, for example, as is required in clock-work, where the hour-hand must perform one exact revolution while the minute-hand revolves exactly twelve times, bands are inapplicable; for, supposing it practicable to make the pullies in so precise a manner that their diameter should bear the exact proportion required, which it is not; this liability to slip would be fatal.

But in all that large class of machinery in which an exact ratio is not required to be maintained in the communication of rotation, endless bands are always employed, and are capable of transmitting very great forces.

179. Bands may be either *round* or *flat*, and the *materials* of which they are formed are various. The best but most expensive is *catgut;* but its durability and elasticity ought to recommend it in every case where it can be obtained of sufficient strength. It acquires by use a hard polished

surface, and it may be procured of any size, from half an inch diameter to the thickness of a sewing needle.

The ends of a catgut band may either be united by splicing or by a peculiar kind of hook and eye which is made for that purpose. Both hook and eye have a screwed socket into which the ends of the gut are forced by twisting, having been previously dipped into a little rosin. The hook and eye may be warmed to keep the rosin fluid while the band is being forced in, and the ends of the band that come out through the socket may, for further security, be seared with a hot wire.

Hempen ropes are only used in coarse machinery, but in the cotton factories a kind of cord is prepared, of the cotton-waste, for endless bands, which is tolerably elastic and soft, and is peculiarly adapted for driving a great quantity of spindles. Also the soft plaited rope, termed patent sash-line, answers very well for these purposes. All these bands must have their ends neatly spliced together, so as to avoid as much as possible the increased diameter at the place of junction, because the periodic passage over the pullies of the lump or knot so formed gives rise to a series of jerks, that interfere with the smooth action of the mechanism*.

Common *iron chains* are also used, but only in very rough and slow-moving mechanism.

Flat leather belts appear to unite cheapness with utility in the highest degree, and are at any rate by far the most universally employed of all the kinds. This they owe partly to the superior convenience of the form of pully which they require, over that which is employed for round bands and chains. Belts vary in width from less than one inch up to fifteen inches, and their extremities may be united by

* Vide Transactions of Society of Arts, Vol. XLIX. Part I. p. 99, for some practical directions.

buckles, but are best joined by simply overlapping the ends and stitching them together with strips of leather passed through a range of holes prepared for the purpose, or they may be glued or cemented at the ends; in which case, by carefully paring and adjusting the parts that overlap, they will be perfectly uniform in thickness throughout; but they thus lose the power of being adjusted in length, and must therefore be provided with stretching pullies.

Belts, on account of their silent and quiet action, are very much employed for machinery in London, to avoid nuisance to neighbours. It appears also from a recent work*, that the use of belts is greatly extended in the American factories. In Great Britain the motion is conveyed from the first moving power, to the different buildings and apartments of a factory, by means of long shafts and toothed wheels; but in America, by large belts moving rapidly, of the breadth of 9, 12, or 15 inches, according to the force they have to exert.

Of late, both flat belts and round bands have been manufactured of *caoutchouc* interwoven with fibrous substances, in various ways; and under peculiar management may be made to answer very well. But changes of temperature occasion great variations of length and elasticity in this material; nevertheless in this latter quality it is greatly superior to catgut, and, like that substance, it requires no stretching pullies, which must always be employed for rope-bands. Belts are also made of *woollen felt*, and round bands are cut out of thick *leather*. In small machinery an endless band may even be cut out, in one piece, of a skin of leather, so as to avoid the necessity of joining the ends, and thus the jerks occasioned by the passage of the knot over the pully are entirely avoided.

* Cotton Manufacture of America, by J. Montgomery, 1840, p. 19.

180. *The form of the pully* upon which an endless band is to act is of importance, as the adhesion of the band

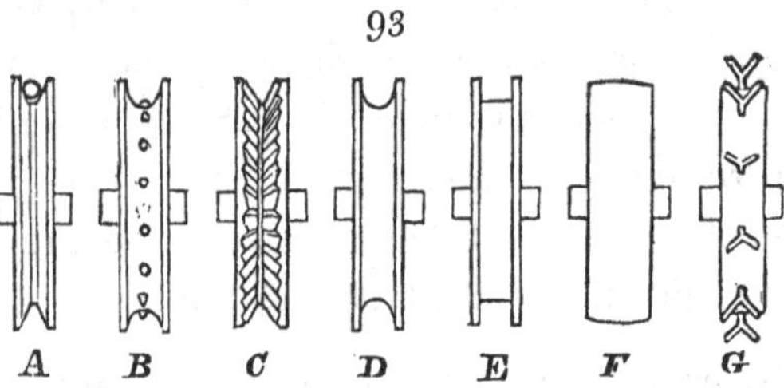

is greatly influenced thereby. Fig. 93 exhibits the principal forms. Round bands of catgut, rope, or other material, or even chains, require an angular groove (as *A*), into which their own tension wedges them, and thereby enables them to grasp more firmly the edge of the pully.

But when ropes or soft bands are used, the bottom of the groove is sometimes furnished with short sharp spikes, (as *B*), or else its sides are cut into angular teeth, (as *C*), which help to prevent the band from slipping, but at the same time are apt gradually to wear it out.

A pully for chains is sometimes formed by fixing Y formed irons at equal distances in the circumference of a cylindrical disk, as at *G*.

When the pully over which the band passes is used merely as a guide-pully (Art. 186), there is no need to provide against slipping, and the groove or *gorge* is made simply of a semicircular section, as *D*, to keep the band in its place.

181. If a tight flat belt run on a revolving cone, it will advance gradually towards the base of the cone, instead of sliding towards its point, as might be expected at first sight.

The reason of this is, that the edge of the belt nearest the base of the cone is tighter, and advances more rapidly than the other, because it is in contact with a portion of the cone of a larger diameter, and consequently moving with a proportionably greater velocity*. Thus the belt is bent into the form *Bb*, shewn in the figure, by which the part which is advancing to the cone is thrown still nearer and nearer to its base. In this manner the belt will gradually make its way from the smaller end of the cone to the larger, where it will remain. Advantage is taken of this curious property in forming the pullies for straps, which are made of the form represented at *F*, fig. 93, a little swelled in the middle. This slight convexity is more effective in retaining the belt than if the pully had been furnished with edges as at *E*; and the form, besides its greater simplicity, enables the belt to be shifted easily off the pully. In fact, when a pully of the latter form *E* is employed, the belt will generally make its way to the top of one of the lateral disks, and remain there, or else be huddled up against one or other of them, but will never remain flat in the center of the rim, if there be the slightest difference of diameter between the two extremities of the cylinder.

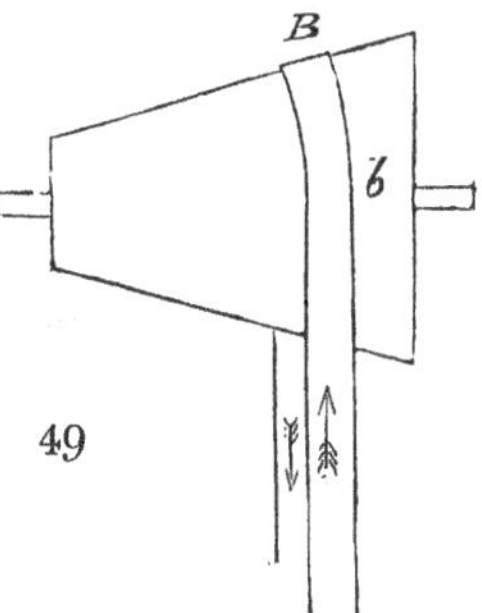

182. In order to bring the belt into contact with as much as possible of the circumference of the pully, it is better to cross it (Art. 177) whenever the nature of the machinery will admit of so doing.

When a strap or other flat belt is crossed, it must be put on to the pullies in the manner represented in fig. 95. Every leather belt has a smooth face and a rough face. Let the rough face be placed in contact with both pullies, by which each straight side of the belt will be twisted half

* Young's Nat. Phil. vol. II. p. 183.

round in the transit from one pully to the other, as shewn in the figure. Now the effect of this is, that at the point where the two sides cross, the belts lie flat against each other; for since the belt at each extremity where it joins the pully is perpendicular to the plane of rotation, and it is twisted half round in its passage, it must be parallel to the plane of rotation half way between the pullies, when the two sides of the belt cross. Hence they pass with very little friction, whereas if this half twist were not employed, the two halves of the belt would pass edgewise, which (in a broad belt especially) would occasion so much friction and displacement as to make the arrangement impracticable.

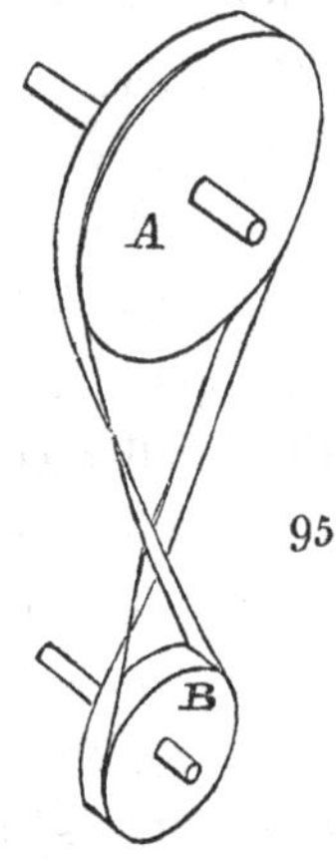

95

183. The band moves with the same velocity as the circumference of the pully with which it is in contact, and consequently the circumferences of the two pullies which it connects move with equal velocities;

$$\therefore \frac{A}{a} = \frac{r}{R},$$

where A, a are the angular velocities, R, r the radii.

But when a thick belt is wrapped over a pully its inside surface is compressed and its outside surface extended, and the center, or nearly so, of the belt alone remains in the same state of tension as its straight sides, and therefore moves with the velocity of the sides Hence the radius of the circle to whose circumference the velocity of the belt is imparted, virtually extends to the center of the belt, and half the thickness of the belt must be added to the radius of the pully, in computing the angular velocities.

Similarly, to find the acting radius of a pully with an angular groove, as at A, fig. 93, the distance of the center of the section of the band from the axis of the pully must be taken, and this in a given pully will be greater the thicker the band employed.

184. An endless band of any kind is easily shifted during the motion to a new position on a cylindrical drum or pully, if the band be pressed in the required direction on its *advancing side*, that is, on the side which is travelling towards the pully; but the same pressure on the retiring side of the belt will produce no effect on its position.

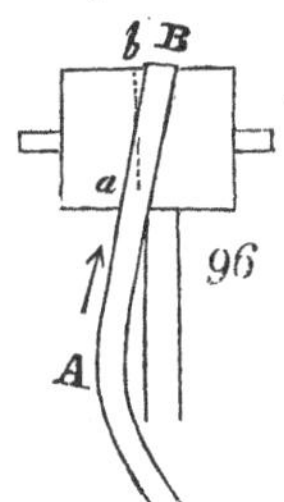

For example, if the belt *AB* has been running over the drum in the position *B*, and this belt be drawn a little aside, as at *A*, those portions of the belt which now come successively into contact with the drum, as at *a*, will begin to touch it at a point to the left of the original position, and in one semi-revolution the whole of the belt in contact with the drum will thus have been laid on to it, point by point, in a new position *ab*, to the left of the original one *B*; but if the direction of the motion were from *B* to *A*, the portions of belt drawn aside are those which are quitting the drum, and therefore produce no effect on its position thereon.

Therefore, to maintain a belt in any required position on a cylindrical drum, it is only *necessary* that the advancing half of the belt should lie in the plane of rotation of that section of the drum upon which it is required to remain, but the retiring side of the belt may be diverted from the plane, if convenient, without affecting its position.

If the machinery be at rest it is very difficult to shift the position of a belt of this kind, on account of the adhesion of its surface; but by attending to the simple principle just explained it becomes very easy to shift the belt by merely turning the drum round, and pressing the advancing side of the belt at the same time. The same principle applies to round bands running on grooved pullies; if it be required to slip them out of the groove, the advancing side of the band must be pressed to one side, so as to make it lay itself over the ridge of the pully, when half a revolution will throw it completely off.

185. Let AM, BN be two shafts, neither parallel nor meeting in a point, and let it be required to connect them by a pair of pullies and an endless band. Recollecting that the advancing side of the band must remain in the plane of rotation of each pully, find the line MN, which is the common perpendicular to the shafts. Fix the pullies upon the respective shafts, so that a line mn parallel to MN* shall be a common tangent to them, which is done by making the distance AM of the upper pully from the point M equal to the radius Bn of the lower pully, and *vice versâ*,

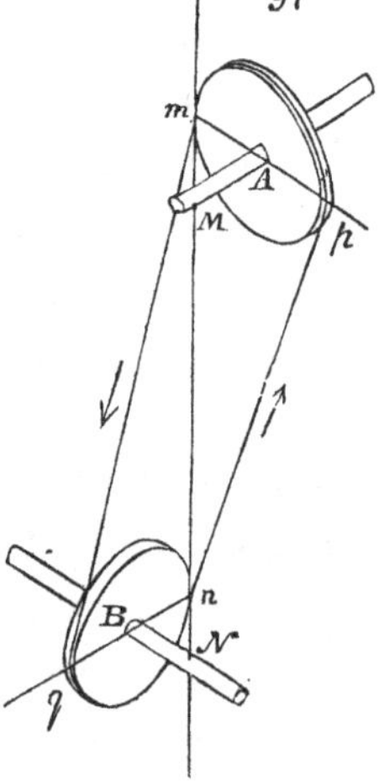

$$BN = mA.$$

Arrange the belt in the manner shewn in the figure, the arrows indicating the direction of motion; then the portion np which is advancing to the upper pully is plainly in the plane of rotation of that pully, and will therefore retain its position thereon, and similarly, the portion mq which is advancing to the lower pully, is also in the plane of rotation of the latter.

If, however, the motion be reversed the belt will immediately fall off the pullies, for in that case the portion pn

* The lines MN, mn are confounded into one in the figure, but it will be easily seen that mn lies considerably behind MN.

will advance towards the lower pully in a plane pn, making an angle with that of the pully. The belt will therefore begin to shift itself towards N, and, by so doing, will be thrown off the pully, and a similar action will take place between the belt qm and the upper pully.

The appearance of this arrangement in practice is very curious; for the retiring belts being twisted at a very considerable angle from the planes of the pullies, at the moment of quitting them appear as if they were slipping off at every instant, which however they never do. The only fault is, that this violent twist at m and n is apt to wear out the leather, especially if the shafts are pretty close together. For which reason it may be better to employ guide pullies to conduct the belt from one wheel to the other, as in Art. 187.

If it be required to cross the belts, the arrangement for so doing will be found by drawing a figure similar to 97, but in which qm shall be the intersection of the planes of rotation, mn the descending belt, and a common tangent from p towards q the ascending belt.

186. Pullies are sometimes employed for the purpose of altering the course or path of a band, in which case they are termed *guide pullies*. Their position and number may be determined in the following manner:

A band moving in the line Ab is required to have its path diverted into the direction bB by guide pullies.

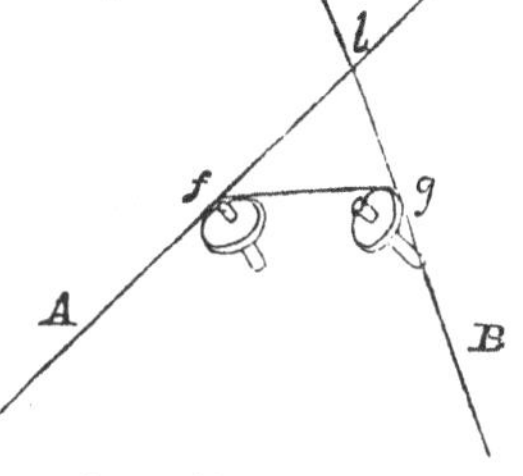

If these lines meet in the point b, one pully is sufficient; the axis of which must be placed perpendicularly to the plane which contains the two lines Ab, bB, and its mean diameter adjusted so that it may touch these lines. If this diameter be too great for convenience, or the point of intersection *b* too remote, or if the lines do not meet in a point, then two pullies are required, whose positions are thus determined.

Draw a third line *fg*, meeting the two former lines in any convenient points *f* and *g* respectively, and let this line be the path of the band in its passage from one line of direction to the other. Place, as before, one guide pully at the intersection *f*, and the other at the intersection *g*, the axes of these pullies being respectively perpendicular to the plane that contains the two directions of the band*.

187. Let *A*, *B* be two pullies whose axes are neither parallel nor meeting in direction, as in Art. 185, and let the line *cd* be the intersection of the two planes of these pullies.

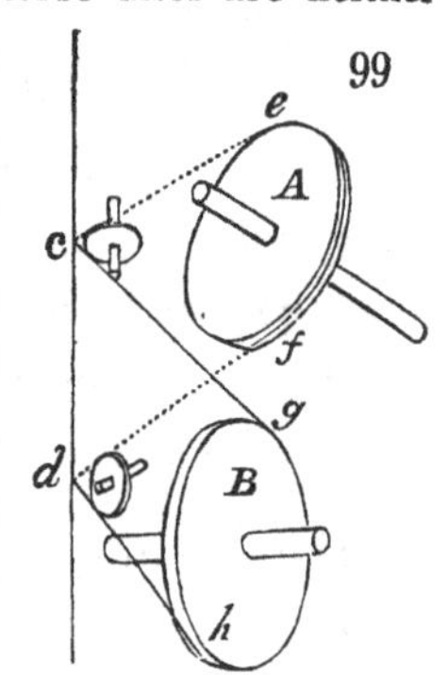

In this line assume any two convenient points *c* and *d*; and in the plane of *A* draw *ce*, *df*, tangents to the opposite sides of this pully; also in the plane of *B* draw *cg*, *dh*, similarly tangents to the pully *B*.

This process gives the path of an endless band *ecghdf*, in which it may be retained by a guide pully at *c* in the plane *ecg*, and another at *d* in the plane *fdh*. In this band both the retiring and advancing sides lie in the planes of each pully. The pullies will therefore turn in either direction at pleasure, and the band is not liable to the twisting wear already deprecated in the arrangement of fig. 97.

In other cases that may present themselves, the position and least number of the requisite guide pullies may be determined by similar methods.

* Poncelet, Mec. Ind. Part III. Art. 24.

188. If the bands are not made of elastic substances they require *stretching pullies;* that is, pullies resembling guide pullies, whose axes can be shifted in position, so as to increase the tension of the band as required; or else their axes are mounted in frames so that a weight or spring may act upon them, to retain the band in the proper state of tension; but as the operation of these contrivances involve considerations of force, they do not fall under the plan of this portion of the present work. Neither do certain arrangements by which the quantity of circumference embraced by the bands are increased or multiplied, for the purpose of improving the adhesion.

189. We have seen that a common iron-chain with oval links may be employed as an endless band; using the form of groove *A*, fig. 93. If the chain be formed with care, and the wheels between which it works be provided with teeth, the spaces between which are accurately adapted to receive the successive links, then the chain will take a secure hold of the circumference of each wheel; and its action upon these teeth will resemble that of one toothed wheel upon another, or rather of a rack upon a toothed wheel, the successive links falling upon and quitting the teeth without shocks or vibration, so that the motion of one toothed circumference will be conveyed to the other without loss from slipping. A chain of this kind is termed a *geering chain*, and various forms have been given to its links to ensure smoothness of action. But these chains are expensive and troublesome, and are not much in use, as, generally speaking, the communication of motion to a distance can be as completely effected by a long shaft with bevil-wheels at each end; and the geering chain, in all its forms, is liable to stretch, by which the spacing or pitch of its links is increased, so that they no longer fit the teeth of the wheels.

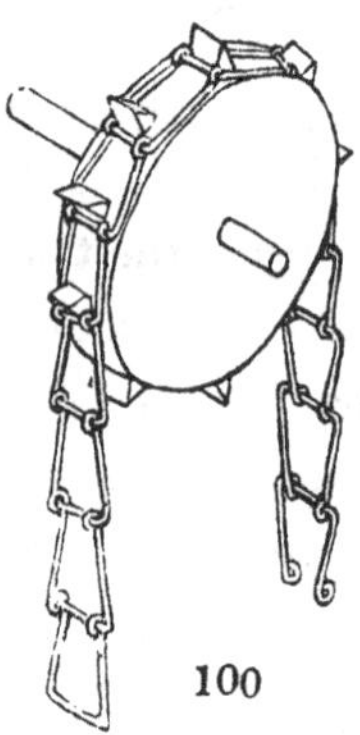

Fig. 100 shews the geering chain which was proposed by the celebrated Vaucanson, about 1750. The links of the chain are made of iron-wire, and adapted to lay hold of the teeth of a wheel in the manner shewn by the figure*.

Geering chains had been, however, employed long before this period, as for example, by Ramelli in 1588†; and the very chain of Vaucanson is represented by Agricola, in 1546, as an endless chain, to carry buckets in a machine for raising water from a mine.

Fig. 101 is another form, from Hachette, in which the

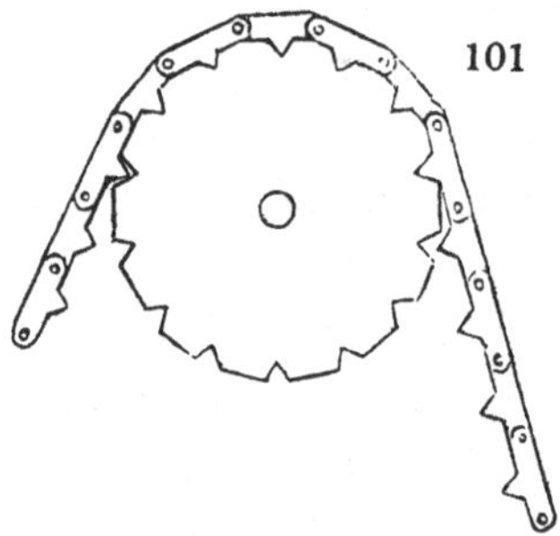

links are made of plates rivetted together, somewhat after the manner of a watch-chain; and 102 is a third modification‡, in which a plate-chain is also employed; but the teeth of the wheel are much better disposed for grasping the successive links. Nevertheless, in all these cases, when the rivets enlarge the holes by wearing, the pitch of the chain is increased, and each link enters its receptacle on the wheel with a jerk, producing vibration and accelerated deterioration.

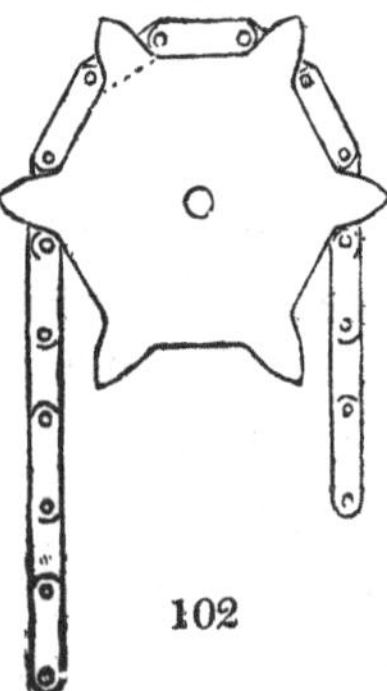

* Vide Encyc. Method. Manufactures, tom. II. p. 132.

† Figs. XXXIX. and XCIII.

‡ Used in Morton's patent slip.

190. If the axes be required to make only a limited number of rotations in each direction, the slipping of the band may be entirely prevented by fixing each end of it to one of the pullies or rollers, and allowing it to coil over them as many times as may be required; as in fig. 103, where rotation is conveyed from one roller *A* to the other *B* by the cord *a*, one end of which is fastened to the surface of *A*, and the other end to that of *B*. To enable the motion to be conveyed in both directions a similar cord *b* may be coiled in the opposite direction round each roller, so that while *b* coils itself round *A*, *a* will uncoil itself, and *vice versa*.

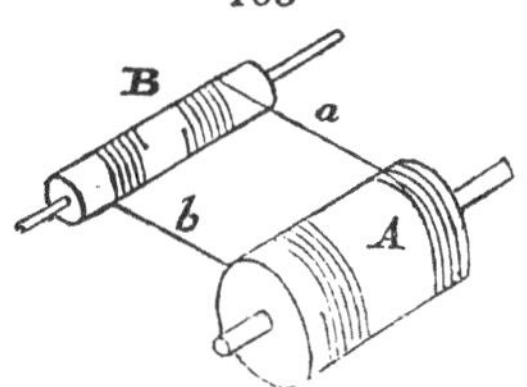

The carriage *B*, fig. 104, runs back and forwards upon the rollers *f*, *e*, and derives its motion from the roller or barrel *A*, which is mounted on an axis above it. A cord *c* is tied to one end of *B*, and another cord *d* to the other end; these cords are passed as many times round the roller as is necessary, in opposite directions, and their ends fastened to its surface. When the roller revolves the carriage will travel along its path, preserving a constant velocity ratio, provided the circumference of the roller nearly touch the line *dc*. Otherwise the variation of the angle *Acd*, during the motion of the carriage, will cause the velocity ratio to change*. If, however, pullies be fixed to the frame of the machine beyond *d* and *c*, and the cords be carried from the barrel over these pullies and then brought back again to *d* and *c*, the axis *A* may be fixed at any required height above *B*. Either piece may be the driver.

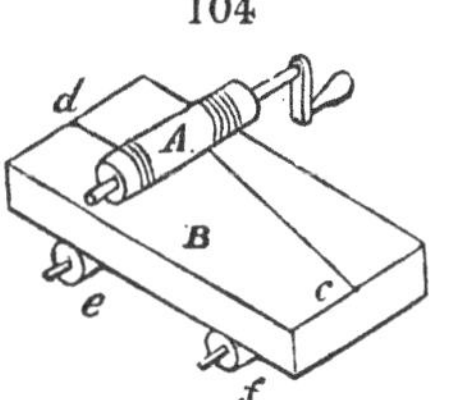

* For the line *Ac* acts as a link jointed at *c*, and therefore;

vel. of *Ac* : vel. of *B* :: cos *Acd* : 1. (Art. 32. Cor. 3.)

Sometimes a single line is employed, which being fastened at *d* is coiled three or four times round the roller, and then carried on to *c*; the coiling is sufficient to enable the cord to lay hold of the roller in most cases, as for example, in the common drill and bow.

191. But the constancy of the ratio is interfered with in both these contrivances, by the varying obliquity of the straight parts of the cords which connect the pieces, as well as by the tendency to heap up the successive coils in layers upon each other, thereby increasing the effective diameter of the rollers. This is remedied by cutting a screw upon the surface of each roller, which guides the cord in equidistant coils as it rolls itself upon the cylinder.

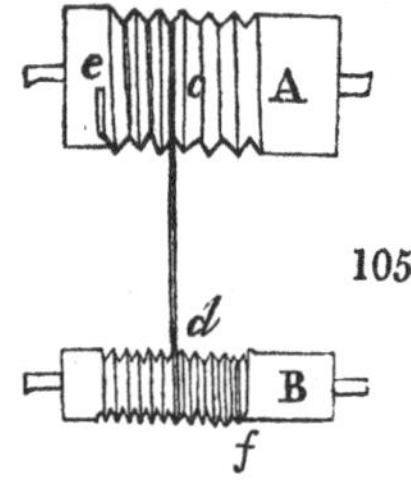

Thus, fig. 105, let *A* give motion to *B* by a cord *cd*, in the manner already shown in fig. 103, but let screws be cut upon the surface of the rollers; then during the motion of *A* the extremity *c* of the straight portion of the cord will be gradually carried to the right as it is wound up, and *vice versa*; and this motion will be constantly proportional to the rotation, and at the rate of one pitch of the screw to each complete turn of the cylinder.

To cause the straight portion *cd* to move parallel to itself, the screw cut upon *B* must be of such a pitch that the endlong motion of *d* may be the same as that of *c*. Now since the velocity of the surfaces of the two cylinders are equal, and every revolution of either screw carries the cord endlong through the space of one pitch, let $m \times$ circumferences of $A = n \times$ circumferences of B, and let C, c be the respective pitches of their screws; R, r their radii, then we must have $mC = nc$,

$$\text{or } \frac{C}{c} = \frac{R}{r}.$$

192. In the combination of fig. 104, the screw roller will prevent the irregular heaping up of the cord on the band, but will not correct the varying obliquity of the cord. This may be got rid of thus.

Let *B*, fig. 106, be the sliding carriage, *CD*, *HK* the

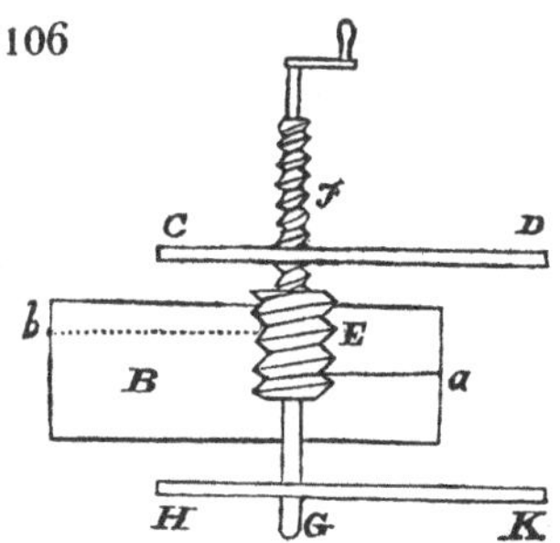

sides of the frame which supports the roller, *E* the roller formed into a screw. This roller has a screw *F* cut on its axis, of the same pitch as that of *E*, and passing through a nut in the frame *CD*; the other extremity of the roller is supported by a long plain axis *G*, passing through a hole in the frame *HK*; the cord being tied at *b* to the carriage, and at the other end to the screw-barrel *E*; it follows, that when the latter is turned round, it will travel at the same time endlong by means of the screw and nut *F*, exactly at the same rate, but in the opposite direction, as the end of the cord is carried along the barrel by its coiling; consequently the one motion exactly corrects the other, and the cord *b* will always remain parallel to the path of the slide *B**.

A similar and contrary cord being employed to connect the other end of the slide with the band, will enable the roller to move the slide in either direction.

193. A well made chain of the common form, with oval links, will coil itself with great regularity upon a re-

* From a machine by Mr. Holtzapfel.

volving barrel, if a spiral groove be formed upon the surface, of a width just sufficient to receive the thickness of the links. As shewn in fig. 107, the links will alternately place themselves edgewise in the groove and flat upon the surface of the barrel.

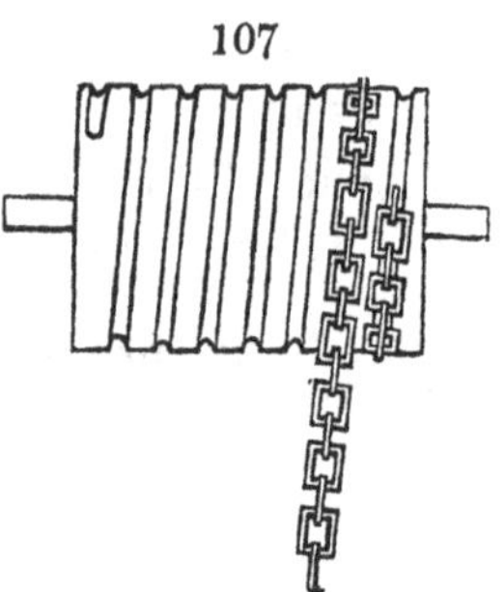
107

194. When the revolving piece is required to move only through a fraction of a revolution, the combination is made more simple.

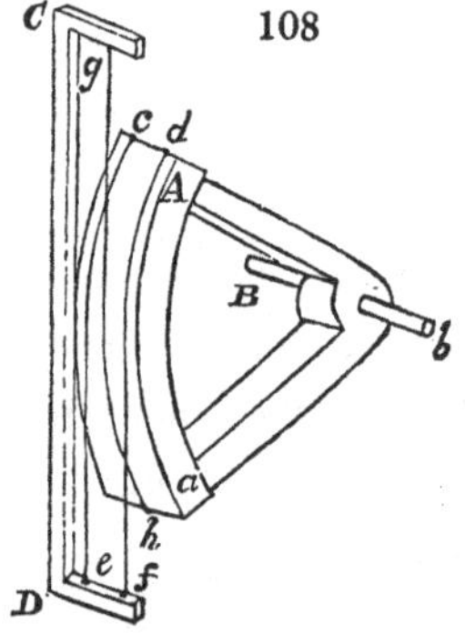

108

Thus let *A* represent a revolving piece or quadrant, whose axis is *B*, *b*, and whose edge is made concentric to it, and let *CD* be the sliding piece represented as an open frame for clearness only, but supposed to be guided so as to move in either direction along the line *CD* produced. If cords or chains be attached at *c*, *d*, to the quadrant and at *e*, *f*, to the sliding frame; and a third cord be attached contrariwise to the quadrant at *h* and the frame at *g*, then either the motion of the quadrant or the frame will communicate motion to the other in a constant ratio, and in either direction at pleasure.

CHAPTER V.

ELEMENTARY COMBINATIONS.

CLASS A. { DIRECTIONAL RELATION CONSTANT.
VELOCITY RATIO CONSTANT.

DIVISION D. COMMUNICATION OF MOTION BY LINK-WORK.

---

195. WHEN two arms revolving in the same plane are connected by a link (Art. 32), their angular velocities are inversely as the segments into which the link divides the line of centers. This relation is constantly changing, as the arms revolve, unless the point of intersection *T* (fig. 6), be thrown to an infinite distance, by making *PQ* parallel to *AB*, in all positions, which can only be effected by making the arms equal, and the link equal in length to the distance between the centers. In this case the angular velocities will become equal, and their ratio consequently constant.

196. This produces the arrangement of fig. 109. *D*, *B* are centers of motion, *Bd* = *Df* the arms, *df* (= *BD*) the link. If *Bd* be carried round the circle, *BdfD* will always be a parallelogram, and consequently the angular distances of *Bd* and *Df* from the line of centers the same, and their angular velocity the same.

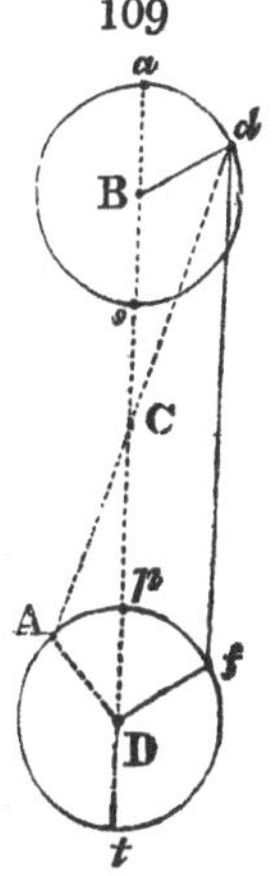

But in any given position of one of the arms *Bd*, there are two possible corresponding positions of the arm *Df*, for with center *d*, and radius *df*, describe an arc which will necessarily cut the circular path of *f* round *D* in two points *f* and *A*; therefore *AD* is also a position of the arm corresponding to *Bd*, in which the link *dA* intersects the line of centers in a point *C*; and if *Bd* be moved, the point *C* will shift its place, and consequently the angular velocity of *AD* will not preserve a constant ratio to that of *Bd*.

It appears, then, that this system is capable of two arrangements, one in which the angular velocity ratio is constant, and the other in which it is variable, according as the link is placed parallel to the line of centers, or across it.

But if the motion of this system in either state be followed round the circle, it will be found that when the extremity *d* of the arm *Bd* comes to the line of centers, either above or below, at *a* or *s*, the extremity of the other arm will also coincide with that line, since the link is equal to *BD*, and therefore to *ap* or *st*. In these two phases (Art. 17) of its motion the two positions *fd*, *Ad* of the link coincide, and at starting from either of these phases, the link has the choice of the two positions. If, for example, the arms be at *Ba* and *Dp*, then as *a* moves towards *d*, *p* may either move towards *f*, in which case the link will remain parallel to *BD*, until the semicircle is completed, or else *p* may move towards *A*, and then the link will lie across *BD*, until the semicircle is completed by *d* coming to *s*, when a new choice is possible. But in any given position of *Bd* intermediate between *Ba* and *Bs*, it is impossible to shift the link from one position to the other without bending it.

The two phases in which the arms coincide with the line of centers, are termed the *dead points* of the system.

197. When this contrivance is employed to communicate a constant velocity ratio, some provision must be made to prevent the link from shifting out of the parallel position into the cross position, when the arms reach the dead points.

There are three ways of passing the link parallel to itself across the line of centers. First, by introducing a third arm, as at c, of the same length as the others, with its center placed on the line of centers, and its extremity jointed to the link, so as to divide the latter in the same proportion as the line of centers is divided by the center of the new arm. This new arm may be placed either between or beyond the others, and plainly renders any position of the link, except that of parallelism to the line of centers, impossible. It is not even necessary that the centers of the three equal arms shall lie in one line, for if the three joint-holes, a, b, c, of the link, be the points of an equal and similarly placed triangle to that formed by the three centers of motion, the arms will all revolve alike.

198. The second way requires only two axes of motion, but has two sets of arms.

Aa, Bb, fig. 111, are the two parallel axes. At one end of each are fixed the equal arms AP, BQ, connected as before by a link $PQ = AB$; at the other end of each are fixed arms ap, bq, also connected by a link, $pq = ab$.

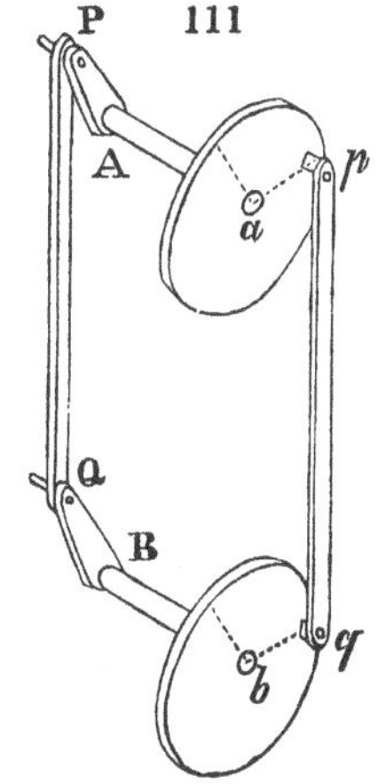

Now since the separate effect of each of these systems is to produce equal rotation in the axes, it is plain that the action of the second will conspire with that of the first to produce this effect, whatever be the angle which AP makes with ap. Let ap then be set at right angles, or nearly so, to AP; therefore when either system arrives at the dead points, the other will be half way between them, and by communicating at that moment the equal rotation to the axes, will thus carry the link of the former system over the dead points, without allowing it the choice of the second set of positions; which second set of positions is besides rendered geometrically impossible by this combination of the two sets of arms.

199. The form of the piece to which the joint-pin is fixed is indifferent; thus (fig. 111) the pin P is carried by an arm AP, and the pin p by a disk; but the motion produced by each is precisely the same; the effective length of the arm being in every case measured in the plane of rotation in a right line from the center of the pin to the center of motion of the piece which carries it, whatever be the form given to the latter.

However, if either axis be carried across the plane of motion of the link, the latter will strike against it, and thus prevent the completion of a single revolution. If the axes be required to revolve continually in the same direction, either the piece which carries the pin must be fixed to the extremity of the axis, as in fig. 111, or else the axis must be bent into a loop or *crank*, as it is termed, as in fig. 112, by

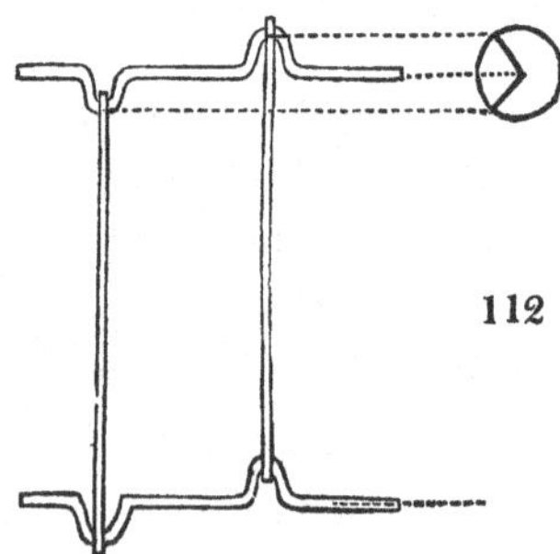
112

which the axis is also removed from the plane of rotation of the link; but the axis may thus be extended indefinitely on either side.

200. The third method of passing the links over the dead points consists, like the latter, in employing two or more sets of arms and links, so disposed as that only one set shall be passing the dead point at the same moment. But in this method, fig. 113, the axes *Aa*, *Bb* are parallel but not oppo-

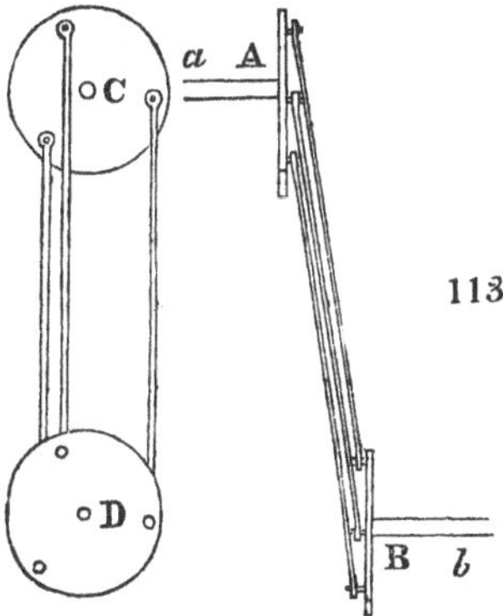

site, and a disk of any convenient form, as *C*, *D*, being attached to the free end of each, pins are fixed in the faces of the disks at equal distances from the centers of motion, and at equal angular distances from each other respectively, and links each equal to the distance of the centers are jointed to them in order, as shewn in the figure.

The planes of rotation of these disks are removed from each other by a distance sufficient to throw the connecting links into a slightly oblique position, which enables them each to clear the others, during the rotation, by passing alternately above and below them.

The number of the links is indifferent. Two are sufficient, as in the former case, and the radii of their pins must be nearly at right angles; but if three or more be employed, the pins may be at equal angular distances round the circle; and it is hardly necessary to add, that in determining the length of the links allowance must be made for the oblique position into which they are thrown by the nature of the contrivance*.

201. It appears (Art. 195), that by Link-work, rotation in a constant velocity ratio can only be communicated between two axes when they are parallel, move in the same direction, and revolve in equal times. If, however, only a motion through a small angle is required, it may be communicated with an approximately constant velocity ratio, whatever be the magnitude of that ratio, the relative position of the axes, or the directional relation.

For if the axes be parallel, it is shewn in Art. 133, that if a pair of arms *AP*, *BQ*, fig. 114, be connected by a link

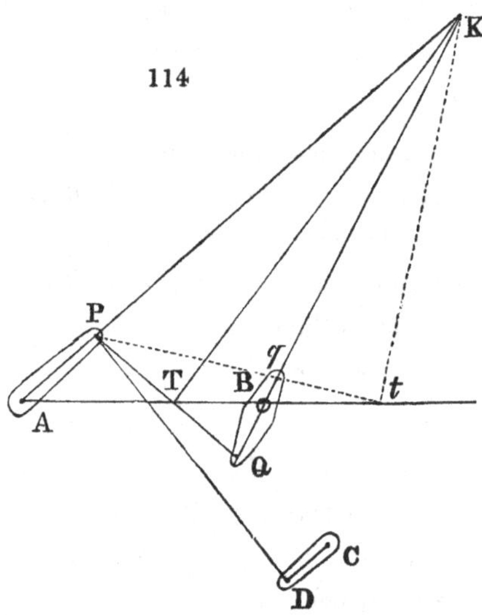

PQ, and placed in such a position that the intersection *T* of the link and line of centers shall coincide with the perpendicular *KT* upon the link from the intersection of the arms produced, then will the angular velocity be momentarily constant, and will be sufficiently near to constancy, if

* By T. Bœhm, of Bavaria, communicated to Soc. Arts. vol. L. p. 83.

the motion of the links be confined to a small angle on each side of the mean position.

Now the arms *AP*, *BQ* will revolve in opposite directions; but if they be required to revolve in the same direction, the centers of motion must lie on the same side of the link. *AP*, *Bq*, are a pair of arms connected by a link *Pq*, which will fulfil this latter condition, and *Kt* the corresponding perpendicular upon the link produced, and intersecting it in *t* in the line of centers produced.

The angular velocities of the arms have been shewn to be inversely as the segments *AT*, *BT*, or *At*, *Bt*.

The simplest mode of arranging the proportions is to make the link perpendicular to the arms in the mean position, as shewn in *AP*, *CD*; *PD* being the link; and in this case, the angular velocities are inversely as the length of the arms themselves, (Art. 137).

202. If the axes be not parallel, let *Ae*, *Bf* (fig. 115), be the axes whose directions do not meet, find their common

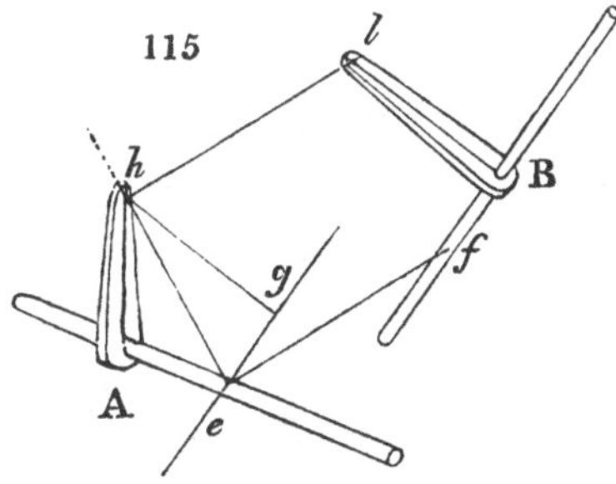

perpendicular *ef*, and draw *eg* parallel to *fB*. In the plane *Aeg* draw *eh* dividing the angle *Aeg* into two, *Aeh*, *heg*; whose sines are inversely as the angular velocities of the axes *Ae*, *Bf* respectively (Art. 44). From any point *h* drop perpendiculars *hA*, *hg*, upon *Ae* and *eg*; make *fB* equal to *eg*, draw *Bl* equal and parallel to *gh*, and join *hl*; which being parallel to *ef*, is plainly perpendicular both to *Ah* and to *Bl*.

If *Ah*, *Bl* be arms, and *hl* the link, then by the construction the link is perpendicular to the arms; and if the angular motion be small and the figure represent the mean position, the angular velocity ratio of the axes will not differ sensibly from that which would be communicated if the axes were parallel, and the arms and link in one plane, and will therefore be nearly constant, and equal to the inverse ratio of the length of the arms.

If the axes be required to revolve with the opposite directional relation to that shewn in the figure, one of the arms must be placed on the opposite side of the axis. In fact, as each arm admits of two positions (thus *h* may be above the axis or below it), so there are four ways in which these arms may be combined, two of which will make the axes revolve one way with respect to each other, and the other two the opposite way.

203. The mechanism of Organs, Pedal-Harps, Bell-hanging, and various other portions of machinery, generally called *bell-crank work*, fall under this class of small sensibly equable angular motions. The same kind of mechanism requires the change of the line of direction of these small motions. This may generally be effected by a single axis with two arms; and by the same combination the velocities may be changed in any required ratio, whether the motions be in the same or in different planes, as follows.

204. If the motions be in one plane, let *ab*, *da* (fig. 116) be the lines of direction of the motions meeting in *a*. Draw *Ca* dividing the angle *bad* into two, whose sines are in the ratio of the given velocities in *ab*, *da* (vide the construction in Art. 44). In *aC* take any convenient point *C* for

a center of motion, from which drop perpendiculars Cb, Cd upon the respective directions. If these be taken for arms moving round C, and links be jointed to them in the lines of direction ab, da, then a small motion given to ab will turn the two-armed piece bCd round its axis C, but will not remove its extremities sensibly from the directions ab, da, which are the tangents to the circles described by those extremities in the mean position of the axes. But these extremities will move with velocities which are directly as the length of the arms. (Art. 11).

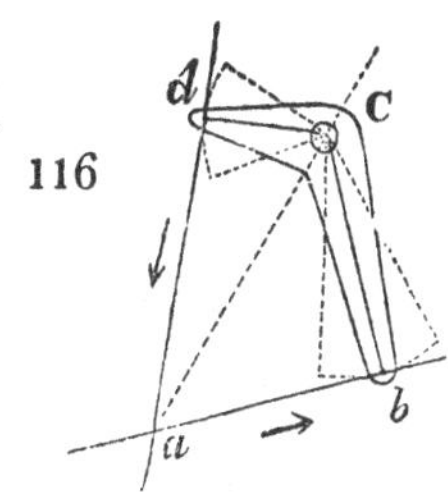

In practice it is better to make the lines ab and ad bisect the versines of the arcs of excursion, in which case each link will be carried to the right and left of its mean position, instead of deviating wholly towards the center of motion, as in the figure.

205. Since the arcs of excursion of the extremities d, b are given, we can by removing the center C to a sufficient distance from a, reduce the angular motion of the piece as much as we please, and thereby diminish the deviations of a, b from the mean positions*.

A two-armed piece or bent lever of this kind is termed a *crank*, or more properly a *bell-crank*, to distinguish it from the looped axis to which the term crank is also applied (Art. 199), but which differs from it considerably; the object of the

* If the links be not perpendicular to the arms in the mean position, but if the angle adC made by one link with its arm be equal to the supplement of the angle abC made by the other link with its arm, then it can be shewn that during a small angular motion of the system the ratio of the velocities of the links will still remain constant, and be equal to the ratio of the respective perpendiculars from C upon the links. This, however, supposes that the links in their deviations are not sensibly removed from parallelism to the mean positions, and it would rarely be of any practical service.

former being to change the direction of motion of a link when that motion is limited in extent; whereas the latter is expressly formed to allow of unlimited rotation in the same direction. The bell-crank is analogous to the guide pullies of wrapping bands (Art. 186), and accordingly these are sometimes employed in lieu of bell-cranks, to change the direction of motion of a link, by inserting at the place where the motion is diverted a piece of chain which passes over a guide pully.

206. If the given directions of motion intersect, as in fig. 116, we obtain four angles round the point of intersection, in two of which the directions of motion both approach the point, in another they both recede from it; and in the two remaining angles one motion approaches and the other recedes. The axis *C* may be placed in either of the two latter angles. If the directions of motion are parallel and opposite, the axis will lie between them, and if parallel and similar, the axis will lie beyond them, on one side or the other, but if also equal, then the axis is removed to an infinite distance, and the crank becomes practically impossible; but the change of motion may be effected by the next Article.

207. If the two directions of motion be not in one plane, let *ad*, *cb*, fig. 117, be these lines; find their common perpendicular *dc*; draw *ce* parallel to *ad*, and in the plane *bce* construct the required crank, as in Art. 204, of which let *B* be the center, *Bb*, *Be* the arms respectively perpendicular to *bc* and *ce*.

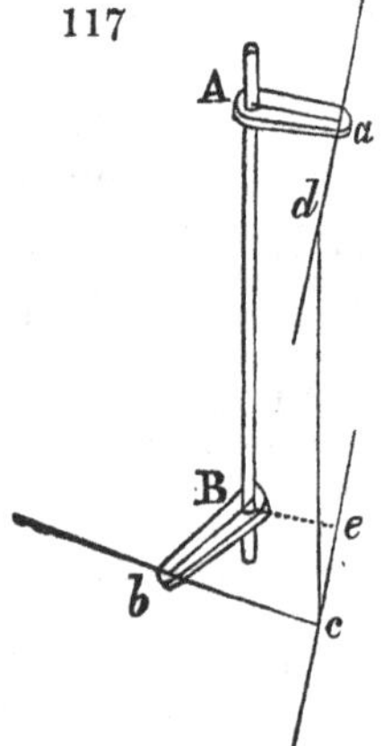

Draw *BA* a common perpendicular to *Bb* and *Be*, and equal to *dc*. Draw *Aa* parallel and necessarily equal to

Be, then will *AB* be the axis, *Aa* and *Bb* the arms required to change the small motion in *ad* into the required motion in *cb*.

By a similar construction we can effect the change of a small motion in a given direction, into another *equal* motion in the same direction parallel to the first; which has been shewn to be impossible by the bell-crank in one plane, although the motions themselves are in one plane.

In the mechanism of organs, in which the transmission of such small motions is of frequent occurrence, the crank is termed a *back-fall* when its arms are in one horizontal straight line, and a *square* when they are at right angles. An armed axis, like fig. 117, is a *roller*, and the links are *stickers* when they act by compression, and *trackers* when by tension.

CHAPTER VI.

ELEMENTARY COMBINATIONS.

CLASS A. { DIRECTIONAL RELATION CONSTANT.
VELOCITY RATIO CONSTANT.

DIVISION E. COMMUNICATION OF MOTION BY REDUPLICATION.

---

208. THE mechanism which results from this principle forms a class which is already separated by common practice from all others under the name of *Tackle*, and is principally employed on shore for raising weights, but in the rigging of ships is used to give motion to the sails, in order either to place them in the requisite positions for receiving the action of the wind, or to furl and unfurl them.

209. If an inextensible string $AfgB$ be passed over any number of fixed pins, as f and g, and if the extremities A, B of the string be compelled to move each in the direction of its own portion, Af, gB of the string, then the motion of one of these extremities will evidently be communicated unaltered to the other, and every intermediate portion of the string will move with the same velocity. This is unaffected by the form of the pins over which the string passes, and they may therefore be fixed cylinders or pullies, that is to say, wheels mounted on revolving axes, which are generally substituted for fixed pins, for the purpose of reducing the friction of the string in passing over them.

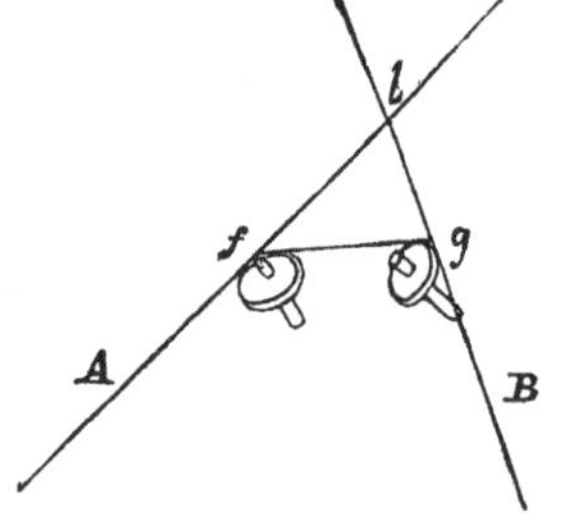

210. If, however, the pins or axes of the pullies be not fixed, then the principle of *reduplication* (Art. 30) is introduced, by which the velocity of the string and its extremities is greatly modified.

Thus let the string be attached to a fixed point M, and then doubled over P, and returned to Q, PQ being parallel

to MP; also let P be capable of moving in a path parallel to MP, then if Q be moved to q, P will travel to p; and it has been shewn in Art. 30, that $qQ = 2pP$. Also, the portion of string Mp is at rest while every point of PQ travels with a velocity equal to that of its extremity Q.

Now let the string be wound back and forwards, beginning with the fixed extremity at M, and passing alternately over P and M, finally ending at Q; and let the number of strings at P be n, which will manifestly be an even number; then if a motion be communicated to Q, which carries it to q, P will be moved to p; and as the string is inextensible, its total length in both positions will be the same; that is,

$$\overline{n-1} \,.\, MP + PQ = \overline{n-1} \,.\, Mp + pq,$$

$$\text{or } \overline{n-1}\,(Mp + pP) + Pp + pQ = \overline{n-1}\, Mp + pQ + Qq;$$

$$\therefore n \,.\, Pp = Qq.$$

If the extremity Q were once more passed over the pin M, and carried into any convenient direction, the velocity of its extremity along that direction would be plainly unaltered.

But if the end of the string were first tied to the moveable pin P, and then wound in the same manner back and forwards over the two pins P and M, finally ending at Q, then the number n of strings at the moveable pin would be an odd number, and we should find in the same manner, $n \,.\, Pp = Qq$, also the velocity of the extremity Q would be unaltered by passing it at last over the fixed pin, and carrying it from that in any convenient direction.

211. In practice these fixed and moveable pins are replaced by *blocks*, each of which contains as many mortises as the reduplication of the string requires, and in each mortise is a friction-pully or *sheave**, having a groove in its circumference round which the string or cord passes. The entire assemblage, consisting of a fixed and moveable block with the cord, is termed a *Tackle*†. The pullies may be arranged in various ways in the block, which are represented in the ordinary treatises on mechanics. As however the diameter of the pullies has been shewn to produce no effect upon the velocity ratios of the combination, it is most convenient to represent the sheaves as in fig. 118, where they are shewn as concentric, but of different diameters, and for the purpose of exhibiting the course of the string with more clearness.

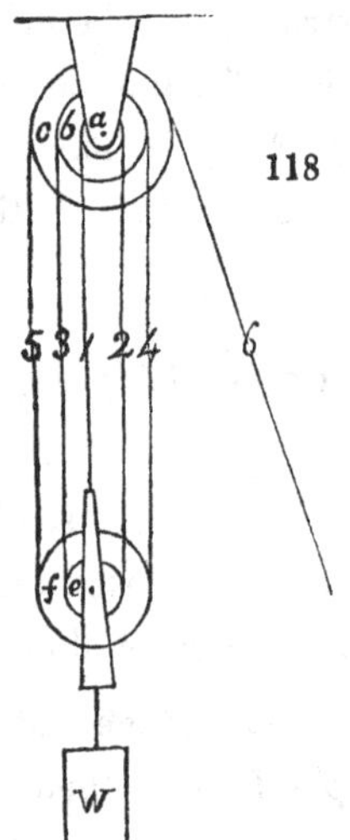

118

212. In this figure the string is attached to the lower or moveable block, and as there are five strings at this block, we have $n = 5$, and the velocity of the extremity $6 = 5 \times$ velocity of W, by Art. 210.

The upper pully being fixed, it is plain that the strings 1 & 2, 3 & 4, 5 & 6, move respectively with the same velocity, but in opposite directions, 1, 3, and 5 ascending, and 2, 4, 6 descending, if W be supposed to move upward, and *vice versa*. Also the velocities of each of these pairs of

* From *Scheibe*. Germ.

† This term appears to have been derived thus: τροχαλια, Gr.; *Trochlea*, Lat.; *Taglia*, Ital.; *Taakel*, Dutch. In French, *Mouffle* is used either for the block alone, or for the block and its sheaves; and *Pully* (Eng.), as well as *Poulie* (Fr.), is used either for the sheave or for the complete block and its sheaves.

the string are different, for the velocity of 1 is equal to that of the lower block; and if 3 were the extremity of the string, 1, 2, 3 would with their sheaves form a tackle in which $n = 3$; and therefore the velocity of 3 is triple that of the lower block; similarly, the strings 1 to 5 form a tackle in which $n = 5$; and thus, whatever odd number of strings are at the lower block, the velocities of these strings, beginning from the center, will be 1, 2, 3, ... an arithmetical series of the odd numbers, in which the velocity of *W* is supposed unity; but if one end of the string be tied to the fixed block, and consequently the number of strings at the moveable block be even, then the series of velocities can be similarly shewn to be 0, 2, 4, 6, ...

213. In figure 118 the sheaves *a*, *b*, *c*,... are supposed to revolve separately, although upon the same axis; but since the perimetral velocity of a wheel varies directly as the radius, and the strings of the tackle have been shewn to move with velocities increasing in an arithmetical progression, it follows that if the lengths of the radii of the sheaves *a*, *b*, *c*... form the same progression as the velocities of the strings, the sheaves will all revolve with the same angular velocity, and may consequently be all made in one piece. Blocks so fitted up form what is termed White's Tackle, from the name of its inventor.*

214. The free portion of rope (as 6) is termed the *fall*, and when the other extremity (1) is tied to the fixed block, and therefore, as we have seen, has no velocity; this is termed the *standing part*. In nautical phraseology, the following terms are applied to *Tackles*. If $n = 1$, the tackle is a *Whip;* if $n = 2$, it is a *Gun-Tackle;* if $n = 3$, a *Luff-Tackle;* the fall being in all cases supposed to be taken

* White's Century of Inventions, p. 33.

from the fixed block. It may however be observed, that in any given tackle the velocity ratio is different according as one block or the other is made the fixed block. Thus in fig. 118 the block from which the *fall* proceeds is made the fixed block, and $n = 5$; but if this block were employed as the moveable block, we should have $n = 6$. The number of sheaves is always less by one than the number of strings at the *fall-block*.

The fall-block is usually fixed, because this allows the fall-rope to be drawn in any direction, whereas if the fall proceed from the moveable block, it must be drawn as nearly as possible in a direction parallel to the path of the moving body, and therefore to the strings of the tackle*.

215. Several tackles may be combined, as shewn in fig. 119. Thus let A be the fixed block, a the moveable

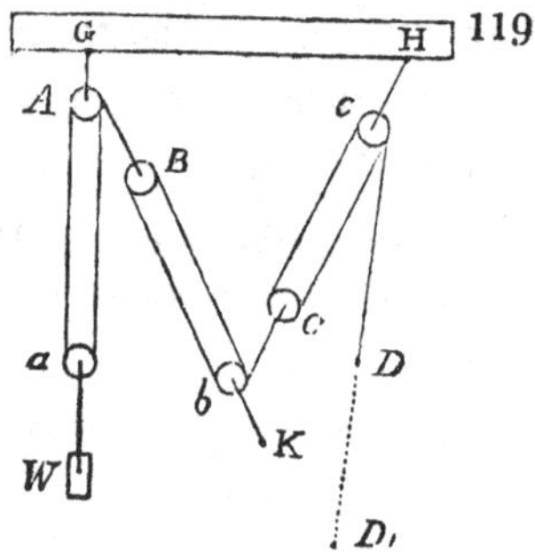

block of a tackle in which there are n_1 strings at a, and of which AB is the fall; let the extremity of this fall be tied to the moveable block B of a second tackle of which b is the fixed block, and n_2 the number of strings at B. Also, let the fall bc of the second tackle be tied to the moveable block C of a third tackle of which c is the fixed block, and cD the fall, and n_3 the strings at C; let a velocity V_4 be given to

* Vide Reduplication, in Chap. VIII.

D, and let V_1, V_2, V_3, be the velocities of W, B and C respectively;

$$\text{then } V_4 = n_3 V_3 = n_3 n_2 V_2 = n_3 n_2 n_1 V_1.$$

If there be m tackles in this series *or train*, and they have all the same number of strings, we should find in a similar way $V_{m+1} = n\,.^{m} V_1$.

Now the total number of strings in this combination $= n \times m$; whence the following problem.

216. Given the velocity ratio $\frac{V_{m+1}}{V_1} = n^m$ of the train of tackles, to find the number and nature of the separate tackles that will require the fewest strings.

$$\text{Here } n^m = \text{constant} = C \text{ suppose;}$$

$$\therefore\ m = \frac{1\,C}{1\,n} \text{ and the number of strings}$$

$$= mn = \frac{n\,.\,1\,C}{1\,n},$$

which is at a minimum when hyp. log $n = 1$, and $n = 2.72$; the nearest whole number to which being 3, it appears that a series of Luff-tackles will produce a given velocity ratio with fewer strings than any single tackle or combination of equal tackles. In fact, sailors combine two Luff-tackles in this manner, which they term *Luff upon Luff*.

If however instead of attaching each tackle to a fall from the fixed block of the previous one, it be tied to a fall from the moveable block, one sheave will be saved out of each tackle without altering the velocity ratio, and the total number of sheaves will be $(n - 1)\,.\,m$; which will be at a minimum when $n - 1\,. = 2.72$, and $\therefore\ n = 3.72$.

A combination of this kind in which $n = 2$, and therefore each pully hangs by a separate string, is commonly represented in mechanical treatises.

CHAPTER VII.

TRAINS OF ELEMENTARY COMBINATIONS.

CLASS A. {DIRECTIONAL RELATION CONSTANT.
VELOCITY RATIO CONSTANT.

---

217. THE elementary combinations which have been the subject of the preceding chapters consist, for the most part, of two principal pieces only, a driver and a follower; and we have shewn how to connect these so as to produce any required constant velocity ratio, or constant directional relation, whatever may be the relative position of the axes of rotation. There are many cases however in which, although *theoretically* possible, it may be *practically* inconvenient, or even impossible, to effect the required communication of motion by a single combination; in which case a series or train of such combinations must be employed, in which the follower of the first combination of the train is carried by the same axis or sliding piece to which the driver of the second is attached; the follower of the second is similarly connected to the driver of the third, and so on.

218. In all the combinations hitherto considered the principal pieces either revolve or travel in right lines. In a train of revolving pieces, the first follower and second driver being fixed to the same axis, revolve with the same angular velocity; and this is true for the second follower and third driver, and generally for the mth follower and $\overline{m+1}|^{\text{th}}$ driver, which will also, if the piece which carries them travel in a right line, move with the same linear velocity.

But, for simplicity, let us consider all the pieces in the train to revolve (Art. 39), and let the synchronal rotations of the axes of the train in order be

$$L_1, L_2, L_3, L_4, \&c. \ldots\ldots L_m,$$

m being the number of axes;

$$\therefore \frac{L_1}{L_2} \times \frac{L_2}{L_3} \times \frac{L_3}{L_4} \ldots\ldots\ldots \frac{L_{m-1}}{L_m} = \frac{L_1}{L_m};$$

that is; *the ratio of the synchronal rotations of the extreme axes of the train is found by multiplying together the separate synchronal ratios of the successive pairs of axes.* Also, if $A_1 A_2 \ldots A_m$ be the angular velocities of the axes, we have

$$\frac{A_1}{A_2} \times \frac{A_2}{A_3} \ldots\ldots \frac{A_{m-1}}{A_m} = \frac{A_1}{A_m} = \frac{L_1}{L_m} \text{ (Art. 20).}$$

219. And since the values of any one of these separate ratios will be unaffected by the substitution of any pair of numbers that are in the same proportion, we may substitute indifferently in any one the numbers of teeth (N), the diameters (D), or radii (R), of rolling wheels, pitch-circles, or pullies, the periods (P) in uniform motion; or express the value of the ratio in any other equivalents that may be most easily obtained from the given machine or train whose motions we wish to calculate, recollecting that

$$\frac{L}{l} = \frac{A}{a} = \frac{n}{N} = \frac{r}{R} = \frac{p}{P}, \text{ (Art. 69).}$$

220. Ex. 1. In a train of wheel-work let the first axis carry a wheel of N_1 teeth driving a wheel of n_2 teeth on the second axis; let the second axis carry also a wheel of N_2 teeth driving a wheel of n_3 teeth on the third axis, and so on.

$$\frac{A_m}{A_1} \text{ or } \frac{L_m}{L_1} = \frac{N_1}{n_2} \times \frac{N_2}{n_3} \times \dots\dots\dots \frac{N_{m-1}}{n_m},$$

that is, *to find the ratio of the synchronal rotations, or angular velocity of the last axis in a given train of wheel-work to those of the first, multiply the numbers of all the drivers for a numerator, and of all the followers for a denominator.*

It is scarcely necessary to remark that the number of drivers and of followers in a train of this kind is less by one than the number of axes.

221. Ex. 2. The ratios may each be expressed in a different manner: thus in a train of five axes, let the first revolve once while the second revolves three times;

$$\therefore \frac{L_1}{L_2} = \frac{1}{3}.$$

Let the second carry a wheel of 60 teeth driving a pinion of 20 on the third;

$$\therefore \frac{N_2}{n_3} = \frac{60}{20}.$$

Let the third axis drive the fourth by a belt and pair of pullies of 18 and 6 inches diameter respectively;

$$\therefore \frac{D_3}{d_4} = \frac{18}{6}.$$

And let the fourth perform a revolution in ten seconds, and the last in two, when the machinery revolves uniformly;

$$\therefore \frac{P_4}{p_5} = \frac{10}{2};$$

therefore we have,

$$\frac{L_1}{L_5} = \frac{1}{3} \times \frac{20}{60} \times \frac{6}{18} \times \frac{2}{10} = \frac{1}{135};$$

that is to say, that the first axis will perform one revolution while the last revolves 135 times.

222. In this manner the synchronal rotations of the extreme axes in any given machine may be calculated; their directional relation may also be found, by examining in order the connexion of the axes, and by help of the few remarks which follow.

In a train of wheel-work consisting solely of spur-wheels or pinions with parallel axes, the direction of rotation will be alternately to right and left. If therefore the train consist of an even number of axes, the extreme axes will revolve in opposite directions, but if of an odd number of axes, then in the same direction. If an annular wheel be employed, its axis revolves the same way as that of the pinion (Art. 58).

223. If a wheel A (fig. 17, page 42) be placed between two other wheels C and B, it will not affect the velocity ratio of these wheels, which is the same as if the teeth of B were immediately engaged with those of C, but it does affect the directional relation; for if B and C were in contact, they would revolve in opposite directions, but in consequence of the introduction of the intermediate axis of A, B and C will revolve in the same direction. Such an intermediate wheel is termed an *idle wheel.*

224. When the shafts of two wheels A and B, fig. 120, lie so close together that the wheels cannot be placed in the same plane without making them inconveniently small, they may be fixed as here shewn, so as to lie one behind the other, and be connected by an idle wheel C, of rather more than double the thickness of the wheels it connects. Such a thick idle wheel is termed a *Marlborough wheel*, in

120

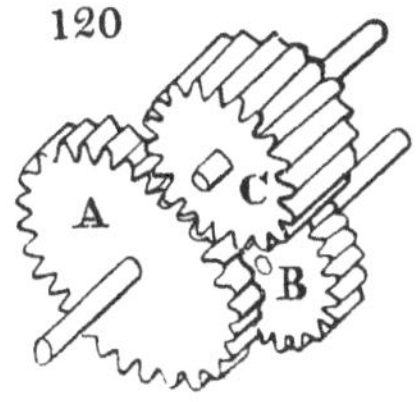

some districts. It is employed in the roller frames of spinning machinery.

225. When the axes in a train are not parallel, the directional relation of the extreme axes can only be ascertained by tracing the separate directional relations of each contiguous pair of axes in order.

By intermediate bevil-wheels parallel axes may be made to revolve either in the same or opposite directions according to the relative positions of the wheels; for example, in fig. 121 the wheel *A* drives *B*, upon whose shaft is fixed the wheel *E*. Now if the wheel *C* be fixed on the same side of the intermediate axis as *A*, the parallel axes of *A* and *C* will revolve in opposite directions; but if the wheel be fixed as at *D*, on the opposite side of the intermediate axis, then the axes of *A* and *D* will revolve in the same direction, the same number of wheels being employed in both cases.

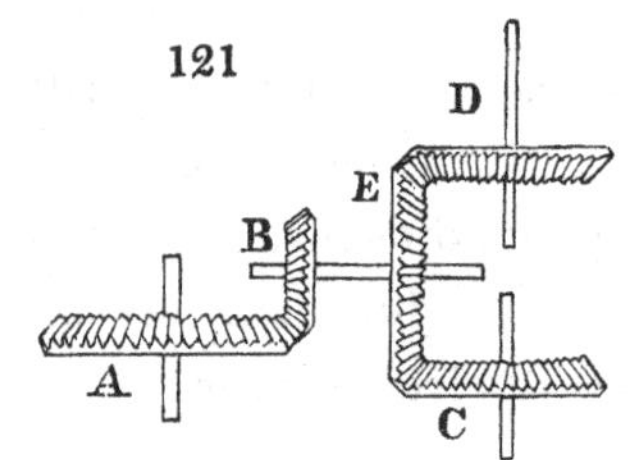

Endless screws may be represented in calculation by a pinion of one or more leaves, according to the number of their threads, (Art. 175), but their effect upon the directional relation of rotation will be different, according as they are right-handed or left-handed screws. (Art. 169).

226. Two separate wheels or pieces in a train may revolve concentrically about the same axes, as for example, the hands of a clock. Also, in fig. 122, the wheel *B* is fixed to an axis *Cc*, and the wheel *A* to a tube *d* or *cannon*, which turns freely upon *Cc*. If these wheels may revolve in op-

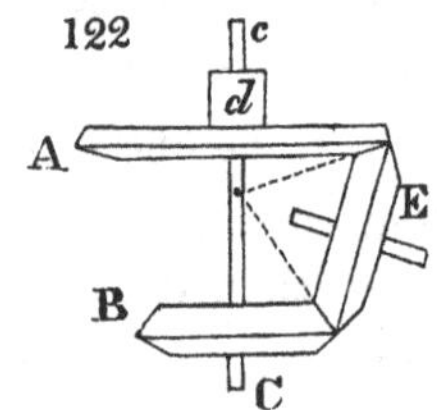

posite directions, a single bevil-wheel E will serve to connect them, if the three cones have a common apex as in the figure; and since E is an idle wheel (Art. 223), the velocity ratio of B to A will depend solely upon the radii of their own frusta.

But if the wheels B, A are to revolve in the same direction, they must be made in the form of spur-wheels, and connected by means of two other spur-wheels fixed to an axis parallel to Cc.

227. Millwrights imagine that in a given pair of toothed wheels it is desirable that the individual teeth of one wheel should come into contact with the same teeth of the other wheel as seldom as possible, on the ground that the irregularities of their figure are more likely to be ground down and removed by continually bringing different pairs of teeth into action.

This is a very old idea, and is stated nearly in the above words by De la Hire. It has also been acted upon up to the present time. Thus Oliver Evans tells us, that "great care should be taken in matching or coupling the wheels of a mill, that their number of cogs be not such that the same cogs will often meet; because if two soft ones meet often, they will both wear away faster than the rest, and destroy the regularity of the pitch; whereas if they are continually changing they will wear regular, even if they be at first a little irregular*."

The clockmakers on the other hand, think that the wearing down of irregularities will be the best effected by bringing the same pair of teeth into contact as often as possible†.

* O. Evans, Young Millwright's Guide, Philadelphia, 1834, p. 193. Vide also Buchanan's Essays, by Rennie, p. 117.

† Francœur, Mécanique Elémentaire, p. 143.

Let a wheel of M teeth drive a wheel of N teeth, and let $\frac{M}{N} = \frac{m}{n}$ when m and n are the least numbers in that ratio;

$$\therefore\ nM = mN,$$

and n is the least whole number of circumferences of the wheel M that are equal to a whole number of circumferences of the wheel N.

If, therefore, we begin to reckon the circumferences of each wheel that pass the line of centers, after a given pair of teeth are in contact, it is clear that after n revolutions of M, and m of N, the same two teeth will be again in contact. Neither can they have met before; for as the entire circumference of one wheel applies itself to the entire circumference of the other tooth by tooth, and as the numbers m and n are the least multiples of the respective circumferences that are equal, it follows that it is only after these respective lengths of circumferences have rolled past each other that the beginnings of each can again meet.

If we act on the watchmaker's principle, by which the contacts of the same pair are to take place very often, the numbers of the wheels M and N must be so adjusted that m and n may be the smallest possible, without materially altering the ratio $\frac{M}{N}$; and this will be effected by making the least of the two numbers m, n equal to unity, and therefore M a multiple of N.

But if the millwright's principle be adopted, m and n must be as large as possible, that is, equal to M and N, or in other words, M and N must be prime to each other. The millwrights employ a *hunting cog* for this purpose. Suppose, for example, that a shaft is required to revolve about three times as fast as its driving shaft, 72 and 24 are a pair of numbers for teeth that would produce this effect and would suit a watchmaker, one being a multiple of the other; but the millwright would add one tooth to the wheel (the hunting cog), and thus obtain 73 and 24, which are prime to each other, and very nearly in the desired ratio*.

228. Sometimes also the nature of the mechanism requires that the wheels shall come as seldom as possible into the same relative positions, and in that case the principle may be applied to a train of several axes. For example, in a train of three axes, in which the drivers have each 22 teeth, and the followers 25 and 35 teeth, we have

$$\frac{L_1}{L_3} = \frac{25 \times 35}{22 \times 22} = \frac{484}{875};$$

which numbers are prime to each other, and therefore the extreme wheels of the train will not return to the same relative position, until one has made 484, and the other 875 revolutions. These are the numbers of the old Piemont silk-reel (1724), which is an excellent example of this principle†.

229. We are now able to calculate the relative motions of the parts in a given machine in which the velocity ratios are constant. The inverse problem is one of considerable importance in the contrivance of mechanism; namely, *Given the velocity ratio of the extreme axes or pieces of a train, to determine the number of intermediate axes, and the proportions of the wheels, or numbers of their teeth.* For simplicity we may suppose the train to consist of toothed wheels only; for a mixed train, consisting of wheels, pullies, link-work, and sliding pieces, can be calculated upon the same principles. Let the synchronal rotations of the first and last axes of the train be

* In a pair of wheels whose numbers are so obtained, any two teeth which meet in the first revolution are distant by one in the second, by two in the third, and so on; so that one tooth may be said to *hunt* the other, whence the phrase, a hunting cog.

† Encycl. Méthodique, Manufactures et Arts, tome II. p. 20.

L_1 and L_m respectively, and let $N_1\ N_2 \ldots$ &c. be the numbers of teeth in the drivers, and $n_1\ n_2 \ldots$ in the followers: then by Art. 220,

$$\frac{L_m}{L_1} = \frac{N_1 . N_2 . N_3 \ldots}{n_1 . n_2 . n_3 \ldots}$$

and by hypothesis the value of $\frac{L_m}{L_1}$ is given, and we have to find an equal fraction whose numerator and denominator shall admit of being divided into the same number of factors of a convenient magnitude for the number of teeth of a wheel. Also to find the value of m.

Synchronal rotations are preferred to angular velocities in stating the question, because it is generally in this form that the data are supplied.

230. In any given train of wheel-work the drivers may be placed in any order upon the axes as well as the followers; for the value of the fraction $\frac{N_1 . N_2 . N_3 \ldots}{n_1 . n_2 . n_3 \ldots}$ will be unaffected by any change of order in the factors, and therefore N_1 may be placed either upon the first, second, or third axes; and similarly for the others.

231. Let w be the greatest number of teeth that can be conveniently assigned to a wheel, and p the least that can be given to a pinion. The train may be either required for the purpose of reducing or increasing velocity. In the first case, L_m will be less than L_1, and the pinions the drivers; but in the second case, L_m will be greater than L_1, and the wheels the drivers.

Let $\therefore \frac{L_1}{L_m}$ or $\frac{L_m}{L_1} = \left(\frac{w}{p}\right)^k$ where k may be a whole number, or a fraction. Take m equal to $k+1$ (Art. 220) if a whole number, or to the next greatest whole number to $k+1$ if a fraction. This will plainly be the least value that can be given to m.

For m must be a whole number, and if it be taken less than $k+1$ then the values of $\frac{w}{p}$ will be greater; that is, either w will become a greater number than can be assigned to a wheel, or p a less than can be given to a pinion, which is absurd.

No general rule can be given for determining the values of w and p, which are governed by considerations that vary according to the nature of the proposed machine; also, it will rarely happen that the fraction will admit of being divided into factors so nearly equal as to limit the number of axes to the smallest value so assigned.

The discussion of a few examples will best explain the mode of proceeding in particular cases.

232. Fig. 123 is a diagram to represent the arrangement of the wheel-work of a clock of the simplest kind, for the purpose of illustrating what follows upon trains of wheel-work in general.

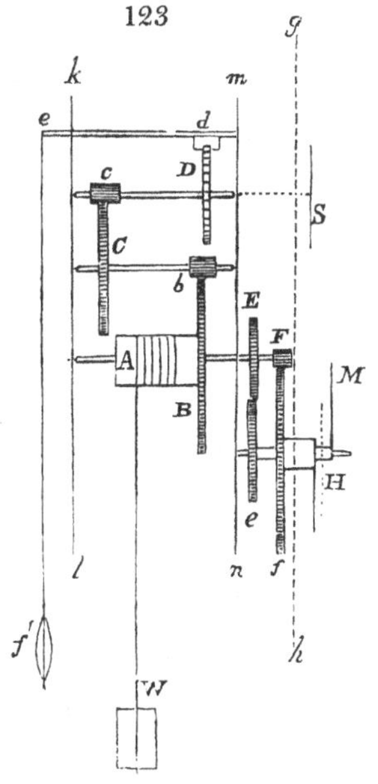

The weight W is attached to the end of a cord, which is coiled round the barrel A. Upon the same axis or *arbor** as the barrel is fixed a toothed wheel B, and this wheel drives a pinion b, which is fixed to the second arbor Cb of the train, which also carries a wheel C. This wheel drives a pinion c upon the third arbor, and upon this arbor is also fixed a toothed wheel D of a peculiar

* *Arbor* is the watchmakers' term for an axis; vide Note p. 44.

construction, termed an escapement wheel or swing-wheel. Above this wheel is an arbor *ed* termed the verge, which is connected with the pendulum *ef* of the clock, and vibrates together with it through a small arc. The verge also carries a pair of teeth which are termed pallets, and are engaged with the teeth of the swing-wheel *D* in such a manner, that every vibration of the pendulum and verge allows one tooth of the wheel to *escape* and pass through a space equal to half the pitch. With the nature of this connexion we have at present nothing to do; for, as the motion of the clock-work is our only object, it is sufficient to know that one tooth of the swing-wheel passes the line of centers for every two vibrations of the pendulum.

Let the time of a vibration of the pendulum be t seconds, where t is a whole number or a fraction, and let the swing-wheel have e teeth, then the period or time of a complete rotation of this wheel is $2te$. To take a simple case, let the pendulum be a seconds' pendulum; $\therefore t = 1$, and if $e = 30$, the swing-wheel will revolve in a minute; and if B have 48 teeth, and C 45, and the pinions 6 leaves each, we have for the train

$$\frac{L_3}{L_1} = \frac{48 \times 45}{6 \times 6} = 60;$$

therefore A will revolve in an hour; and supposing the cord to be coiled about sixteen times round the barrel, the weight in its descent will uncoil it and turn the barrel round, communicating motion to the entire train until the cord is completely uncoiled, which it will be after sixteen hours.

This train of wheel-work is solely destined to the purposes of communicating the action of the weight to the pendulum in such a manner as to supply the loss of motion from friction and the resistance of the air. But besides this, the clock is required to indicate the hours and minutes by the rotation of two separate hands, and accordingly two other trains of wheel-work are employed for this purpose.

The train just described is generally contained in a frame consisting of two plates, shewn edgewise at *kl*, *mn*, which are kept parallel and at the proper distance by means of three or four pillars, not shewn in the diagram. Opposite holes are drilled in these plates, which receive the pivots of the axes or arbors already described. But the axis which carries *A* and *B* projects through the plate, and other wheels *E* and *F* are fixed to it.

Below this axis and parallel to it a stout pin or *stud* is fixed to the plate, and a tube revolves upon this stud, to one end of which is fixed the minute-hand *M*, and to the other a wheel *e* engaged with *E*. In our present clock *E* revolves in an hour, consequently the wheels *E* and *e* must be equal.

A second and shorter tube is fitted upon the tube of the minute-hand so as to revolve freely, and this carries at one end the hour-hand *H*, and at the other a wheel *f*, which is driven by the pinion *F*; and because *f* must revolve in twelve hours, it must have twelve times as many teeth as *F*.

233. To exhibit the ramifications of motion in a machine, and the order and nature of the several parts of which the trains are composed, it is convenient to employ a *notation*. This notation should be of such a form as not only to exhibit these particulars, but also to admit of the addition, if necessary, of dimensions and nomenclature, as well as to allow of the necessary calculations by which the velocity ratios may be deduced. To exhibit in this way the actual arrangement of the parts is out of the question; this can only be done by drawings, and the very object of a notation is to unravel the apparent confusion into which the trains of

motion are thrown by the *packing* of the parts into the frame of the machine, and to place them in the order of their successive action.

Clock and watchmakers have long employed a system which consists simply in representing the wheels by the numbers of their teeth, and writing these numbers in successive lines, placing the wheels which are fixed on the same arbor on the same horizontal line, with the sign – interposed, and writing the numbers of the wheels that are in geer vertically over each other. The first driver in the train is always placed at the top of the series.

Thus in the principal train of the clock, fig. 123, if the letters represent the wheels we should write down the train thus:

B
b——*C*
c——*D*;

or, employing the numbers already selected,

48
6——45
6——30,

and adding the names, which is sometimes done,

Great wheel 48,
Pinion 6——45 second-wheel,
Pinion 6——30 swing-wheel*.

* Farey in Rees' Cyclopædia, art. *Clockwork*, calls this the ordinary mechanical method of writing down the numbers. Oughtred in his Opuscula, 1677, proposes another method in which the wheels which are on the same axis are written vertically over one another, and those which are in geer are placed in the same line with the character) between; thus, (the first driver being at the bottom, and all the drivers to the right of the followers):

30
6) 45
6) 48

He employs, however, letters in lieu of figures, and introduces other artifices which are scarcely worth dwelling upon. Derham (Artificial Clockmaker, 1696)

234. This method requires very little addition to make it a very convenient system for mechanism in general. Thus the entire movement of the clock, fig. 123,

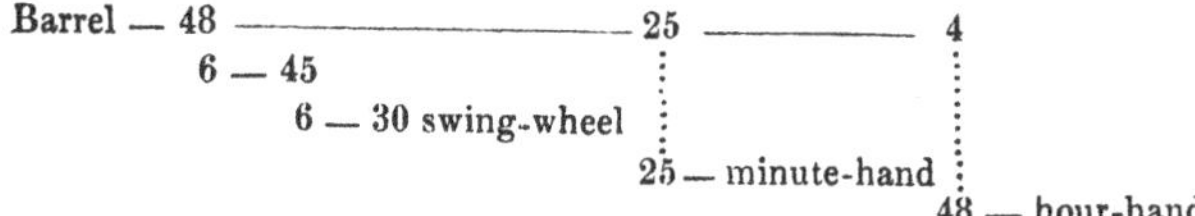

may be thus represented, and by which is shown very clearly the three trains of mechanism from the barrel to the swing-wheel, the minute-hand and the hour-hand; as well as the distinction of the pieces into drivers and followers, and the nature of their connexion; namely, whether they be permanently united by being fixed upon the same axis, or connected by geering. If however other connexions are introduced, as by wrapping-bands, or links, this must be written in the diagram, or expressed by a proper sign. I shall have occasion to return to this subject in a future page*.

235. In the explanation of the clock, fig. 123, I have assumed the numbers of the wheel-work and of the axes; let us now examine whether these are the best for the purpose, or generally how such numbers would be determined.

If the arbor of the swing-wheel revolve in a minute,

follows this method, and also uses another which consists in writing all the numbers in one line, thus, 48) 6 — 45) 6 — 30, where the character) implies that the wheels between which it lies geer together, and — that they are fixed on the same axis. Allexandre, Traité général des Horloges, 1735, writes the numbers thus, 48.6 — 45.6 — 30; and Derham also gives the "usual way of watchmakers in writing down their numbers," thus,

48
45 — 6
30 — 6

which, to use his own words, "though very inconvenient in calculation, representeth a piece of work handsomely enough, and somewhat naturally."

* Mr. Babbage is the only one who has endeavoured to extend Notation to Mechanism in general. His elaborate and complete system is fully explained in his paper on "A method of expressing by signs the action of Machinery," in the Philosophical Transactions, 1826, vide below, Chap. IX.

and that of the barrel in an hour, we have $\frac{L_m}{L_1} = 60$; or if D be the product of all the drivers, and F of the followers, $D = 60 . F$, an indeterminate equation, for the solution of which any numbers may be employed that are proper for the teeth of wheels. Now in common clocks six is the least number of leaves that is ever employed in a pinion, and 60 teeth the greatest number that can be given to a wheel;

$$\therefore \frac{w}{p} = \frac{60}{6} = 10.$$

Now $\frac{L_m}{L_1} = 10^{1 \cdot 8}$, therefore by Art. 231, 3 is the least number of axes; and there will be two pinions of six each, $\therefore D = 60 \times 6^2 = 2160$, which is the product of two wheels.

We are at liberty to divide this into any two suitable factors. The best mode of doing it is to begin by dividing the number into its prime factors, writing it in this form:

$$2160 = 2^4 \times 3^3 \times 5.$$

For this enables us to see clearly the composition of the number; and it is easy to distribute these factors into two groups; as for example,

$$2^4 . 3 \times 3^2 . 5 = 48 \times 45, \text{ or } 2^3 . 5 \times 2 . 3^3 = 40 \times 54,$$
$$\text{or } 2^2 . 3^2 \times 2^2 . 3 . 5 = 36 \times 60.$$

The nearest to equality is the first, 48 and 45; and these will probably be selected for the train, which will stand thus:

$$\frac{D}{F} = \frac{48 \times 45}{6 \times 6}.$$

This is the best form in which to exhibit the numbers for a train when they have been merely divided into proper factors for teeth. If the distribution of the wheels and pinions upon the several axes is also settled, the train may then be written in the form 48

6——45

6.

236. Six is however too small a number of leaves to ensure perfect action in a pinion, for it appears in the Table (p. 93) that a pinion of 6 will only work with a wheel of 20 when the receding arc of action is equal to $\frac{2}{3} \times$ pitch, and that if this arc be greater, the pinion becomes impossible. A pinion of 8 will be better, but 10 or 12 should be employed if a very perfect action is required. If 8 be selected, we have $F = 8^2 = 64$, and $D = 64 \times 60$, which will form a good train.

But in well-made clocks we may allow more than 60 teeth to the wheel: 100 or even 120 is very admissible. If we begin, then, with the wheels, and assume that three arbors are to be employed,

$$\text{let } \frac{D}{F} = \frac{(100)^2}{p^2} = 60; \;\therefore\; p = 13, \text{ nearly.}$$

$$\text{Assume, therefore, } F = 12 \times 14; \;\therefore\; D = 60 \times 12 \times 14$$
$$= 96 \times 105;$$

which gives the train 105
14——96
12

237. In a train of $\overline{k+1}$. axes of which every wheel has w teeth, and every pinion p leaves, we have

$$\frac{L_m}{L_1} = \left(\frac{w}{p}\right)^k = x^k \text{ if } \frac{w}{p} = x.$$

Now $xp \; (= w)$ is the number of teeth in each wheel, and $k\,(p + xp)$ is the entire number of teeth in the train.

$$\text{Let } \left(\frac{w}{p}\right)^k \text{ or } x^k = \text{constant} = C;$$

$$\therefore\; k = \frac{\mathrm{l}\,C}{\mathrm{l}\,x},$$

$$\text{and number of teeth} = \frac{1\,C}{1\,x} \cdot p \cdot (1 + x)$$
$$= \text{a minimum.}$$

Differentiating we obtain in the usual manner,

$$1\,x = \frac{1 + x}{x}\ ; \quad \text{whence } x = 3 \,.\, 59.$$

If therefore a given angular velocity ratio is to be obtained with the least number of teeth, we must make $\frac{w}{p} = 3 \,.\, 59$. This theorem is due to Dr. Young*.

As a practical rule this is not of much value, for it proceeds on the assumption that simplicity is best consulted by reducing the number of teeth *only* as much as possible; but, in fact, it is necessary in doing this to avoid also increasing the number of axes in a train. For example, in our clock $\frac{L_m}{L_1} = 60$, which being greater than the cube of $3 \,.\, 59$ would require for the least number of teeth at least three wheels; and, in fact, if we compute the number of teeth required in the case of one, two, three, and four wheels, assuming the number of leaves in the pinions to be six, we find, putting D for the denominator, and dividing it into convenient factors:

	Wheels.	Total Number of Teeth.
one wheel	$D = 6 \times 60 = 360$	$360 + 6 = 366$
two wheels	$D = 6^2 \times 60 = 45 \times 48$	$45 + 48 + 2 \times 6 = 105$
three wheels	$D = 6^3 \times 60 = 20 \times 27 \times 24$	$20 + 27 + 24 + 3 \times 6 = 89$
four wheels	$D = 6^4 \times 60 = 15 \times 16 \times 18 \times 18$	$15 + 16 + 18 + 18 + 4 \times 6 = 91$
five wheels	$D = 6^5 \times 60 = 12^3 \times 15 \times 18$	$3 \times 12 + 15 + 18 + 5 \times 6 = 99$

So that, as the theorem has already taught us, the least number of teeth, 89, is required when three wheels are employed. But the universal practice is to employ two wheels and pinions only in the train between the hour-arbor and swing-wheel arbor, for, in fact, the increase in the number of

* Young's Nat. Philosophy, vol. II. p. 56.

teeth does not occasion so great a loss of simplicity as the additional arbor with its wheel and pinion would do. Some mechanicians have fallen into the opposite error of supposing that the simplicity of the clock would be still more improved by reducing the train to a single wheel and pinion, and hence increasing inordinately the number of teeth in the wheel. Of this nature are Ferguson's and Franklin's clocks*.

238. If a clock has no seconds' hand there is no necessity for the arbor of the swing-wheel to perform its revolution in a minute, which when the pendulum is short, would become impracticable, from the great number of teeth required. Now from Art. 232, if t be the time of vibration of the pendulum in seconds, and e the number of teeth of the swing-wheel, $2te$ is time of rotation of the swing-wheel.

But the vibrations of small pendulums are commonly expressed by stating the number of them in a minute. Let p be this number, $\therefore \frac{2e}{p}$ is the time of one rotation of the swing-wheel in minutes, and the hour-arbor revolves in 60 minutes; the train between them is represented by $\frac{D}{F} = \frac{30p}{e}$.

Ex. The pendulum of a clock makes 170 vibrations in a minute, and there are 25 teeth in the swing-wheel, and eight leaves are to be given to the pinions; to find the wheels:

$$\frac{D}{64} = \frac{30 \times 170}{25};$$

$$\text{whence } D = 13056 = 128 \times 102.$$

239. In a watch the vibrations of the balance are much more rapid than in any pendulum-clock, varying in different constructions from 270 to 360 in a minute. Also, from the

* Vide Ferguson's Mechanical Exercises, or any Encyclopædia.

small size of the machinery, it becomes impossible to put so many teeth into the wheels. The escapement-wheel, termed in a watch the balance-wheel, has from 13 to 16 teeth, instead of having, as in a clock, from 20 to 40, and the numbers of teeth in the wheels vary from 40 to 80, or in chronometers and larger work are sometimes carried as high as 96, whereas in large clocks, 130 may even be employed. Now as the number of leaves in the pinions do not admit of reduction, the consequence is, that an additional arbor must be employed in watches, and the train of wheel-work between the hour-arbor and the arbor of the balance-wheel consists of 3 wheels and 3 pinions, instead of the two pair employed in a clock.

Ex. The balance of a watch makes 360 vibrations in a minute, and there are 15 teeth in the balance-wheel, and eight leaves in the pinions; to find the wheels:

Here $F = 8 \times 8 \times 8$,

$$\text{and } D = 8^3 \frac{30 \times 360}{15} = 368640 = 80 \times 72 \times 64.$$

240. The examples of clock-trains already given, refer merely to the connexion between the hour-arbor and the swing-wheel, and it has been assumed throughout that the barrel for the weight is carried by the hour-arbor; but in this case the clock will not go for more than sixteen hours, and must therefore be wound up every night and morning. If it be required to go longer the barrel must be fixed to a separate axis, and this connected by wheel-work with the hour-arbor, so that the barrel may revolve much more slowly, and consequently allow the weight to occupy a longer time in its descent.

Now the cord, as we have seen, is wound spirally round the barrel, and by making the barrel of the requisite length, we could of course make it hold as many coils as we please.

But in practice it is found that if more than about sixteen coils are placed on it, it becomes inconveniently long. So that if the clock be required to go for eight days without fresh winding up, each turn of the barrel will occupy twelve hours. As the arbor of the hour-hand revolves in one hour, any pair of wheels whose ratio is 12 will answer the purpose of connecting them; 96 and 8 are the numbers usually employed, which will produce this train:—

Train for Eight-day Clock.	Periods.
96	12^h
8 — 105	1^h
8 — 96	
8 — 30....	$1'$

241. If the clock be required to go a month, or 32 days, without winding, then supposing the barrel, as before, to have sixteen turns, each turn of the barrel will occupy 48 hours, and the train from the barrel to the hour-arbor $= \frac{D}{F} = 48$, which is too great a number for a single pair, but will do very well for two. If pinions of nine are employed,

$$D = 9 \times 9 \times 48 = 72 \times 54;$$

which numbers being small we are at liberty to employ larger pinions; for example, if we take twelve and sixteen,

$$D = 12 \times 16 \times 48 = 96 \times 96;$$

whence the following train:—

Train for Month-Clock.	Periods.
96	48^h
16 — 96	...
12 — 105	1^h
8 — 96......	...
8 — 30	1′

242. Now in the clock (fig. 123), the arbor of A is made to revolve in an hour, because the wheels E and e are equal. By making these wheels of different numbers, we get rid of the necessity of providing an arbor in the principal train that shall revolve in an hour, and may by that means, in an eight-day clock, or month-clock, distribute the wheels more equally. For example, in an eight-day clock let the swing-wheel revolve in a minute; and let the train from the barrel-arbor to this minute-arbor be $\frac{108 \times 108 \times 100}{12 \times 12 \times 10} = 810$, in which the barrel will revolve in 810 minutes or thirteen hours and a half, and consequently fourteen or fifteen coils of the cord will be sufficient.

The second wheel in this train, which in fig. 123 corresponds to B, will revolve in $\frac{12}{108} \times 810$ minutes, or an hour and a half, and on its arbor must be fixed, as in the figure, the two wheels E and F for the minute and hour-hands; consequently, the ratio of

$$\frac{F}{f} = \frac{1}{8}, \text{ and } \frac{E}{e} = \frac{3}{2}.$$

It is convenient that the size or pitch of the teeth in these two pairs should be about the same. To effect this, let x be the multiplier of the first ratio, and y of the second;

so that x and $8x$ are the numbers of teeth in the first pair, and $3y$, $2y$ in the second. Then, if the teeth of the two pairs be of the same pitch, we have

$$x + 8x = 3y + 2y, \text{ or } 9x = 5y; \therefore\ x = \frac{5y}{9}.$$

$$\text{Let } y = 9z; \therefore\ x = 5z;$$

and if $z = 1$, $y = 9$, $x = 5$, numbers are $\frac{5}{40}$ and $\frac{27}{18}$

$z = 2$, $y = 18$, $x = 10$, ... $\frac{10}{80}$ and $\frac{54}{36}$;

either of which may be adopted.

Train of Eight-day Clock.	Periods.
108	810′
12 — 108 —————— 54 ———— 10	90′
12 — 110	...
10 — 30	1′
36 — minute-hand	60′
80 — hour-hand	720′

I have confined the above examples to clock-work, because its action is more generally intelligible than that of other machines; but the principles and methods are universally applicable, or at least require very slight modifications to adapt them to particular cases.

TO OBTAIN APPROXIMATE NUMBERS FOR TRAINS.

243. If $\frac{L_m}{L_1} = a$ when a is a prime number, or one whose prime factors are too large to be conveniently employed in wheel-work, an approximation may be resorted to. For example, assume $\frac{L_m}{L_1} = a \pm E$. This will introduce an error

of $\pm E$ revolutions of the last axis, during one of the first, and the nature of the machinery in question can alone determine whether this is too great a liberty.

But we may obtain a better approximation than this, without unnecessarily increasing the number of axes in the train; for determine in the manner already explained the least number m of axes that would be necessary if a were decomposable, and the number of leaves that the nature of the machine makes it expedient to bestow on the pinions, and let F be the product of the pinions so determined;

$$\therefore \frac{L_m}{L_1} \text{ or } \frac{D}{F} = \frac{Fa}{F}, \text{ supposing the wheels to drive.}$$

$$\text{Assume } \frac{D}{F} = \frac{Fa \pm E}{F};$$

where E must be taken as small as possible, but so as to obtain for $Fa \pm E$ a numerical value decomposable into factors. There will be in this case an error of $\pm E$ rotations in the last axis during F of the first, or of $\frac{\pm E}{F}$ rotations during one of the first.

If the pinions be the drivers, then in the same manner assume $\frac{L_1}{L_m} = \frac{Da \pm E}{D}$; and there will be an error of $\frac{\pm E}{D}$ rotations in the first axis during one of the last.

244. Ex. Let it be required to make $\frac{L_m}{L_1} = 269$ nearly. Now if the nearest whole number 270 be taken, a train may be formed, but with an error of one revolution in 270. But suppose that from the nature of the machine, a ratio of $\frac{1}{8}$ is the greatest that can be allowed between wheel and pinion,

then since 269 lies between 8^2 and 8^3, it appears that three pair of wheels and pinions are necessary.

If pinions of 10 are employed, $\frac{D}{F} = \frac{269000}{1000}$,

and $\frac{269001}{1000} = \frac{3^8 \times 41}{10^3}$, will make a very good train, with an error of $\frac{1}{1000}$ of a revolution only in 269.

245. Ex. 2. Let it be required to find a train that shall connect the twelve hour-wheel of a clock with a wheel revolving in a lunation, $= 29^d. 12^h. 44'$ nearly, for the purpose of shewing the Moon's age upon a dial. Reducing the periods to minutes, we have

$$\frac{S_1}{S_m} = \frac{42524}{720},$$

of which the denominator $(= 2^2 \times 10631)$ contains a large prime, but

$$\frac{42524 + 1}{720} = \frac{945}{16} = \frac{3^3 . 5 . 7}{2^4},$$

is well adapted to form a train of wheel-work, with an error of one minute in a lunation.

246. This method is sufficient for ordinary purposes, but if greater accuracy be required, or if the terms of the fraction, although divisible into proper factors, should require so many wheels and pinions, as to make it necessary to find a fraction which shall approximate to the value in smaller terms, then *continued fractions* must be resorted to.

$\frac{S_1}{S_m}$ being given in the form of a fraction with large terms, must be treated in the usual manner* to obtain the

* Vide Euler's Algebra, Barlow on Numbers, or Bonnycastle's Algebra, &c.

series of principal and intermediate fractions, which must be separately examined until one is found that will admit of a convenient division into factors, and at the same time approximate with sufficient accuracy.

247. Ex. *To find an annual train.*

Let it be required to find a train of wheel-work for a clock, by means of which a wheel may be made to revolve in an exact year, that is, in 365 days, 5 hours, 48 minutes, 48 seconds*.

If the hours, minutes, and seconds, be reduced to decimals of a day, the period becomes $365.24\dot{2}$ days; and supposing the pinion from which the motion is to be derived to revolve in one day, the required ratio becomes $\frac{365.24\dot{2}}{1.000}$, which by the common rule for circulating decimals is equal to

$$\frac{365242 - 36524}{900} = \frac{328718}{900} = \frac{164359}{450},$$

when in its lowest terms.

Now as the nearest whole number to this is 365, it appears that three axes, at least, would be required to produce this variation of motion, and therefore the fraction itself would not be in terms too great, provided it were manageable. Now

$$\frac{164359}{450} = \frac{269 \times 47 \times 13}{10 \times 9 \times 5};$$

which has an inconveniently large number, 269, but has been actually employed to form a train, in Mr. Pearson's Orrery for Equated Motions†, in this form,

$$\frac{269 \times 26 \times 94}{10 \times 10 \times 18}.$$

* The length of the year determined by different astronomers varies in the number of seconds from 47″.95 to 51″.6; the mean of five results is 49″.77.

† Rees' Cyclopædia, art. Orrery.

If the ratio be treated by the method of continued fractions, we obtain in the usual manner,

Quotients.			365	4	7	1	3	1	2
Principal Fractions.	$\frac{0}{1}$	$\frac{1}{0}$	$\frac{365}{1}$	$\frac{1461}{4}$	$\frac{10592}{29}$	$\frac{12053}{33}$	$\frac{46751}{128}$	(B) $\frac{58804}{161}$	(A) $\frac{164359}{450}$
Intermediate Fractions.							$\frac{34698}{95}$ $\frac{22645}{62}$		(C) $\frac{105555}{289}$

The whole of these fractions will be found unmanageable, from containing large primes, with the exception of those marked *A*, *B* and *C*, of which *A* is the original fraction.

$$(B) = \frac{241 \times 61 \times 4}{7 \times 23} = \frac{241 \times 61 \times 52}{23 \times 13 \times 7}$$

corresponds to a period of $365^d. 5^h. 48'. 49''.19218$; this has been employed by Janvier*.

$$(C) = \frac{105555}{289} = \frac{227 \times 31 \times 15}{17 \times 17}$$

is equivalent to a period of $365^d. 5^h. 48'. 47''. 3$, and is rather more accurate than the last; but as they each include a large wheel, it appears that the original fraction is quite as convenient.

248. If, as in the example just cited, the series of fractions obtained will not give a sufficiently convenient result, the more general method which follows may be employed, which however requires the calculation of the continued fractions, at least of the principal fractions, as they are called, and which, therefore, will not supersede the

* Rees' Cyclopædia, art. Planetary Machines.

method just explained, but may be used after it, should it be found to fail.

To find a fraction $\frac{x}{y}$ very near to $\frac{a}{b}$, we have their difference $= \frac{a}{b} - \frac{x}{y} = \frac{ay - bx}{by} = \frac{k}{by}$, suppose: k will be by the supposition a very small integer, compared with by, and either positive or negative; to find k, we have the indeterminate equation $ay - bx = k$. Let the fraction $\frac{a}{b}$ be converted into a series of principal converging fractions, and let $\frac{p}{q}$ be the last but one, then it can be shewn* that the following expressions will include all the solutions of this equation that are possible in integer numbers: $x = pk + ma$, $y = qk + mb$,

$$\text{and } \frac{x}{y} = \frac{pk + ma}{qk + mb}$$

will be the approximate fraction required, in which m may be any whole number, positive or negative, as well as k, but k must be small with respect to by or ax. Thus a multitude of values of $\frac{x}{y}$ may be obtained, from whence the one may be chosen that best admits of decomposition into factors. The only part of this process which is left to choice is the selection of values for k and m. The numbers obtained from them for x and y must necessarily be small, for we are seeking numbers less than a and b, and therefore k and m must have different signs, but even with this limit there is an infinite latitude given to the choice.

Assume $k = 0, -1, +1, -2$, and so on; and in each case take such values of m as will make the values of x and y,

* Euler's Algebra, p. 530. Barlow on Numbers, p. 317. Francœur, Cours de Mathematiques, Art. 565. Par. 1819.

not too great for the purpose, trying always whether the pair of results are decomposable into factors, and if they be, then proceeding to calculate the consequent error. In this way a pair of numbers will at last be found, that will give sufficient exactness without employing too much wheel-work*. Tables of factors will greatly assist in these operations†.

249. For example, to find a fraction $\frac{x}{y}$ very near to $\frac{45}{14}$, (Art. 251.) the last fraction but one of the series of principal converging fractions, is $\frac{16}{5}$, and putting these numbers in the expression for $\frac{x}{y}$, we have

$$\frac{16\,k + m\,45}{5\,k + m\,14}.$$

$$\text{Let } m = 1\ k = -1, \quad \therefore \frac{x}{y} = \frac{29}{9}.$$

$$m = 1\ k = -2 \qquad \frac{x}{y} = \frac{13}{4}.$$

$$m = 2\ k = -3 \qquad \frac{x}{y} = \frac{42}{13}.$$

Two of these have already been obtained from the series of converging fractions, but the third $\frac{42}{13}$ is a new one. In fact, since the expression $\frac{m\,a + pk}{mb + qk}$ includes the whole of the principal and secondary converging fractions, as well as many other approximate values of the original fraction, it must be expected that some assumed values of m and k will reproduce these already calculated approximations.

* Francœur, Dict. Technologique, tom. XIV. p. 423, and Traité de Mécanique, p. 146.

† Such as Barlow's New Mathematical Tables, 1814. Chernac. Cribrum Arithmeticum, Davent. 1811. Burckhardt, Table des Diviseurs. Par. 1817.

Barlow's Table extends only to 10000, Chernac's to 1019999, and Burckhardt's to 3035999.

But the coexisting values of m and p that belong to th converging fractions, may be obtained at once, to save thi useless trouble. For this purpose, write the quotients obtained from the original fraction in a *reverse order*, and proceed to deduce converging fractions from them in the usual manner, both principal and intermediate. Then will the numerator and denominator of each fraction of this new set be the coexisting values of m and k, that belong to a corresponding fraction in the first set, supposing it to be represented by the formula $\frac{ma - pk}{mb - qk}$, the principal fractions in one set corresponding reversely to those of the other set, and likewise the intermediates to the intermediates. It is useless therefore to try a pair of values of m and k so obtained, but any other pair will give new fractions.

250. For in the series of converging fractions,

$$\frac{A}{A_1}, \quad \frac{B}{B_1}, \quad \frac{C}{C_1}, \quad \frac{D}{D_1}, \quad \frac{E}{E_1},$$

in which $\alpha, \beta, \gamma, \delta, \epsilon$ are the quotients, it is known that

$$A = \alpha,$$

$$B = \beta A + 1 = \alpha\beta + 1,$$

$$C = \gamma B + A = (\alpha\beta + 1)\gamma + \alpha,$$

$$D = \delta C + B = \{(\alpha\beta + 1)\gamma + \alpha\}\delta + \alpha\beta + 1,$$

$$E = \epsilon D + C = \&c.\ldots$$

$$A_1 = 1,$$

$$B_1 = \beta,$$

$$C_1 = \beta\gamma + 1,$$

$$D_1 = (\beta\gamma + 1)\delta + \beta, \quad \&c.$$

(Euler's Algebra, . 476.

whence we obtain

$$C = E - \epsilon D,$$
$$-B = \delta E - (\delta\epsilon + 1) \cdot D,$$
$$A = (\gamma\delta + 1)E - \{(\delta\epsilon + 1)\gamma + \epsilon)\}D.$$

In which the coexisting values of the coefficients of E and D, the last and last but one of the series of numerators, are 1 and ϵ, δ and $\delta\epsilon + 1$, $\gamma\delta + 1$ and $(\delta\epsilon + 1)\gamma + \epsilon$, and so on, which manifestly follow the same law as the corresponding values of A_1 and A, B_1 and B, &c., if we substitute $\epsilon\delta\gamma\beta a$ for $a\beta\gamma\delta\epsilon$ respectively. Also the same may be similarly shewn for the denominators A_1, B_1, C_1,... &c., as well as for the intermediate fractions. The coefficients of E and D will therefore be obtained from these quotients, if we treat them in this reverse order in the same manner as when we obtain from them the values of the successive converging fractions. And since E and D correspond to a and p, their coefficients are the values of m and k in the formula $\frac{ma - pk}{mb - qk}$, which belong to the continued fractions.

251. To shew this more clearly take this example, $\frac{45}{14}$, which treated in the usual manner gives the following set of quotients and converging fractions.

Quotients.		3	4	1	2	
Principal Fractions.	(a) $\frac{0}{1}$	(b) $\frac{1}{0}$	(c) $\frac{3}{1}$	(d) $\frac{13}{4}$	(e) $\frac{16}{5}$	(f) $\frac{45}{14}$
Intermediate Fractions.			(b′) $\frac{2}{1}$	(c′) $\frac{10}{3}$		(e′) $\frac{29}{9}$
			(b″) $\frac{1}{1}$	(c″) $\frac{7}{2}$		
				(c‴) $\frac{4}{1}$		

Writing the quotients in the reverse order and proceeding as before, we obtain the following set.

Quotients.		2	1	4	3	
Principal Fractions.	(f) $\frac{0}{1}$	(e) $\frac{1}{0}$	(d) $\frac{2}{1}$	(c) $\frac{3}{1}$	(b) $\frac{14}{5}$	(a) $\frac{45}{16}$
Intermediate Fractions.			(e′) $\frac{1}{1}$		(c‴) $\frac{11}{4}$	(b″) $\frac{31}{11}$
					(c″) $\frac{8}{3}$	(b′) $\frac{17}{6}$
					(c′) $\frac{5}{2}$	

Now every one of the fractions in the last set consist of the value of $\frac{m}{k}$ that belongs to one of the fractions of the first set, as shown by the corresponding letters of reference; the fractions of the first set being supposed to be represented by the formula

$$\frac{m \times 45 - k \times 16}{m \times 14 - k \times 5}.$$

This is shown in the following table:

	$\frac{m}{k}$	Principal Fractions.	$\frac{m}{k}$	Intermediate Fractions.	
f	$\frac{0}{1}$	$\frac{1 \times 45 - 0 \times 16}{1 \times 14 - 0 \times 5} = \frac{45}{14}$	$\frac{1}{1}$	$\frac{1 \times 45 - 1 \times 16}{1 \times 14 - 1 \times 5} = \frac{29}{9}$	e′
e	$\frac{1}{0}$	$\frac{0 \times 45 - 1 \times 16}{0 \times 14 - 1 \times 5} = \frac{16}{5}$	$\frac{11}{4}$	$\frac{4 \times 45 - 11 \times 16}{4 \times 14 - 11 \times 5} = \frac{4}{1}$	c‴
d	$\frac{2}{1}$	$\frac{1 \times 45 - 2 \times 16}{1 \times 14 - 2 \times 5} = \frac{13}{4}$	$\frac{8}{3}$	$\frac{3 \times 45 - 8 \times 16}{3 \times 14 - 8 \times 5} = \frac{7}{2}$	c″
c	$\frac{3}{1}$	$\frac{1 \times 45 - 3 \times 16}{1 \times 14 - 3 \times 5} = \frac{3}{1}$	$\frac{5}{2}$	$\frac{2 \times 45 - 5 \times 16}{2 \times 14 - 5 \times 5} = \frac{10}{3}$	c′
b	$\frac{14}{5}$	$\frac{5 \times 45 - 14 \times 16}{5 \times 14 - 14 \times 5} = \frac{1}{0}$	$\frac{31}{11}$	$\frac{11 \times 45 - 31 \times 16}{11 \times 14 - 31 \times 5} = \frac{1}{1}$	b″
a	$\frac{45}{16}$	$\frac{16 \times 45 - 45 \times 16}{16 \times 14 - 45 \times 5} = \frac{0}{1}$	$\frac{17}{6}$	$\frac{6 \times 45 - 17 \times 16}{6 \times 14 - 17 \times 5} = \frac{2}{1}$	b′

Any other integrals substituted for m and k will give new approximate fractions; as for example,

$$\frac{2 \times 45 - 3 \times 16}{2 \times 14 - 3 \times 5} = \frac{42}{13} = 3.230,$$

$$\frac{3 \times 45 - 7 \times 16}{3 \times 14 - 7 \times 5} = \frac{23}{7} = 3.285,$$

the decimals serve to show the closeness of the approximation for the original fraction, $\frac{45}{14} = 3.2\dot{1}\dot{4}$.

252. If we apply this method to the example (Art. 247) of an annual movement, the approximate fraction becomes

$$\frac{164359 \times k - m \times 58804}{450 \times k - m \times 161},$$

in which k and m may have any values; for example,

$$\frac{7 \times 164359 - 22 \times 58804}{7 \times 450 - 22 \times 161} = \frac{143175}{392} = \frac{25 \times 69 \times 83}{8 \times 7 \times 7},$$

corresponding to a period of $365^d. 5^h. 48'. 58''.6944$. (error $10''.69$). This is the annual train which has been calculated by a different method by P. Allexandre, in 1734, and afterwards by Camus and Ferguson.

However, the expression

$$\frac{3 \times 164359 - 10 \times 58804}{3 \times 450 - 10 \times 161} = \frac{94963}{260} = \frac{11 \times 89 \times 97}{2^2 \times 5 \times 13},$$

which corresponds to a period of $365^d. 5^h. 48'. 55''.38$, is quite as convenient, and rather more accurate.

In a train of this kind one or more endless screws may be introduced, by way of saving teeth; for example, in the fraction last cited the numerator does not admit of being divided into less than three wheels; but the denominator may be distributed between two pinions and an endless screw, (remembering that the latter is equivalent to a pinion of one leaf) thus, 1 × 20 × 13, or 1 × 10 × 26. If the endless screw be not convenient, then the terms of the fraction must be multiplied by 4, to make the numbers of the denominator large enough for three pinions, and the train will stand thus,

$$\frac{44 \times 89 \times 97}{8 \times 10 \times 13}.$$

253. Ex. *To find a Lunar train that shall derive its motion from the twelve-hour arbor of a clock.*

The mean synodic period of the Moon is $29^d. 12^h. 44'. 2''.8032$, which is exactly equal to $29^d.530588$, or nearly $29^d.5306$, and since twelve hours is equal to $0^d.5$, the ratio will be $\frac{295306}{5000}$, or, dividing each term by 2, $\frac{147653}{2500}$; from which the following quotients and fractions may be obtained.

Quotients.	59	16	2	1	16	3	
Principal Fractions.		$\frac{59}{1}$	(A) $\frac{945}{16}$	$\frac{1949}{33}$	$\frac{2894}{49}$	$\frac{48253}{817}$	$\frac{147653}{2500}$
Secondary Fractions.					(C) $\frac{4843}{82}$ (B) $\frac{19313}{327}$	(D) $\frac{99400}{1683}$	

Now as the whole number nearest to the original fraction is 59, which is less than 8^2, it is clear that two pair of wheels should suffice. The whole of the secondary fractions which would not admit of reduction, are omitted. The principal fractions are refractory, with the exception of (A), $\frac{945}{16} = \frac{3^2 . 5 . 7}{4^2}$, which has been employed by Ferguson and by Mr. Pearson; it corresponds to a period of $29^d. 12^h. 45'$ exactly, and has an error in excess of 57″.2; as it is a multiple of seven, it may be introduced into a clock which has a weekly arbor.

This fraction has been already obtained by a coarser method in (Art. 245.).

$(B) = \frac{19313}{327} = \frac{7 \times 31 \times 89}{3 \times 109}$ has an error in defect of 0″.6 in each lunation.

$(C) = \frac{4843}{82} = \frac{29 \times 167}{2 \times 41}$ has an error of $-$ 8″.6.

$(D) = \frac{99400}{1683} = \frac{2^3 \times 5^2 \times 7 \times 71}{3^2 \times 11 \times 17}$ has an error of + 1″.03.

Other results may be obtained from the expression,

$\frac{147653 \times k - m \times 48253}{2500 \times k - m \times 817}$, as in the following Table.

	Values of k	Values of m	$\frac{D}{F}$	$\frac{D}{F}$ in Factors.	Error in a Lunation.
a	12	59	$\frac{41520}{703}$	$\frac{5 \times 48 \times 173}{19 \times 37}$	$-$ 0″.4
	31	97	$\frac{103298}{1749}$	$\frac{2 \times 13 \times 29 \times 137}{3 \times 11 \times 53}$	+ 0″.08
b	29	89	$\frac{12580}{213}$	$\frac{2^2 \times 5 \times 17 \times 37}{3 \times 71}$	$-$ 6″.18
c	76	233	$\frac{21321}{361}$	$\frac{103 \times 23 \times 9}{19 \times 19}$	$-$ 9″.84
d	29	92	$\frac{157339}{2664}$	$\frac{7^2 \times 13^2 \times 19}{2^3 \times 3^2 \times 37}$	+ 0″.44
e	11	35	$\frac{64672}{1095}$	$\frac{2^5 \times 43 \times 47}{3 \times 5 \times 73}$	+ 0″.48
	1633	5000	$\frac{147651}{2500}$	$\frac{3 \times 7 \times 79 \times 89}{2^2 \times 5^4}$	$-$ 33″.5

Of these *a* is a train given by Francœur, *b* and *c* by Allexandre, *d* by Camus, *e* by Mr Pearson; each of these writers having arrived at his result by a method of his own*.

* Vide Francœur, Mécanique Elementaire, p. 146. Allexandre, Traité Général des Horloges, p. 188. Camus on the Teeth of Wheels. Rees' Cyclopædia, art. Planetary Numbers.

254. The early mechanists were content with much more humble approximations, and employed a great number of unnecessary wheels. In the annual movement of the planetary clock, by Orontius Finæus (about 1700), the following annual train is employed, from a wheel which revolves in three days*.

12——48
36——180
48——48
24——146 $= \frac{365}{1}$.

A train of half the number of wheels would do as well,

$$\text{thus } \frac{60 \times 73}{6 \times 6}, \quad \text{or } \frac{146 \times 180}{12 \times 18}.$$

Again Oughtred†, in 1677, is satisfied to represent the synodic period of the Moon by $29\frac{1}{2}$ days, and employs the train $\frac{40 \times 59}{10 \times 4}$. Huyghens employed for the first time continued fractions in the calculation of this kind of wheel-work ‡.

255. Let it be required to connect an arbor with the hour arbor of an ordinary clock, in such a manner that it may revolve in a sidereal day; so as to indicate sidereal time upon a dial, while the ordinary hands of the clock shew mean time upon their own dial.

Twenty-four hours of sidereal time are equivalent to $23^h. 56'. 4''.0906$ of mean solar. Neglecting the decimals and reducing to seconds, we obtain 86400″ of sidereal time equivalent to 86164″ of mean time, and therefore one wheel must make 86400 turns while the other makes 86164, or dividing by the common factor 4, we get

$$\frac{S_1}{S_m} = \frac{21600}{21541}, \text{ an unmanageable fraction.}$$

* Allexandre, p. 167. † Oughtred, Opuscula. ‡ Hugenii Op. posth. 1703.

Approximating as before, we obtain the expression

$$\frac{3651\,k + 21541.m}{3661\,k + 21600.m},$$

in which $k = -4$, $m = 7$, gives

$$\frac{1096}{1099} = \frac{8 \times 137}{7 \times 157},$$

with a daily sidereal error of $0''.0586$, or $21''\frac{1}{2}$ in the year*.

256. Another mode of indicating sidereal and solar time in the same clock, consists in placing behind the ordinary hour hand a moveable dial concentric with and smaller than the fixed dial†. Both dials must in this case be divided into twenty-four hours. The hand of the clock performs a revolution in twenty-four solar hours, and therefore indicates mean solar time upon the fixed dial as usual, but a slow retrograde motion is given to the moveable dial, so that the same hand shall point upon the latter to the sidereal time, which corresponds to the solar time shewn upon the fixed dial. For this purpose it is evident that during each revolution of the hour hand, the moving dial must retrograde through an angle corresponding to the quantity which sidereal time has gained upon solar time in twenty-four hours; which is $3'.56''.555 = 236''.555$, and as the entire circumference of the dial contains $86400''$, we have

$$\frac{\text{Ang. vel. of hour hand}}{\text{Ang. vel. of dial}} = \frac{86400000}{236555} = 60 \times \frac{288000}{47311}.$$

From this fraction approximate numbers may be obtained, by which the proper wheel-work for the motion of the dial can be set out.

* This is Francœur's result.

† This method is due to Mr Margett, the details of his mechanism may be found in Rees' Cyclopædia, art. Dialwork.

The fraction $\frac{288000}{47311}$ reduced to continued fractions gives

Quotients.	6	11	2	3	1	152	
Fractions.		$\frac{6}{1}$	$\frac{67}{11}$	$\frac{140}{33}$	$\frac{487}{80}$ (A)	$\frac{627}{103}$ (B)	&c.

(*A*) contains a large prime 487, but is employed by Mr Margett. (*B*) $= \frac{3 \times 11 \times 19}{103}$ contains a smaller number, and is a better approximation.

CHAPTER VIII.

ELEMENTARY COMBINATIONS.

CLASS B. {DIRECTIONAL RELATION CONSTANT. / VELOCITY RATIO VARYING.}

---

257. THE elementary combinations which are the subject of the preceding chapters, include those which are employed in all the largest and most important machines; for the parts of heavy machinery are always made to move with uniform velocity, if possible; and consequently with a constant velocity ratio and directional relation to each other. In the combinations which remain to be considered, either the velocity ratio, or directional relation, or both, vary; but as the arrangement of them is for the most part derived from some one or other of the previous contrivances, it will no longer be necessary to enter so much at large into the explanation of principles and of various forms, as a reference to the preceding chapters will for the most part suffice, at least for the less important machines. For this reason I have not thought it necessary to assign a separate chapter to each division of the classes *B* and *C*, as in class *A*, but shall include these classes each in a single chapter.

CLASS B. DIVISION A.

COMMUNICATION OF MOTION BY ROLLING CONTACT.

258. It has been already shewn, in Art. 35, that when a pair of curves revolving in the same plane in contact are of such a form as to roll together, the point of contact remains in the line of centers. The two radii of contact

coincide therefore with this line, and the tangents of the angles made by the common tangent of the curves at the point of contact with their radii respectively are the same.

259. Ex. 1. In the *logarithmic spiral* the tangent makes a constant angle with the radius vector. Let two equal logarithmic spirals be placed in reverse positions, and made to turn round their respective poles as centers of motion, and let these centers be fixed at any distance that will permit the curves to be in contact. Then in every position of contact the common tangent will make the same angle with the radius vector of one curve that it makes on the opposite side with the radius vector of the other. The two radii of contact will therefore be in one line, and coincide with the line of centers, and hence, *equal logarithmic spirals are rolling curves.*

Ex. 2. Let aPm, APM be two similar and equal ellipses of which s, h; S, H are the foci, and let them be

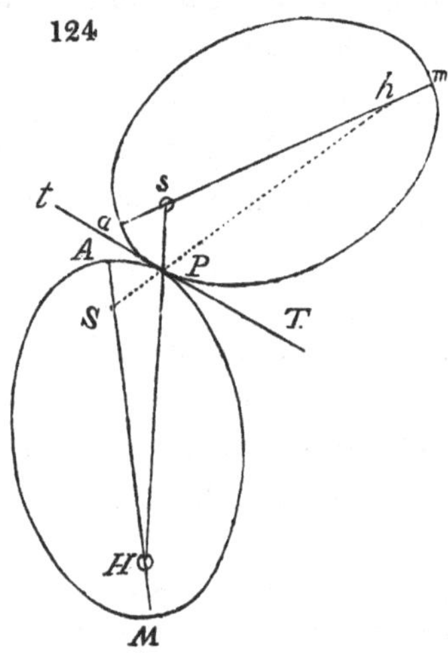

placed in contact at any point P situated at equal distances aP, AP from the extremities of their major axes, and draw tPT the common tangent at P.

Now by the property of the ellipse the tangent makes equal angles with the radii sP, Ph; and because $aP = AP$, and the ellipses are equal, the tangent makes the same angle with the radii $SP\ PH$; whence $tPs = TPH$, and sPH is a right line. Also $sP = SP$; $\therefore sP + PH = SP + PH = AM$ is a constant distance, whatever be the distance of the point of contact P from the extremity of the axes major. If, therefore, the foci s, H be made centers of motion, and their distance equal to the major axes of the ellipses, the curves will roll together.

The logarithmic spiral and ellipse round the focus appear to be the only two rolling curves that admit of simple independent demonstrations of their possessing this property.

260. Suppose fig. 124 to represent any pair of rolling curves, and let $r = sP$ be the distance of their point of contact P from the center of rotation s of the first curve, and $\theta = asP$ the angle made by r with a fixed radius sa, and let $r_{,} = PH$, $\theta_{,} = PHA$, be the corresponding quantities in the second curve, and c the distance sH of the centers; then since r and r are in the same straight line,

$$r + r_{,} = c, \quad \therefore dr = -dr_{,};$$

also the lengths of those parts of the curves aP, AP, that have been in contact are equal;

$$\therefore \int\sqrt{dr^2 + r^2 d\theta^2} = \int\sqrt{dr^2_{,} + r^2_{,} d\theta^2_{,}},$$

$$\text{and as } dr = -dr_{,}, \quad \therefore r d\theta = r_{,} d\theta_{,} = \overline{c - r} \,.\, d\theta_{,}.$$

Again, $\dfrac{r d\theta}{dr}$ is the tangent of the angle the first curve makes with r, and $\dfrac{r_{,} d\theta_{,}}{dr_{,}}$ is the tangent of the angle the second curve makes with $r_{,}$, and these angles are the same;

$$\therefore \frac{r d\theta}{dr} = -\frac{r_{,} d\theta_{,}}{dr_{,}}, \text{ whence } r d\theta = r_{,} d\theta_{,}, \text{ as before.}$$

Hence, if one curve be given by an equation between r and θ, the other is determined by the equations

$$r_{,} = c - r, \text{ and } \theta_{,} = \int \frac{r d\theta}{c - r}.$$

Ex. Let the first curve be the logarithmic spiral, (Art. 259) and let ϕ be the constant angle between the radius vector and the curve, $\therefore \theta = \phi \log \frac{r}{b}$ is its equation;

$$\therefore d\theta = \phi \frac{dr}{r}, \quad \theta_{,} = \int \frac{r d\theta}{c - r} = \phi \int \frac{dr}{c - r} = C - \phi \log r.$$

Now when $\theta_{,}$ vanishes, $r = c - b$; $\therefore 0 = C - \phi \log \overline{c - b}$;

$\therefore -\theta_{,} = \phi \log \frac{r}{c - b}$ is the equation to the second curve, which is the same logarithmic spiral in the reverse position.

The general equation of this article is given by Euler, in the fifth volume of the Acta Petropolitana, but it is not easy to obtain many convenient results in this manner. The properties of one class of rolling curves have been investigated in the most complete and able manner, in a paper in the Cambridge Philosophical Transactions, by the Rev. H. Holditch, to which I must refer those of my readers who are desirous of following out the subject I have substituted in the next article a simpler but more limited investigation, for which I am indebted to the author of the paper in question.

261. Let each rolling curve enter into itself, in which case it must have greater and less apsidal distances, a and b. And as in revolving the greater apsidal distance of one must come into contact with the less apsidal distance of the other,

$$a + b = c = r + r_{,}.$$

Let p be the perpendicular from the center upon the tangent to the point of contact; then since $\frac{p}{r}$ is the sine of the angle the tangent makes with the radius vector, and this angle must be the same from the very nature of rolling curves, when $a + b - r$ is substituted for r; it is plain that if we assume

$$\frac{r^n}{p^n} = \frac{A + Br + Cr^2 + \&c.}{A_{,} + B_{,}r + C_{,}r^2 + \&c.}$$

that this equals

$$\frac{A + B\,.\,\overline{a + b - r}\,.+ C\,.\,\overline{a + b - r}|^2 + \&c.}{A_{,} + B_{,}.\overline{a + b - r} + C_{,}.\,\overline{a + b - r}|^2 + \&c.}$$

Let $r = c - z$, $2c = a + b$;

$$\therefore \frac{A + B\,.\,\overline{c - z} + C\,.\,\overline{c - z}|^2 + \&c.}{A_{,} + B_{,}.\,\overline{c - z} + C_{,}\,.\,\overline{c - z}|^2 + \&c.}$$

$$= \frac{A + B\,.\,\overline{c + z} + C\,.\,\overline{c + z}|^2 + \&c\ldots}{A_{,} + B_{,}.\overline{c + z} + C_{,}.\,\overline{c + z}|^2 + \&c\ldots}$$

And as the coefficients of the even powers of z are the same in both fractions, and the coefficients of the odd powers only differ in their signs; if O be the sum of the odd, and E of the even powers in the numerators, and $O_{,}$ of the odd and $E_{,}$ of the even in the denominators, then

$$\frac{E - O}{E_{,} - O_{,}} = \frac{E + O}{E_{,} + O_{,}}, \quad \therefore \frac{E}{E_{,}} = \frac{O}{O_{,}};$$

$$\therefore \frac{r^n}{p^n} = \frac{E - O}{E_{,} - O_{,}} = \frac{E}{E_{,}} = \frac{\alpha + \beta z^2 + \gamma z^4 + \ldots}{\alpha_{,} - \beta_{,} z^2 + \gamma_{,} z^4 + \ldots},$$

$$\text{or } \frac{r^n}{p^n} = \frac{A + A_2\left(r - \frac{a+b}{2}\right)^2 + A_4\left(r - \frac{a+b}{2}\right)^4 + \ldots}{B + B_2\left(r - \frac{a+b}{2}\right)^2 + B_4\left(r - \frac{a+b}{2}\right)^4 + \ldots},$$

which is a more convenient notation.

Now at the apses $r = a$ or b, and either of these values will give the equation

$$1 = \frac{A + A_2\left(\frac{a-b}{2}\right)^2 + A_4\left(\frac{a-b}{2}\right)^4 + \&c\ldots}{B + B_2\left(\frac{a-b}{2}\right)^2 + B_4\left(\frac{a-b}{2}\right)^4 + \&c\ldots},$$

which being substracted from the former, we have

$$\frac{r^n - p^n}{p^n} = \frac{A + A_2\left(r - \frac{a+b}{2}\right)^2 + \ldots}{B + B_2\left(r - \frac{a+b}{2}\right)^2 + \ldots} - \frac{A + A_2\left(\frac{a-b}{2}\right)^2 + \ldots}{B + B_2\left(\frac{a-b}{2}\right)^2 + \ldots}.$$

If these fractions be reduced to a common denominator, the general term in the numerator is

$$A_{2m} \,.\, B_{2n}\left(r - \frac{a+b}{2}\right)^{2m}\left(\frac{a-b}{2}\right)^{2n}$$

$$- A_{2m}\, B_{2n}\left(r - \frac{a+b}{2}\right)^{2n}\left(\frac{a-b}{2}\right)^{2m}$$

$$= A_{2m} \,.\, B_{2n}\left(r - \frac{a+b}{2}\right)^{2m}\left(\frac{a-b}{2}\right)^{2n}$$

$$\left\{\left(r - \frac{a+b}{2}\right)^{2m-2n} - \left(\frac{a-b}{2}\right)^{2m-2n}\right\},$$

which is divisible by

$$\left(r - \frac{a+b}{2}\right)^2 - \left(\frac{a-b}{2}\right)^2, \text{ or by } (a-r)\,.\,(r-b);$$

$$\therefore \frac{r^n - p^n}{p^n} = (a-r)\,.\,(r-b)\,\frac{C_1 + C_2\left(r - \frac{a+b}{2}\right)^2 + \ldots}{D_1 + D_2\left(r - \frac{a+b}{2}\right)^2 + \ldots}, \quad (1)$$

in which equation C_1, $C_2\ldots$ D_1, $D_2\ldots$ are arbitrary constants. If then, for the sake of simplicity, we limit the investigation to the first term by taking $C_2 \ldots D_2 \ldots = 0$, and make $n = 2$, we have*

$$\frac{p}{\sqrt{r^2 - p^2}} \ (= \tan\theta) = \frac{r\,d\theta}{dr} = \frac{k}{\sqrt{\overline{a-r}\,.\,\overline{r-b}}}:$$

to integrate this, let

$$\sqrt{\overline{a-r}\,.\,\overline{r-b}} = \overline{a-r}\,.\,z;\ \therefore\ r - b = \overline{a-r}\,.\,z^2;$$

$$\therefore\ r = \frac{b + z^2.\,a}{1+z^2},\ dr = \frac{2\,\overline{a-b}\,.\,z\,dz}{\overline{1+z^2}|^2},\ a - r = \frac{a-b}{1+z^2};$$

$$\therefore\ d\theta = \frac{2k\,dz}{b + az^2} = \frac{2k}{\sqrt{ab}}\ \frac{\sqrt{\frac{a}{b}}\,.\,dz}{1 + \frac{az^2}{b}}:$$

$$\text{let } \frac{k}{\sqrt{ab}} = \frac{1}{n};$$

$$\therefore\ \frac{n\theta}{2} = \tan^{-1} z\sqrt{\frac{a}{b}} = \tan^{-1}\sqrt{\frac{a}{b}}\sqrt{\frac{r-b}{a-r}};$$

$$\therefore\ \frac{ar - ab}{ab - br} = \tan^2\frac{n\theta}{2},$$

$$\text{and } r = \frac{ab}{a\cos^2\frac{n\theta}{2} + b\sin^2\frac{n\theta}{2}} = \frac{2ab}{a + b + \overline{a-b}\,.\cos n\theta},$$

one equation amongst many others† that may be obtained

* Otherwise; the solution of the equation $\frac{r\,d\theta}{dr} = \frac{r_{,}d\theta_{,}}{dr}$ is $\frac{r\,d\theta}{dr} = u(r\,.\,\overline{c-r})$ where $u(r\,.\,\overline{c-r})$ is any symmetric function of r and $\overline{c-r}$.

Now $\overline{a-r}\,.\,\overline{r-b} = r\,.\,\overline{a+b} - ab - r^2 = r(a+b-r) - ab = r\,.\,\overline{c-r} - ab$ is a symmetric function of r and $\overline{c-r}$, therefore we may assume

$$\frac{r\,d\theta}{dr} = \frac{k}{\sqrt{a-r}\sqrt{r-b}},\ \text{as in the text.}$$

† For example, the equation $\frac{r\,d\theta}{dr} = \frac{k_{,} + k\,.\left(r - \frac{a+b}{2}\right)^2}{\sqrt{\overline{a-r}\,.\,\overline{r-b}}}$, which gives a much greater variety of rolling curves than the one in the text. This equation has been fully discussed in the paper already referred to.

from the expression (1). This however includes a variety of curves, according as different values are taken for ab and n.

262. Now if c be the distance of the centers of the two curves, it is evident from what has been said before, that if $\frac{r d\theta}{dr} = f(r)$ (any function of r) be the differential equation of one curve, then $\frac{r_{,} d\theta_{,}}{dr_{,}} = f(c - r_{,})$ (the same function of $\overline{c - r_{,}}$) will be the differential equation of the other.

Since therefore we have taken $\frac{r d\theta}{dr} = \frac{k}{\sqrt{\overline{a - r} \cdot \overline{r - b}}}$ for the equation of the first curve, that of the second will be

$$\frac{r_{,} d\theta_{,}}{dr_{,}} = \frac{k}{\sqrt{\overline{a - c + r_{,}} \cdot \overline{c - r_{,} - b}}},$$

$$= \frac{k}{\sqrt{\overline{a_{,} - r} \cdot \overline{r_{,} - b_{,}}}}, \quad (\text{if } c - b = a_{,}, \text{ and } c - a = b_{,})$$

which is the same form as the differential equation of the first curve; and being solved in the same manner we have these equations to a pair of rolling curves:

$$r = \frac{2ab}{a + b + \overline{a - b} \cdot \cos n\theta} \qquad (1),$$

$$r_{,} = \frac{2a_{,}b_{,}}{a_{,} + b_{,} + \overline{a_{,} - b_{,}} \cdot \cos n_{,}\theta_{,}} \qquad (2).$$

263. Let $n\theta = 2m\pi$;

$$\therefore \theta = \frac{2m}{n} \cdot \pi, \quad \text{and } r = b,$$

which shews that the minor distances recur at equal angles, the major in like manner correspond to

$$\theta = \frac{2m+1}{n}\pi,$$

and therefore bisect the angles between the minor apsidal distances. If the portion of curve between two minor distances, including as they do a major distance between them, be called a Lobe, then $\frac{2\pi}{n}$ is the angle which contains a lobe, and there are therefore n lobes in one revolution. In order that the curve may return to itself, and so be capable of successive revolutions, n must be an integer.

264. The constants in the pair of equations (1), (2), may be assumed at pleasure, subject to the conditions

$$\frac{ab}{n^2} = \frac{a_{\prime}b_{\prime}}{n_{\prime}^2}, \quad \text{since } k = \frac{n}{\sqrt{ab}} = \frac{n_{\prime}}{\sqrt{a_{\prime}b_{\prime}}},$$

$$\text{and } a - b = a_{\prime} - b_{\prime}.$$

For the greater apsidal distance of one curve corresponds to the less of the other,

$$\text{that is, } c - a = b_{\prime}, \text{ and } c - b = a_{\prime}.$$

A system of wheels or curves thus found will roll together in pairs or in any combinations.

Let $a - b = l$, then since $\frac{ab}{n^2} = k^2$, $b^2 + bl = n^2k^2$;

$$\therefore\ b = \sqrt{n^2k^2 + \frac{l^2}{4}} - \frac{l}{2},$$

$$a = \sqrt{n^2k^2 + \frac{l^2}{4}} + \frac{l}{2};$$

and if n be taken successively equal to 1, 2, 3, &c. we have thus the major and minor distances of a system of wheels of one two three &c. lobes, which will work together, where k and l may have any assignable values, but must be the same for the same system.

265. Describe therefore a circle whose diameter is l and draw a tangent at any point A, (fig. 125,) in which take $AC = k$, and $AE = nk$, and draw EG through the center, then the apsidal distances for a wheel of n lobes are EG and EF;

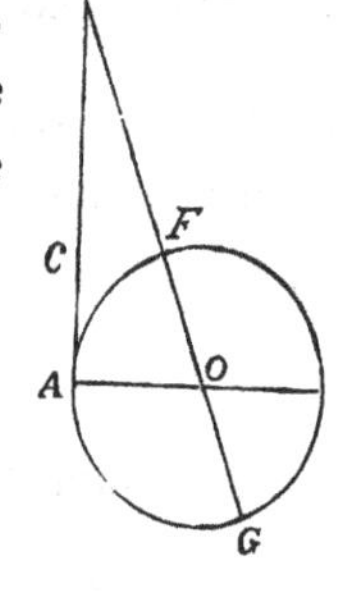

for $EF = EO - FO = \sqrt{n^2k^2 + \frac{l^2}{4}} - \frac{l}{2} = b$,

and $EG = EO + OG = \sqrt{n^2k^2 + \frac{l^2}{4}} + \frac{l}{2} = a$,

Ex. If $k^2 = \frac{2}{9}$, and $l = 1$, then if $n = .1, 3, 4\ldots$

we have $b = .56, 1, 1.45\ldots$

$a = 1.56, 2, 2.45\ldots$

the figures will roll together or in any pairs, or two similar ones will roll together.

Substituting these values of a and b in (1) (2), the equation to a curve of n lobes will be

$$r = \frac{2n^2k^2}{2\sqrt{n^2k^2 + \frac{l^2}{4}} + l.\cos n\theta} \quad (3).$$

266. In the equation (3) let $n = 1$;

$$\therefore r = \frac{2k^2}{2\sqrt{k^2 + \frac{l^2}{4}} + l.\cos\theta},$$

the equation to an ellipse round the focus, of which the major axis

$$= 2\sqrt{k^2 + \frac{l^2}{4}} = a + b,$$

and $l = a - b =$ the distance between the foci.

The curve of one lobe in the system defined by the equation (3), is therefore always an ellipse round the focus, which has been already shewn to be capable of rolling with another equal and similar ellipse; and this equation will also give curves of any number of lobes capable of rolling with it.

267. These curves may be set out practically as follows. Having determined the values of a and b for a curve in a system of any given number of lobes, describe an ellipse whose axis major is $a + b$, and $a - b$ the distance between its foci.

Draw straight lines from one of the foci to the elliptic circumference, making equal angles with each other. Divide the base of each lobe into as many equal parts as there are equal angles round the focus, then the distances from the center to the several points of the lobe are easily shewn to be equal to the elliptic distances, and may therefore be set off from them.

268. Thus, let it be required to construct a set of three rolling curves of one, three, and four lobes respectively, in a system of which the constants l and k are given.

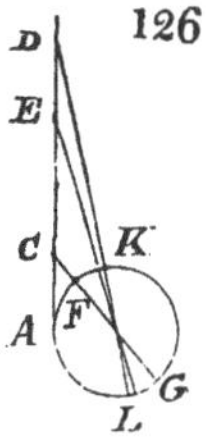

Describe the circle AKG, fig. 126, with a diameter $= l$, and upon the tangent AD set off $AC = k$, $AE = 3k$, and $AD = 4k$. Draw through the center CG, EL, and DL.

126

The curve of one lobe will be an ellipse round the focus M, fig. 127, whose apsidal distances are CF and CG, and major axis consequently $= CF + CG$.

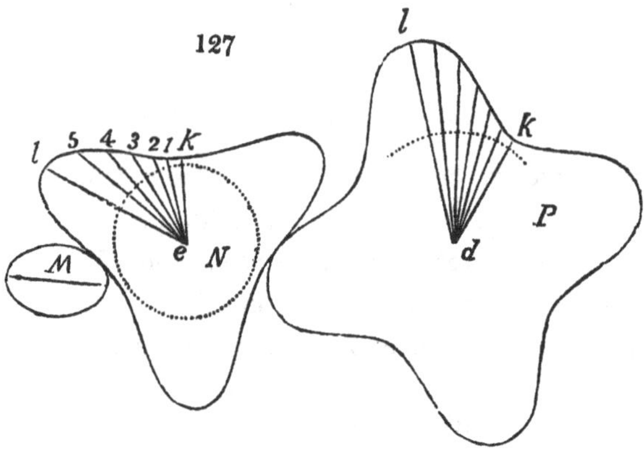

For the curve of three lobes describe a semi-ellipse *Q*, fig. 128, with apsidal distances *ek*, *el* respectively equal to *EK*, *EL*; and from *e* draw a sufficient number of radii *e*1, *e*2, *e*3, &c....at equal angular distances.

To construct the three-lobed curve *N*, describe a circle round its center *e*, which divide into six equal sectors, each one of which will contain half a lobe. Divide this into as many equal angles as those of the semi-ellipse *Q*, and draw radii, upon which set off in order distances equal to the radii of the semi-ellipse, as indicated by the corresponding letters and figures. Through the points thus obtained draw the curved edge of the semi-lobe, and this curve repeated to right and left alternately will complete the three-lobed curve.

To describe the four-lobed curve *P*, draw an ellipse whose apsidal distances are *DK*, *DL*, and major axis *DK* + *DL*, and proceed in a precisely similar manner to divide it and transfer its radial distances from the focus to the semi-lobe *dkl* of the four-lobed curve *P*.

Any two of these curves will roll together, or if two of them be made alike, the pair so obtained will roll together.

I cannot conclude these extracts without strongly recommending a perusal of the original paper, in which the forms and properties of a great variety of these curves are completely worked out.

269. If, however, a curve be given, and another be required to roll with it, which has been shewn (Art. 260) to be a problem that admits of solution, this in practice can only be solved by tentative methods, which will readily occur, but require some patience in application.

270. By Art. 35, the angular velocities in rolling contact are inversely as the segments into which the point of contact divides the line of centers.

In a pair of rolling ellipses, let A, $A_{,}$ be the angular velocities of the driver and follower respectively, r, $r_{,}$ their radii,

$$\text{then } \frac{A_{,}}{A} = \frac{r}{r_{,}} = \frac{r}{a+b-r}.$$

This is at a maximum when $r = a$; $\therefore \frac{A_{,}}{A} = \frac{a}{b}$,

and at a minimum when $r = b$; $\therefore \frac{A_{,}}{A} = \frac{b}{a}$.

Let the ratio of the maximum to the minimum $= m$;

$$\therefore m = \frac{a^2}{b^2}.$$

$$\text{But, } r = \frac{2ab}{a+b+(a-b).\cos\theta};$$

$$\therefore \frac{A_{,}}{A} = \frac{2ab}{a^2+b^2+(a^2-b^2)\cos\theta} = \frac{2.m^{\frac{1}{2}}}{m+1+(m-1).\cos\theta};$$

which will also apply to a pair of equal rolling curves (as in Art. 262) of any number of lobes; but if they have different numbers of lobes, and a, $a_{\prime}$, b, $b_{\prime}$, be the respective apsidal distances, we should find $m = \frac{a a_{\prime}}{b b_{\prime}}$.

271. *To employ rolling curves in practice.* In fig. 124, let the upper curve be the driver, and let it revolve in the direction from T to t. Then since the radius of contact sP increases by this motion, and the corresponding radius PH decreases, the edge of the driver will press against that of the follower, and so communicate a motion to it of which the angular velocity ratio will be $\frac{PH}{sP}$. But when the point m has reached M, the radii of contact in the driver will begin to diminish, and its edge to retire from that of the follower, so that the communication of motion will cease.

To maintain it, it is necessary to provide the retreating edge with teeth, as in fig. 129, which will engage with similar teeth upon the corresponding edge of the follower, and thus maintain the communication of motion until the point a has reached A, when the advancing side of the driver will come into operation, and the teeth be no longer necessary.

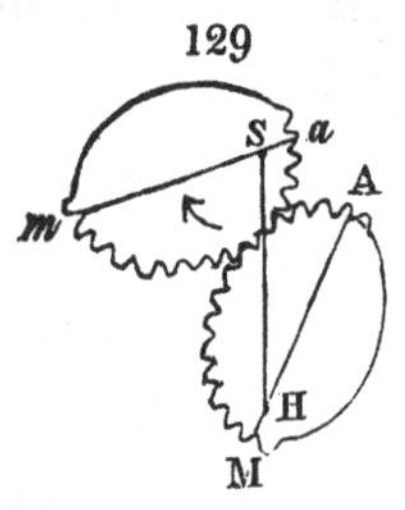

These teeth, however, necessarily destroy the advantage of no friction, and another practical difficulty is introduced. If the curves be not very accurately executed, it may happen that the first pair of teeth and spaces that ought to come together at M, m in each revolution, may not accurately meet, and that either the tooth may get into the wrong space, or become jammed against another tooth, by which the machinery may be broken.

272. To prevent this accident, a curved guide-plate n (fig. 130) may be fixed to one of the wheels, and a pin p to

the other. The edge of this plate must be made of such a form that the pin p may be certain of engaging with it, even if the wheels are not exactly in their proper relative position. When the pin has fairly entered the fork of the plate, it will press either on the right or left side, and so correct the position, and guide the first pair of teeth into contact. It is easy to see that the edge of this plate should be the epicycloid that would be described by p, if the lower plate were taken as a fixed base, and the upper made to roll upon it; but the outer edge of the plate must be sloped away from the true form, to ensure the entrance of the pin into the fork.

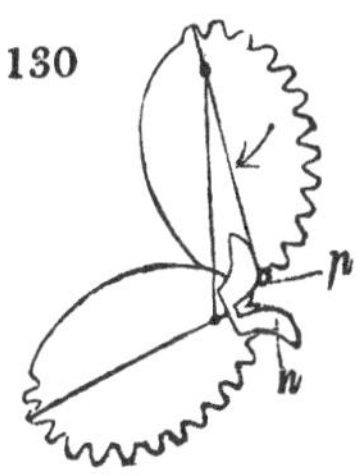

273. Another method is to carry the teeth all round the two plates, which effectually prevents them from getting entangled in the above manner, but at the same time entirely destroys the rolling action. This method, however, is the one always adopted in practice, as for example, in the Cometarium, and in the silk-mills, and is an excellent method of obtaining a varying velocity ratio. Fig. 131 represents a pair of such wheels that were employed by Messrs. Bacon and Donkin in a printing machine.

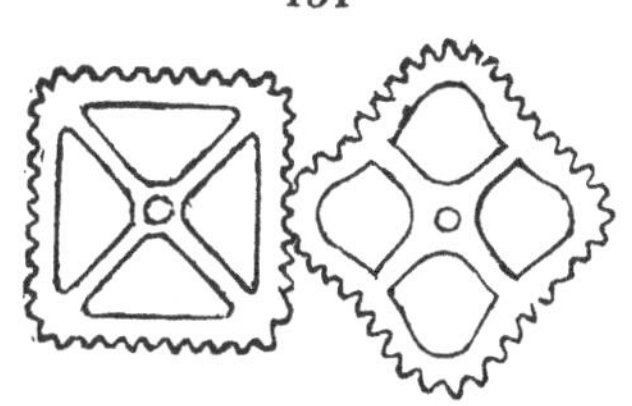

274. The forms of the teeth to be applied to these rolling curves may be obtained by a slight extension of the general solution in Art. 82. For calling the rolling curves pitch curves, it can be shewn for them, precisely in the same manner as it is there shewn for pitch circles, that if any given circle or curve be assumed as a describing curve, and if it be made to roll on the inside of one of these pitch curves, and on the outside of the corresponding portion of

the other pitch curve, that the motion communicated by the pressure and sliding contact of one of the curved teeth so traced upon the other, will be exactly the same as that effected by the rolling contact of the original pitch curves.

275. *The Cometarium* is a machine which has two parallel axes of motion carrying indices or clock-hands; one of which axes is the center of a circle, and the other the focus of an ellipse, which represents the orbit of a comet. The two axes must be connected by mechanism, so that when the first revolves uniformly, the second shall revolve with an angular velocity that will make it describe equal areas of its ellipse in equal times, and thus *represent the motion of a comet round the sun**, for which purpose the machine is constructed. Now, according to what is termed Seth Ward's hypothesis, if one radius vector HP of an ellipse (fig. 124), revolve uniformly round the focus H, the other SP will describe equal areas round the focus S. This, although a very coarse approximation, is considered sufficient for the mechanical representation of planetary or cometary motions in this instrument, and is accordingly obtained by connecting the two axes with a pair of rolling ellipses, as in fig. 124. For by Art. 259, it appears that $HP = hP$, and the angle $SHP = shP$. The motion therefore of HP and hP with respect to the axis major of their respective ellipses is the same, and the ratio of the angular velocities of sP and hP round their foci s and h is the same as those of SP and HP round S and H Also, since the corresponding radii sP, PH have been shewn to

* In any ellipse APM (fig. 124), we have

$$\frac{\text{Angular velocity of } SP \text{ round } S}{\text{Angular velocity of } HP \text{ round } H} = \frac{HP}{SP} = \frac{SP \,.\, HP}{SP^2} = \frac{CD^2}{SP^2},$$

where CD is the conjugate diameter of the ellipse. If the ellipse be nearly a circle, CD *may be supposed constant*, in which case if the angular velocity of HP be uniform, that of SP will vary as $\frac{1}{SP^2}$, which is the law of motion of the radius vector of a planet. This is termed Seth Ward's hypothesis, but is a very coarse approximation.

coincide with the fixed line of centers, it follows that the angular velocities of SH and sa round the centers H and s are respectively the same as those of HP and sP, that is, of HP and SP with respect to the major axes of the ellipses.

276. This machine was first introduced by Dr Desaguliers, and may be considered as the first attempt to employ rolling curves in machinery. He did not however furnish his ellipses with teeth, but connected them by means of an endless band of catgut, which embraced the circumference of each ellipse, lying in a groove in the circumference. The addition of teeth was a subsequent improvement.

277. When the required periodic variation in the ratio of angular velocity is not very great, a pair of equal common spur-wheels, with their centers of motion a little excentric, may be substituted for the equal ellipses revolving round their foci; but in this method the action of the teeth will become very irregular, unless the excentricity be very small.

278. The difficulty of forming a *pair* of rolling curves is sometimes evaded in the manner represented by fig. 132. A is a curved plate revolving round the center B, and having its edge cut into teeth. C a pinion with teeth of the same pitch. The center of this pinion is not fixed, but is carried by an arm or frame, which revolves on a center D. So that as A revolves, the frame rises and falls to enable the pinion to remain in geer with the curved plate, notwithstanding the variation of its radius of contact. To maintain the teeth at a proper distance for their action, the wheel A has a plate attached to it which extends beyond it, and is furnished with a groove de, the central line of which is at a

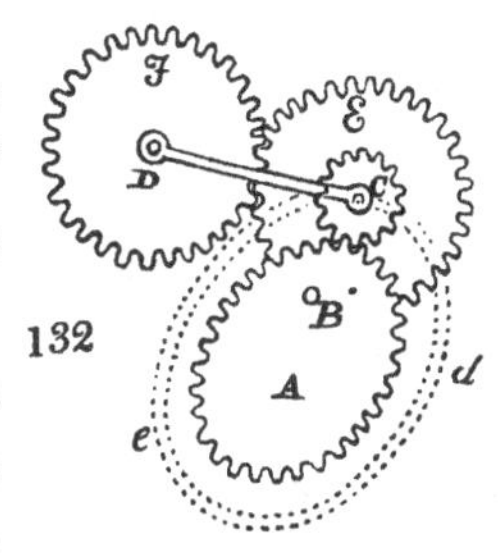

132

* Vide Rees' Cyclopædia, art. Cometarium; or Ferguson's Astronomy.

constant normal distance from the pitch line of the teeth equal to the pitch radius of the pinion. A pin or small roller attached to the swinging frame *D* and concentric with the pinion *C* rests in this groove. So that as the wheel *A* revolves, the groove and pin act together, and maintain the pitch lines of the wheel and pinion in contact, and at the same time prevent the teeth from getting entangled, or from escaping altogether.

Let *R* be the radius of *C*, *r* the radius of contact of *A*, ϕ the angle between *R* and *r*; then it can be easily shewn that $$\frac{\text{ang. vel. of } A}{\text{ang. vel. of } C} = \frac{R}{r} \times \cos \phi.$$
But as the center of motion of *C* continually oscillates, and it is generally necessary to communicate the rotation of *A* to a wheel revolving on a fixed center of motion, a wheel *E* must be fixed to the pinion *C*, and this wheel must geer with a second wheel *D* concentric to the center of the swing-frame. When *A* revolves, the rotation of *C* will be communicated through *E* to *F*, but will also be compounded with the oscillation of the swing-frame, in a manner that will be explained under the head of Aggregate Motions, in the second Part of this work.

279. If for the curved wheel *A* an ordinary spur-wheel *A*, (fig. 133) moving on an *excentric center of motion B*, be substituted, a simple link *AC* connecting the center of the wheel *A* with that of its pinion *C* will maintain the proper pitching of the teeth, in a more simple manner than the groove and pin. The wheel *A* must be of course fixed to the extremity of its axis, to prevent the link from striking it in the course of its revolutions*. This

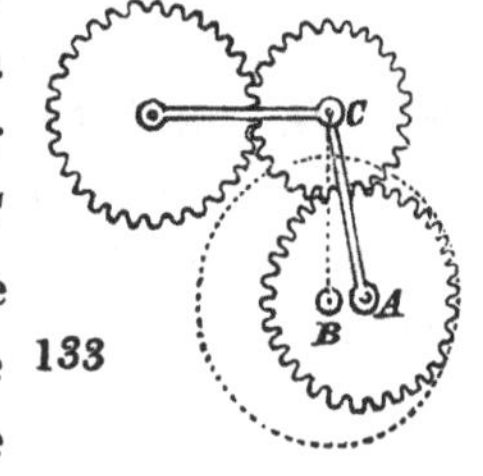

133

* From a machine by Mr Holtzapfel.

combination being wholly formed of spur-wheels, is one of the simplest modes of effecting a varying angular velocity ratio.

280. *On Roëmer's wheels.* These wheels were proposed by the celebrated astronomer Olaus Roëmer*, to effect the varying motion of planetary machines. Aa, Bb are two parallel axes, of which the lower one is provided with a cone C, fluted into regular teeth like those of ordinary bevel-wheels, but occupying the surface of a much thicker frustum of the cone than usual. Opposite to this cone is fixed upon the axis Aa a smooth frustum D, whose apex d is in the reverse direction, and this latter cone is so formed as just to clear the tops of the teeth of C. Upon the surface of D are planted a series of teeth or pins, so arranged as to fall in succession between the teeth of C. By placing these pins at different distances from the apex d, we can obtain any velocity ratio we please between the extremes; for if R, r be the greatest and least radii of D, and $R_{,}, r_{,}$ of C; then the angular velocity ratio of C to D will vary between the limits of $\frac{R}{r_{,}}$ and $\frac{r}{R_{,}}$; the first being obtained by placing the pins close to the large end of D, and the second by fixing them at the small end; and when the pins are fixed in any intermediate position, an intermediate velocity ratio will be obtained.

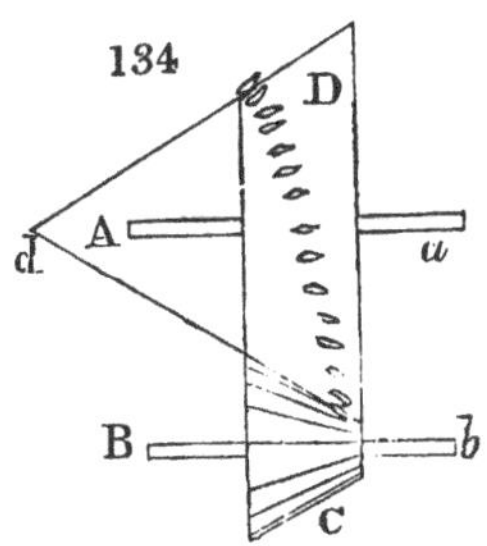

281. If the axes be not parallel, a varying ratio of angular velocity may be obtained by the excentric crown-wheel.

This was invented by Huyghens, for the purpose of representing the motion of the planets in his Planetarium.†

* Machines Approuvées, t. I. † Descriptio Automati Planetarii.

AB is an axis, to the extremity of which is fixed a crown-wheel F, exactly similar to that represented in fig. 26, page 50, only that its center of motion B is excentric to its circumference This wheel is driven by a long cylindrical pinion CD, whose axis meets that of AB in direction, and is at right angles to it. Now since the radius of contact of the pinion is constant, while the radius of contact of the teeth of the hoop varies at different points of the circumference by virtue of its excentricity, it follows that the angular velocity ratio of the axes will vary.

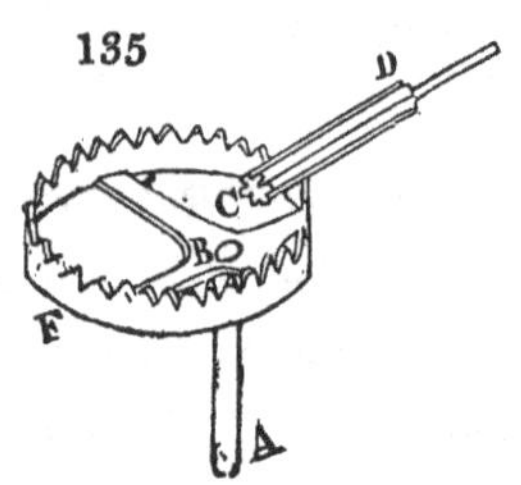

In Huyghen's machine the pinion is the driver, and is supposed to revolve uniformly, but if the contrivance be adopted in other machines, the wheel or pinion may be made the driver, according to the law of velocity required. Also, by making the circumference of the crown-wheel of any other curve than a circle, different laws of velocity may be obtained at pleasure. The action of the teeth however will be irregular, if the excentricity of the hoop be too much increased.

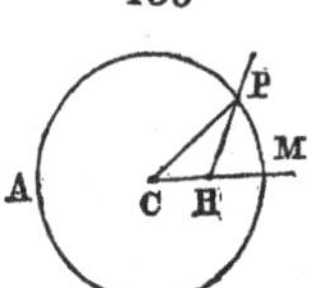

282. Let H, fig. 136, be the center of motion of the crown-wheel, C the center of its circumference,

$CP = R$, $HP = r$, $MHP = \theta$, and $HC = E$.

Then, since the axis of the pinion is directed to H in the line of the excentric radius HP, the perimetral velocity of the pinion will be communicated to this radius in a direction perpendicular to it; and if ρ be the radius of the pinion, we have

$$\frac{\text{angular velocity of pinion}}{\text{angular velocity of crown-wheel}} = \frac{r}{\rho}.$$

$$\text{But } R^2 = r^2 + E^2 \mp 2rE\cos\theta,$$

$$\text{whence } r = \pm E\cos\theta + R \,.\, \sqrt{1 - \frac{E^2}{R^2} \,.\, \sin^2\theta}.$$

Now in planetary machines E is small with respect to R;

$$\therefore\ r = \pm E\cos\theta + R.$$

And since the pinion revolves uniformly, angular velocity of crown-wheel

$$\propto \frac{1}{r} \propto \frac{1}{R \pm E\cos\theta} \propto R \mp E\cos\theta \text{ nearly.}$$

But if MP were the elliptic orbit of a planet, of which C the center, H the focus, HP the radius vector, and $AM\ (= 2R)$ the axis major, we should have angular velocity of HP

$$\propto \frac{1}{HP^2} \propto (R \mp E\cos\theta)^2 \propto R \mp 2E\cos\theta \text{ nearly.}$$

By making therefore the *excentric distance* CH of the crown-wheel equal to the *distance of the foci* of the elliptic orbit, the radius vector HP will revolve with an approximate representation of planetary motion, when the driving pinion revolves uniformly*.

283. Huyghens also proposed another method of obtaining the varying velocity; namely, by varying the pitch of the teeth. If in a pair of ordinary spur-wheels the pitch of one wheel be constant as usual, but in the other it vary so that a given arc of the circumference shall contain N teeth in one part, and an equal arc n teeth in another part of the circumference, and so on; then as every tooth of the first wheel causes one tooth of the other wheel to cross the line of centers, and the driver is supposed to move uni-

* In the article Equation Mechanism, in Rees' Cylopædia, will be found a minute and popular account of the various contrivances employed to represent planetary motion. Those that I have introduced into the text are applicable to machinery in general, and on this account, as well as from the celebrity of their authors, deserve to be studied.

formly, it follows that these equal arcs of the follower will pass the line in times that will be directly as their numbers of teeth N and n, and thus an unequal velocity will be obtained for the follower. But it is evident that this contrivance is but a make-shift, since teeth of unequal pitch will never work well together, although, if the variations from the mean pitch be small, they may be made to act so as to pass tooth for tooth across the line, with a kind of hobbling motion.

Nevertheless, a pair of wheels very similar to these admit of having their teeth formed upon correct geometrical principles; but the difficulty of executing them would be so much greater than those of the rolling curves (Art. 274), that I do not think it worth while to occupy space by developing their theory, which may be easily deduced from the preceding pages.

284. It may happen that the variation of angular velocity in the follower may consist in a sudden change from motion to rest, and *vice versa;* that is, that the follower may be required to move by short trips with intervals of complete rest between, or with an *intermittent* motion.

This may readily be effected with a pair of common spur-wheels, by cutting away the teeth of the driver, as in fig. 137, where the follower *B* is an ordinary spur-wheel, and

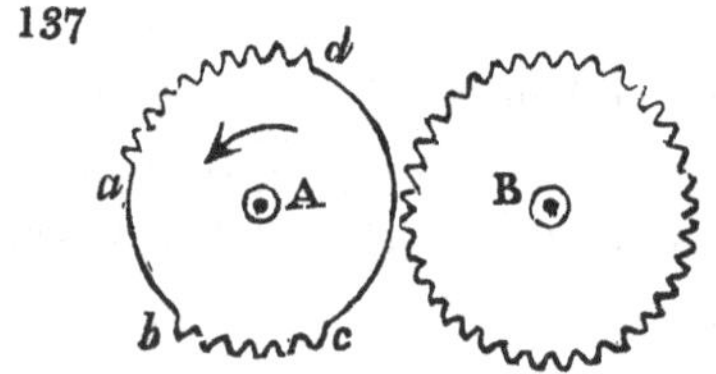

the driver *A* is a wheel of the same pitch whose teeth have been cut away between *a* and *b*, *c* and *d*; consequently, when *A* revolves it will cease to turn *B* while the plain parts of its circumference are passing the line of centers, but will turn it in the usual manner when the teeth come into action. By properly proportioning the plain arcs to those which contain the teeth, we can obtain any desired ratio of rest and motion that can be included within one revolution of the driver.

285. These intermitted teeth are liable to the same objection as those in Art. 271, namely, the chance of the first pair of teeth in each row getting jammed together, and a similar remedy may be employed—a guide-plate and pin. Thus in fig. 138, the wheel A will revolve in the direction of

138

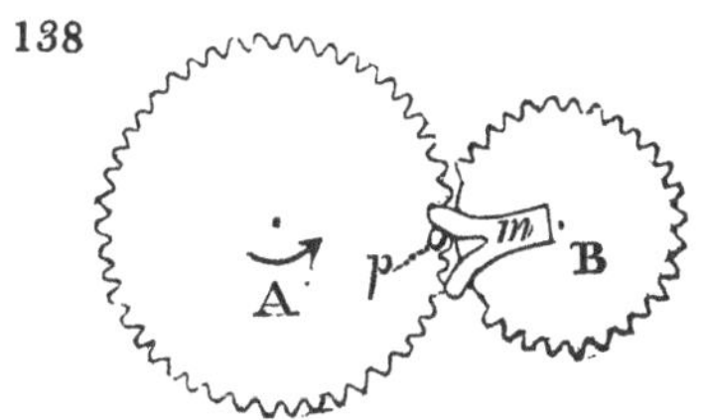

the arrow without communicating any motion to B, until the pin p enters the fork of the guide-plate m, and thus communicates to it a motion which brings the teeth of B into geer with those of A; and A will then continue to turn B until the plate m again reaches the position of the figure, when B will rest until the pin p returns.

In this combination B must make a complete revolution, (unless there be more guide-plates than one) and if R, r be the respective radii of driver and follower, it is easy to see that when A revolves uniformly, the time of B's rest is to the time of its motion as $R - r : r$. Also, several pins may be fixed to A if required, and the intermitted teeth may be given to A instead of to B, or to both.

286. As there is no contrivance in the above to protect B from being displaced during its period of rest, and thereby

preventing the guide-plate from receiving the pin, the action will be rendered more complete by the arrangement of fig. 139.

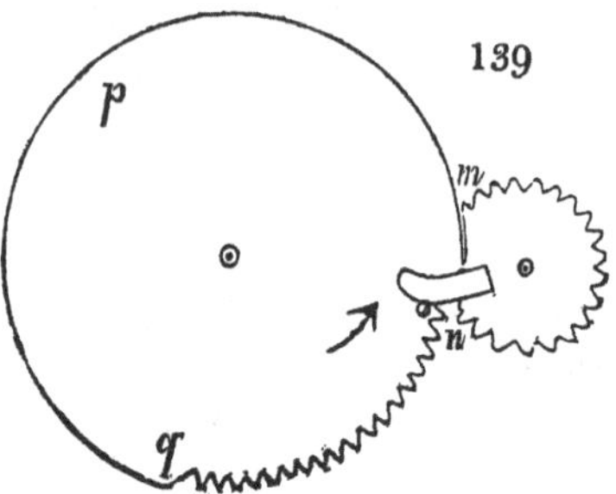

Here the follower has its edge *mn* formed into an arc of a circle whose center is the center of motion of the driver, and the circumference of the driver is a plain disk *npq* of a greater diameter than the pitch circle of the toothed portion *qn*. This plain edge runs past *mn* without touching it, but effectually prevents the follower from being moved out of its position of rest, and therefore ensures the meeting of the pin and guide-plate.

287. Bevil or crown-wheels may be employed if necessary, and the combinations may be thrown into a great many other different forms. The pin and disk of fig. 139 have this advantage, that, when properly formed, they allow the intermittent wheel to begin and end its motion gradually, whereas in fig. 137 the motions begin with a jerk, and the follower is apt to continue its motion through a small space, after the teeth of the driver have quitted it.

288. In many machines a lever is required to move another by the mere contact of their extremities. As the angular motion required is always small, these extremities may be formed into rolling curves, by which the friction will be entirely got rid of, and the small variation in the angular velocity ratio will generally be of little or no con-

sequence. Arcs of the logarithmic spiral or ellipse round the focus will be the most easily described; but since the motion is small, arcs of circles may be substituted as an approximation for the rolling curves, and these may be described as follows.

Let *A*, *B*, fig. 140, be the centers of motion of the levers, *AB* the line of centers divided in *T* in the proportion of the

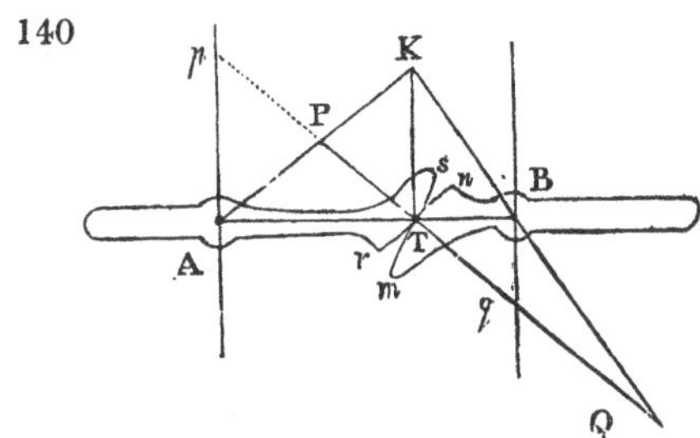

radii in their mean position. Draw *KT* perpendicular to *AT*, and through *T* draw *PTQ* inclined to *AT* at any angle less than a right angle. Assume a point *K* in *KT*. Join *AK* intersecting *PTQ* in *P*, and join *KB*, producing it to meet *PTQ* in *Q*. With center *P* and radius *PT* describe an arc *rTs*, and with center *Q* and radius *QT* describe an arc *mTn*. These arcs will roll together in the mean position of the figure.

For by Art. 33, it appears that the action of these arcs is equivalent to that of a pair of rods *AP*, *BQ*, connected by a link *PQ*. Now during the motion of this system the link may be considered as revolving round a momentary center, which center is always changing its position. But as the extremity *P* of the link begins to move in a direction perpendicular to *AP*, this center must be somewhere in the line *AP* produced; and in like manner, as the extremity *Q* begins to move perpendicularly to *BQ*, the center must be somewhere in *BQ* produced; it must therefore be in *K*, the intersection of *AP* and *BQ*. But since *K* is the momentary center of motion of the link, and KT is perpendicular to AB, it follows that the point of contact T of the arcs rs, mn, will begin to move in the line of centers, and therefore the contact will be *rolling contact.*

289. Since the distance of K from T is arbitrary, let it be supposed infinite, in which case AK, QK become parallel to each other, and perpendicular to the line of centers, as at Ap and Bq, and p, q are now the centers of the arcs. This is a simpler construction.

In practice the angle PTA must be made much greater than in the figure, to avoid oblique action.

Class B. Division B. COMMUNICATION OF MOTION BY SLIDING CONTACT.

290. The simplest mode of obtaining a varying angular velocity ratio, when the rotations are to be continued indefinitely in the same direction, is by the pin and slit, fig. 141, where Aa, Bb are axes parallel in direction, but placed with their ends opposite to each other. Aa is provided with an arm carrying a pin d, which enters and slides freely in a long straight slit formed in a similar arm, which is fixed to the extremity of Bb. If one of these axes revolve, it will communicate a rotation to the other with a varying velocity ratio; for the pin in revolving is continually changing its distance from the axis of Bb.

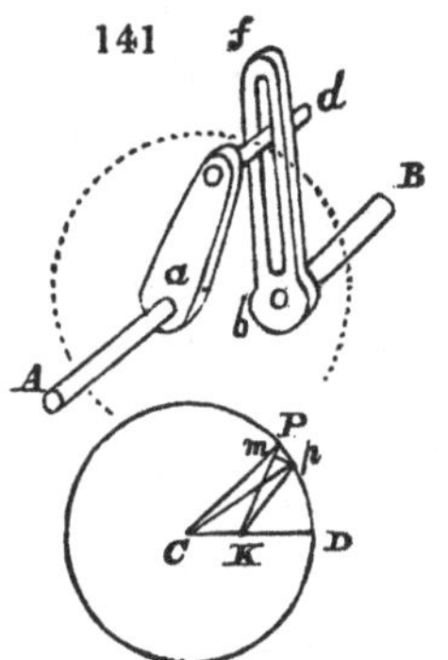

Let C be the center of motion of the pin-arm, K the center of motion of the slit-arm, P the pin, R the constant radius of the pin from C, r the radial distance from K, and let P move to p through a small angle; draw pm perpendicular to CP, then angular velocity of pin : angular velocity of slit

$$:: \frac{Pp}{PC} : \frac{pm}{PK} :: \frac{1}{R} : \frac{\cos CPK}{r}.$$

If CP revolve uniformly, the angular velocity of KP will vary as $\frac{\cos CPK}{r}$, or if CK be small, as $\frac{1}{r}$; therefore when the centers of motion are near, this contrivance produces the same law of motion as that of Art. 282.

If $PCD = \theta$, $PKD = \beta$, $CK = E$, we have

$$R \sin\theta = (R\cos\theta - E)\tan\beta;$$

$$\therefore \tan\beta = \frac{R \cdot \sin\theta}{R\cos\theta - E},$$

will give the position of KP corresponding to any given position of CP.

By altering the direction of the slit, or by making it curvilinear, other laws of motion may be obtained.

291. In the endless screw and wheel (Art. 170), the thread of the screw is inclined to the axis of the cylinder at a constant angle ϕ, and the angular velocity ratio of screw and wheel is constant. If, however, the inclination ϕ of the thread be made to vary at different points of the circumference, as shewn in fig. 142, the angular velocity ratio will vary accordingly. For example, if the threads through half the circumference lie in planes perpendicular to the axis of the screw, the wheel will revolve with an intermittent motion, remaining at rest during the alternate half rotations of the screw. If A, a be the respective angular velocities of the screw and wheel, R, r their pitch-radii, it appears, from Art 166, that $\frac{A}{a} = \frac{r}{R}\tan\phi$.

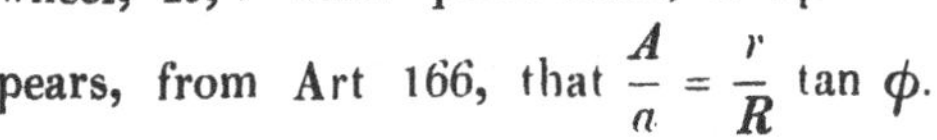

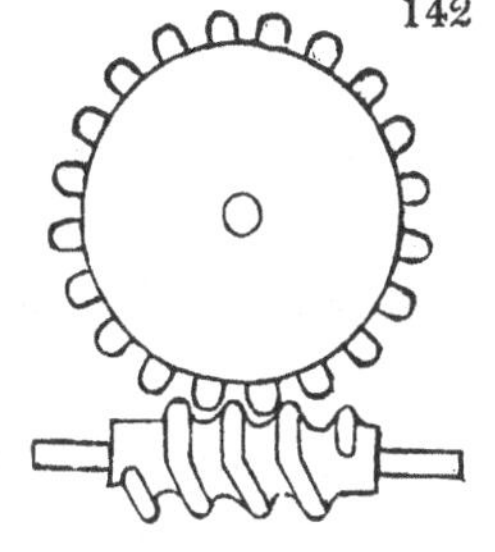
142

But as the inclination ϕ changes, the teeth of the wheel must be made in the form of solids of revolution, having their axes radiating from the center of the wheel.

292. A simple intermittent motion is effected by a pinion of one tooth *A*, fig. 143. This tooth will in each revolution pass a single tooth of the wheel *B* across the line of centers; but during the greatest portion of its rotation will leave the wheel undisturbed. To prevent the wheel *B* from continuing this motion by inertia through a greater space than this one tooth, a *detent* *C* may be employed. This turns freely upon its center, and may be pressed by a weight or spring against the teeth. It will be raised as the inclined side of the tooth passes under it by the action of *A*. and will fall over into the next space, but when *A* quits the wheel, the detent pressing upon the inclined side of the tooth will move it through a short space backwards, until the point *m* rests at the bottom of the nook, as shewn. The detent thus retains the wheel in its position during the absence of the tooth *A*. These detents receive other forms, for which I shall refer to the section on Link-work, in Chap. IX.

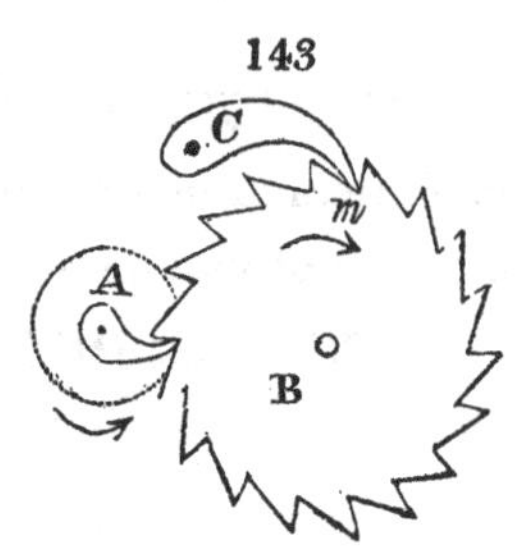

293. A better intermittent motion is produced by a contrivance (fig. 144) which may be termed the *Geneva stop*, as it is introduced into the mechanism of the Geneva watches.

A is the driver which revolves continually in the same direction, *B* the follower, which is to receive from it an intermittent motion, with long intervals of rest. For

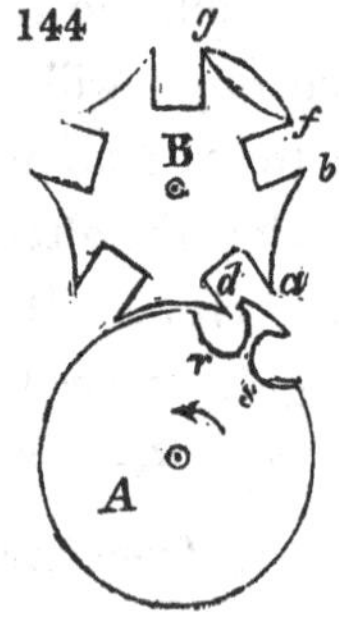

this purpose its circumference is notched alternately into arcs of circles as *ab*, concentric to the center of *A* when placed opposite to it, and into square recesses, as shown in the figure.

The circumference of *A* is a plain circular disk, very nearly of the same radius as the concave tooth which is opposed to it; this disk is provided with a projecting hatchet-shaped tooth, flanked by two hollows *r* and *s*. When it revolves (suppose in the direction of the arrow), no motion will be given to *B* so long as the plain edge is passing the line of centers, but at the same time the concave form of the tooth of *B* will prevent it from being moved (as in fig. 139).

But when the hatchet-shaped tooth has reached the square recess of *B*, its point will strike against the side of the recess at *d*, and carry *B* through the space of one tooth, so as to bring the next concave arc *ab* opposite to the plain edge of the disk, which will retain it until another revolution has brought the hatchet into contact with the side of the next recess *bf*.

The hollow recess at *r* is necessary to make room for the point *d*, which during the motion is necessarily thrown nearer to the center of *A* than the circumference of the plain edge of the latter. The hatchet-tooth being symmetrical will act in either direction.

294. The office of this contrivance in a Geneva watch is to prevent it from being over-wound, whence it is termed *a stop*; and for this purpose one of the teeth is made convex, as shown in dotted lines at *fg*. If *A* be turned round, the hatchet-tooth will pass four notches in order, but after passing the fourth across the line of centers, the convex edge *gf* will prevent further rotation, so that in this state the combination becomes a contrivance to prevent an axis from being turned more than a certain number of times in the same direction.

For the wheel A is attached to the axis which is turned by the key in winding, and the wheel B thus prevents this axis from being turned too far, so as to overstrain the spring. As the watch goes during the day the axis of A revolves slowly in the opposite direction, carrying the stop-wheel with it by a similar intermitting motion.

The late Mr. Oldham applied this kind of mechanism to intermittent motions*, and his arrangement is in some respects superior to that of fig. 144. Instead of the hatchet-tooth he employed a pin carried by a plate fixed to the back of the driver, by which means he was enabled to reduce the size of the square notches of the follower.

295. Any required variation in the ratio of angular velocities may be produced by a cam-plate; but if the directional relation is constant the motion will necessarily be limited, as in fig. 71, (page 153). In this contrivance, by altering the form of the curve we may obtain different velocity ratios at every point of its action; as, for example, if a portion of the edge of the cam-plate be concentric to its axis, the pin or bar which it drives will receive no motion while that part of the edge is sliding past it.

296. The curve for a cam of this kind is generally described by points. The methods of doing this will readily occur in each particular case, but one example may serve to shew the nature of the process. In the combination of fig. 72, page 153, let the angular velocity ratio vary so that when a series of points 1, 2, 3, 4, 5, fig. 145, in the circumference of the circle C 3, 5 shall have reached in order the point C, the pin in the sliding bar shall be moved into the corresponding positions I, II, III, IV, V. To each of the position points in

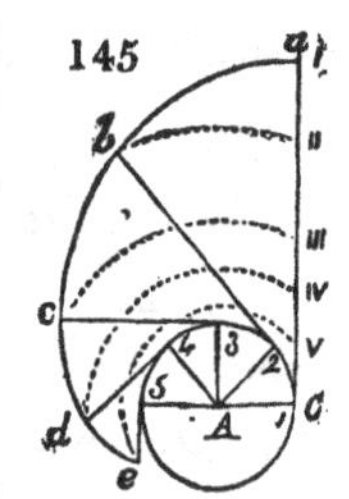

* In the machinery of the Banks of England and Ireland.

the circumference of the circle draw tangents, and with center A draw circular arcs in order, each intersecting one of the position points I, II, III, &c., and the corresponding tangent, as at a, b, c, d, e; thus is obtained a series of points through which, if a curve be drawn, it will be the cam required; for it is manifest, that if any point (as 3) of the circle be brought to C, the corresponding point c of the curve will be moved to III, and thus the pin will be placed in its required position; and so for every other pair of positions.

The curve for a pin of sensible diameter must be obtained from this by the usual method (Art. 88).

Class B. Division C. COMMUNICATION OF MOTION BY WRAPPING CONNECTORS.

297. If an indefinite number of rotations be required to be communicated from one revolving axis to another, an endless band may be employed, as in fig. 146. A is a driving pully, whose edge is shaped to the curve required, and is also grooved or otherwise adapted for the reception of an endless band, (Art. 180). The follower C is a cylindrical pully of the usual form. A stretching pully D (Art. 188) will be required for one side of the band, and if Ap be a perpendicular upon the direction of the other side, and Cq be the radius of the follower pully, we have by Art. 37 and 38,

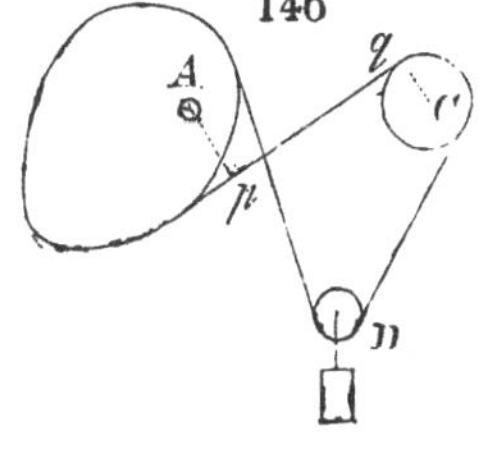

$$\frac{\text{ang. vel. of } A}{\text{ang. vel. of } C} = \frac{Cq}{Ap}.$$

298. If the motion be limited to a small arc the combination assumes the form of fig. 3, (p. 21), but if the limited motion extend to more than a complete revolution, a spiral groove is employed, as in the *fusee* of a watch.

Aa, *Bb*, fig. 147, are parallel axes, one of which carries

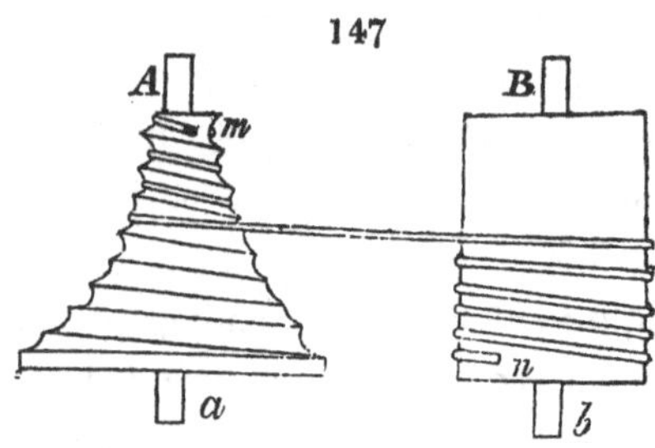

a solid pully, or fusee, as it is termed, upon whose surface is formed a spiral groove, extending in many convolutions from one end to the other. The axis *Bb* carries a plain cylinder ; a band, a cord, or chain, is fastened as at *m* to one end of the fusee, and coiled round it, following the course of the spiral ; the other end of the cord is fixed to the barrel at *n*. If the cord be kept tight by the action of a weight or spring upon one of the axes, the rotation of the other axis will communicate by means of the cord a rotation to the first axis, the velocity ratio of which will vary inversely as the perpendiculars from the axes upon the direction of the cord. And the motion may be continued through as many revolutions as there are convolutions in the spiral.

In like manner a pair of fusees may be employed instead of a fusee and cylinder.

299. If the fusee be required to communicate motion in both directions without the use of the re-acting weight or spring, a double cord will answer the purpose. Thus let it be required to employ the fusee in the manner of the barrel *A*, fig. 104, (p. 181), to give motion to a carriage *B*. The fusee will enable us to obtain a varying velocity ratio between *A* and *B*. In fig. 148, *Aa* is the axis of the fusee, which in this example is made to diminish at both ends. One cord is fas-

tened at *m*, and being coiled round the fusee is carried away at *n*, and attached to the carriage, as at *c*, fig. 104. The other cord is fixed at *p* to the fusee, and being coiled in the opposite direction, leaves the fusee at the same point at which the first cord is carried off. But this cord is taken in the opposite direction, as at *q*, and fixed to the end *d* (fig. 104) of the carriage, (or, which is better, both cords are carried over pullies and brought back to the carriage.)

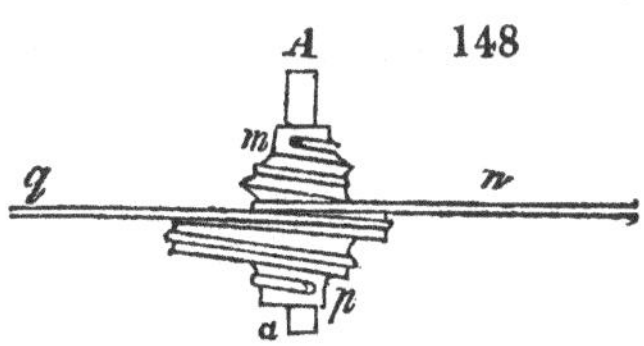

When the axis *Aa* revolves, one cord will unwrap itself from the fusee, while the other wraps upon it, and *vice versa*. But they will always leave its surface in opposite directions at the same point.

Since the fusee (fig. 148) is small at each end and large in the middle, it will, if turned with a uniform angular velocity, have the effect of gradually accelerating the motion of the carriage, till it has reached the middle of its path, and then of gradually retarding it to the end. It is employed in this manner in the self-acting mule of Mr. Roberts, of Manchester.

Class B. Division D. COMMUNICATION OF MOTION BY LINK-WORK.

300. Let *AB*, *CD* be two axes parallel in direction, but not opposite to each other, and let the arms *AP*, *CQ* be fixed to their extremities and connected by a short link *PQ*, jointed to the opposite faces of their arms; then if *AP* and *CQ* be each greater than *AC*, the perpendicular distance of the axes, a continual rotation of one axis will communicate a continual

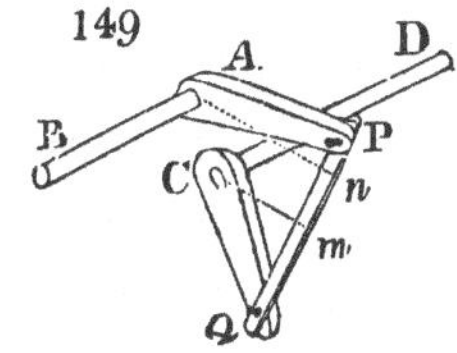

rotation to the other, but with a varying angular velocity ratio; for, if An, Cm be perpendiculars from the centers of motion upon the link, we have

$$\frac{\text{ang. vel. of } AP}{\text{ang. vel. of } CQ} = \frac{Cm}{An},$$

by Art. 32, Cor. 1; which perpendiculars continually change during the motion of the system.

But the properties of this kind of link-work will be more conveniently discussed in the corresponding division of the next Chapter.

301. The combination which is termed Hooke's Joint, however, properly belongs to this division of the subject. This is a method of connecting by link-work two axes whose directions meet in a point, so that the rotation of one shall communicate rotation to the other with a varying angular velocity. It has another use, as an universal joint of flexure, which will be afterwards considered.

150

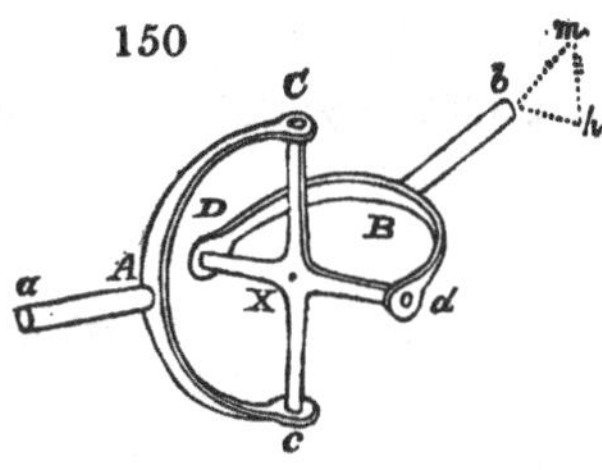

302. This contrivance was invented by Dr. Robert Hooke, and fully described by him in his Cutlerian Lectures*, as well as its properties and the uses to which he intended to apply it, of which however no demonstrations are given. To use his own words, somewhat abridged, "The Universal Joynt consisteth of five several parts. The two first parts are

* No. 2, Animadversions on the Mach. Cælestis, 1674, p. 73. No. 3, Description of Helioscopes, p. 13...1676.

the axes *Aa* and *Bb*, on which the semicircular arms are fastened which are to be joyned together so, as that the motion of one may communicate a motion to the other, according to a proportion which, for distinction's sake, I call elliptical or oblique. The two next parts are the two semicircular arms *CAc* and *DBd*, which are fastned to the ends of those rods, which serve to take hold of the four points of the *ball*, *circle*, *medium*, or *cross* in the middle, X; each of these pair of arms has two center holes, into which the sharp ends of the medium are put, and by which the elliptical or oblique proportion of motion is steadily, exactly, and most easily communicated from the one rod or axis to the other. These center holes I call the *hands*. The fifth and last thing is the ball, round plate, cross, or medium X in the middle, taken hold of by the hands both of one and the other pair of semicircular arms, which, for distinction's sake, I henceforth call the *medium;* and the two points *C*, *c*, taken hold of by the hands of the (driving) axis I call the *points;* and the other two points *D*, *d*, taken hold of by the second pair of arms I call the *pivots*.

"Great care must be had that the pivots and points lie exactly in the same plane, and that each two opposite ones be equally distant from the center, that the middle lines of them cut each other at right angles, and that the axes of the two rods may always cut each other in the center of the medium cross or plate, whatever change may be made in their inclination.

"The shape of this medium may be either a cross, whose four ends hath each of them a cylinder, which is the weakest way; or secondly, it may be made of a thick plate of brass, upon the edge of which are fixed four pivots, which serve for the hands of the arms to take hold of. This is much better than the former, but hath not that strength and stea diness that a large ball hath, which is the way I most approve of, as being strong, steady, and handsome."

To this may be remarked, that at present a stout ring or hoop is generally employed for the medium.

303. *To find the angular velocity ratio of axes connected by a Hooke's joint.* Let C be the intersection of the

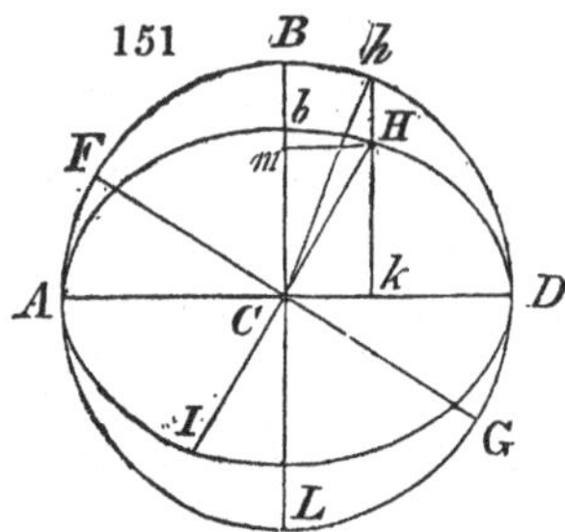

axes, the circle $ABDL$ that described by the extremities of the driver's arms, the plane of the paper being supposed perpendicular to the driving axis. Let the plane which contains the two axes intersect the paper in BCL, and let the ellipse AbD be the projection of the circle described by the extremities of the follower's arms. If θ be the inclination of one axis to the direction of the other produced, we have

$$bC = BC \cdot \cos\theta.$$

Let FCG be that branch of the medium cross which is jointed to the driver; then as this branch is always in the plane of the circle ABD, the projection of the other arm will be perpendicular to it. Draw HCI at right angles to FCG, and passing through the center C, then will this be the projection of that branch of the cross which is jointed to the follower, and H the position of its extremity.

Now in the projected circle AbD all lines parallel to the major axis ACD are unaltered in length by the projection; $\therefore$ Hm, perpendicular upon BC, is the sine of the angle through which the axis of the follower has been moved, if we reckon the motion from bC; and drawing hHk perpendicular to ACD, hCB is that angle. But the corresponding angle, through which the driver has moved, is $ACF = HCB$. Let α be the angle through which the driver has moved, β the corresponding angle of the follower, both being measured as above from that position of the machine in which the arms of the follower lie in the plane which contains the two axes (1),

$$\text{then } \frac{\tan \alpha}{\tan \beta} = \frac{Hk}{hk} = \frac{bC}{BC} = \cos \theta;$$

$$\therefore \tan \beta = \frac{\tan \alpha}{\cos \theta}, \text{ or } \frac{\tan \beta}{\tan \alpha} = constant.$$

The relative positions of the driver and follower are exhibited graphically by means of the ellipse and circle, (fig. 151), where if HCB be the angular distance of any given radius HC of the driver from its position at the beginning of the motion at B, reckoned as above at (1), then will hCB be the corresponding angular distance of the radius hC of the follower, which coincided with it at starting from B.

If we follow these radii round the circle, it appears that they coincide at four points B, D, L, and A; that at starting from B the follower moves slower than the driver at first, and falls behind it, and then accelerates, until it overtakes it at D, beyond which it takes the lead through the next quadrant DL, first moving quicker than the driver and then retarding; so that the driver overtakes it at L, and passes it. The motion through LA is similar to that through BD; and that from A to B the same as that from D to L. The amount of retardation and acceleration depends upon the value of θ; and therefore if a single joint be employed, the axes must be inclined to each other sufficiently to produce the desired variation of velocity.

304. By means of two joints, however, the axes may be parallel or inclined at any angle, and a greater variety of motion be procured.

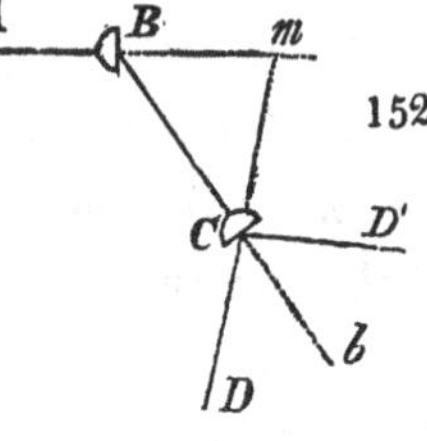

Thus let AB, fig. 152, be the driving axis, and let it be connected to the first following axis BC by a Hooke's joint at B, and let this be similarly jointed to a second axis CD at C. The plane of ABC may be different from that of BCD.

First, let the angular motion of the second joint at C be reckoned like that of the first, from the position in which the fork of the follower lies in the plane of the two axes. Then for the motion of the joint B we have, as before,

$$\tan\beta = \frac{\tan\alpha}{\cos\theta};$$

and if γ be the corresponding angles of the axis CD, and $\theta_{,}$ its inclination to BCb,

$$\tan\gamma = \frac{\tan\beta}{\cos\theta_{,}} = \frac{\tan\alpha}{\cos\theta\,.\,\cos\theta_{,}}.$$

If there be a series of similar axes, whose successive mutual inclinations are θ, $\theta_{,}$, $\theta_{,,}\ldots\theta_{n}$, δ the angular distance of a radius of the last, corresponding to α,

$$\text{then, } \tan\delta = \frac{\tan\alpha}{\cos\theta\,.\cos\theta_{,}\cos\theta_{,,}\ldots\cos\theta_{n}}.$$

In a system of this kind any desired amount of variation may be obtained, and the last follower may be set at any given angle to the first driver, or even in its own direction produced, by three Hooke joints only.

In the system just described the shafts may lie in different planes, but it is supposed that the joints are all so adjusted that when the following arms of the first joint B lie in the plane *ABC* of its two axes, that the following arms of every other joint also lie in the plane of their two axes.

Let there be a system of three axes with two joints, as fig. 152, but let the *driving* arms of the second lie in the plane *BCD*, when the *following* arms of the first lie in the plane *ABC*. The angles of the second are therefore now reckoned from a fixed radius distant one quadrant from those of the first.

If $\tan\beta = \dfrac{\tan\alpha}{\cos\theta}$ be the equation to the first,

$$\tan\gamma = \frac{\tan\left(\dfrac{\pi}{2}+\beta\right)}{\cos\theta_{,}} \text{ is the equation to the second.}$$

$$\text{But } \tan\left(\frac{\pi}{2}+\beta\right) = \frac{1}{\tan\beta};$$

$$\therefore \tan\gamma = \frac{\cos\theta}{\tan\alpha \cdot \cos\theta_{,}}.$$

$$\text{Let } \theta = \theta_{,}; \quad \therefore \tan\gamma = \frac{1}{\tan\alpha};$$

which shews that if the forks be set as above, and if the angles of inclination of the axes be equal, then the variations of motion will counteract each other, and the angular velocity ratio of the extreme axes *AB*, *CD*, remain constant.

305. When the double Hooke's joint is thus employed it is commonly for this purpose of correcting the varying ratio of angular velocity, and the intermediate piece may therefore be made short, as in fig. 153. Care must be taken however that the extreme axes are so placed as to meet in a point, and that the angles they each make with the intermediate piece are the same.

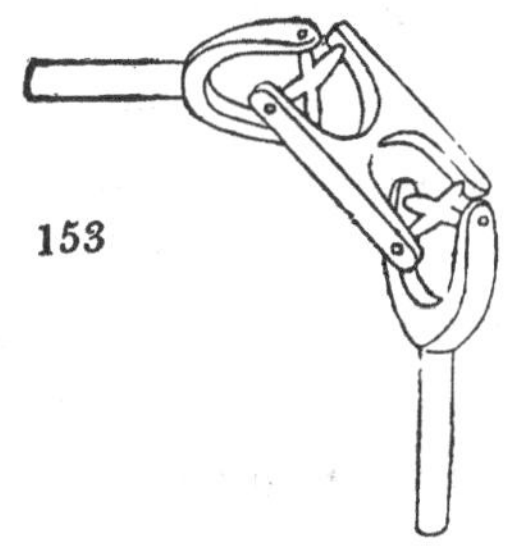

153

In this form the Hooke's joint properly belongs to the Link-work of the previous Chapter. It may be used either to communicate equal rotation between two axes inclined at an angle AmD, fig. 152, or between parallel axes, as AB, CD'.

306. Considering only the elementary form of this contrivance, it is evident that two branches to each axis are necessary only to give greater steadiness to the motion, and that a single pair of arms AC, BD (fig. 150), connected by a straight link from C to D, would produce the same motion; so that in this way we are brought to a form very similar to that of fig. 149. Also, Oldham's joint, in Art. 176, becomes a Hooke's joint, if the axes, instead of being parallel, are set so as to intersect in the center of the cross.

307. More complex relations of angular velocity may be obtained by making the two arms of each axis unequal, as AC greater than Ac; or setting the arms of the cross at different angles. For this purpose in Hooke's figure the arms are provided with the power of adjustment, in length. He proposed to employ this contrivance in the resolution of spherical questions and various other purposes, which however have been long forgotten; but may be found in the original Essay already referred to.

308. One of the applications which Hooke made of this device was to the construction of Sun-dials.

Two axes being connected by this joint, let one of them be fixed parallel to the axis of the Earth, and the other perpendicular to the plane of the required dial; and let the first be furnished with an index, travelling round a common equally-divided hour-circle, and let the other carry a similar index which travels round the circle of the dial. The two indexes being so adjusted that when the first points to noon the other shall coincide with the twelve-o'clock line; then by turning round the upper axis "till you see its index to point at those hours, halves, quarters, or minutes, you have a mind to take notice of in your dial, by the second index you are directed to the true corresponding point in the plane of the dial itself*;" which may be shown as follows.

Let AOE, fig. 154, be a dial-plate inclined at any given angle to the horizon, C its center, PC the style, PO an arc of the circle of declination perpendicular to the plate, therefore CO is the substyle. Let PE be an arc of the meridian, therefore CE is the twelve-o'clock line; and if the sun decline through any angle EPA from the plane of the meridian, we have in the spherical triangle POA, right-angled at O,

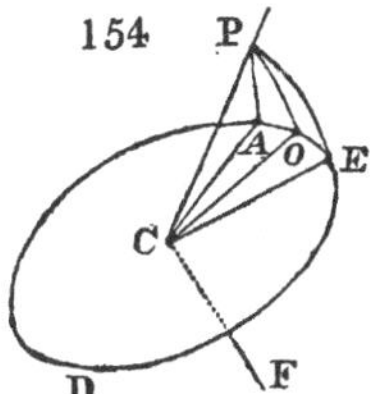

$$\sin PO = \frac{\tan AO}{\tan OPA}$$

If then PC be the position of one axis of the Hooke's joint, and the other axis CF be perpendicular to the dial-plate, $\sin PCO = \cos PCF$. And the expression becomes the same as that we have already found for the synchronal motions of the two axes; AO measuring the motion of the shadow of the style over the plate, and OPA the corresponding rotation of the solar hours.

The axis PC having been fixed in its due situation parallel to the axis of the earth, the meridian line or twelve-o'clock point E may be found by a plummet. And in employing the instrument the actual dial-plate will of course be fixed at the lower extremity of the axis CF, parallel to the circle AOE; but this does not affect the demonstration.

* Hooke, *Desc. of Helioscopes*, p. 14.

309. With respect to the use of Hooke's joint as a universal joint of flexure, let Aa, fig. 150, be a fixed rod, and let it be required to move the extremity b of Bb in any direction bm concentric to X the intersection of the rods. Draw bh, mh meeting in h, and also concentric to X, and in planes respectively perpendicular to the axes of rotation or joints Cc, Dd. Then b can move in the direction bh, by virtue of the joint Cc, and in the direction hm by virtue of the joint Dd; and if it be made to revolve simultaneously round these joints with velocities respectively proportional to bh and hm, it will describe the path bm.

The motion bm is performed round an axis passing through X, which being perpendicular to the plane bmX is also situated in the plane of the joints Cc, Dd; and whatever be the angle between the joints Cc and Dd, that is, between bh and hm, the triangle bhm can be constructed upon bm. Bb is here supposed perpendicular to the plane of the joints, but the same will be true for any piece or rod rigidly connected with Bb, and therefore for a rod making any angle with the plane of the joints. It follows therefore that if two rods or pieces are connected by two joints of flexure whose axes meet at any angle, these pieces are at liberty to turn round any axis of flexure situated in the plane of these two axes, and passing through their point of intersection.

310. Now let the second rod be required to bend with respect to the first round any axis passing through the intersection X.

Let AB, BC, fig. 155, be the rods, B their point of intersection, FBD the plane containing the two axes of flexure, as before, and not necessarily perpendicular to BC, and let BC be required to move about an axis BK passing through B, but not in the plane of FBD. In this plane draw BD, which suppose to be rigidly connected with BC, and let Dn be the direction of motion of D in revolving round the axis BK. Now by virtue of the two axes in the plane FBD, the point D is at liberty to move in the direction Dm perpendicular to that plane, but in no other. But let a third axis of flexure be introduced into the system passing through B in any direction not in the plane FBD, and let mn be the line of motion round this new axis, then the triangle Dmn can always be constructed upon Dn, and thus, as before, Dn be described by the simultaneous motions Dm and mn.

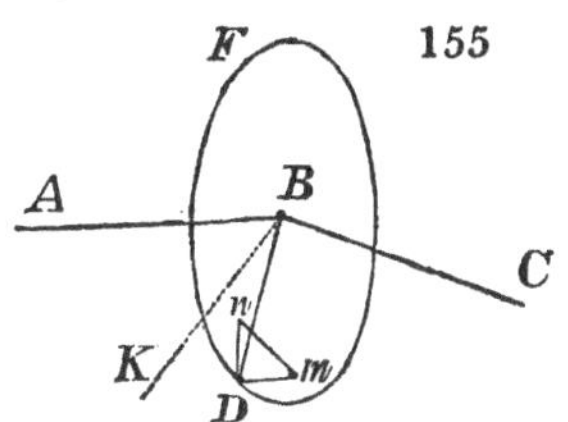

Hence it appears, that if two pieces are connected by three joints of flexure whose axes intersect in a point, and make any angles with each other, but are not in the same plane, these pieces are at liberty to turn about the point of intersection round any axis of flexure whatever which passes through that point.

If the axes of the joints pass each other without meeting in a point, it can be similarly shewn that the moving piece has still the unlimited choice of an axis in direction, and that this axis will lie somewhere between the component axes.

311. The joints by which the members of crustaceous animals and insects are united, furnish many beautiful examples of these principles.

Every separate joint in these animals is a hinge-joint very curiously constructed, but of course possessing but a single axis of flexure; these axes however are grouped so as to produce compound joints having two or three axes of flexure, and therefore forming universal joints, or swivel-joints, in the manner explained in the previous Article.

As an example of this we may take the front claw of the common crab, represented in fig. 156. This consists in fact of five separate pieces, *A*, *B*, *C*, *D*, *E*, not including the moveable jaw *F* of the actual claw; each piece is jointed to the next by a hinge-joint. But upon our principles the entire limb may be considered to consist of two principal members *C* and *E*; of which the first is jointed to the body of the animal by a universal joint of three axes of flexure, and the second to the first by a joint of two axes of flexure, or Hooke's joint.

For the piece *C* is united to the claw *E* by means of an

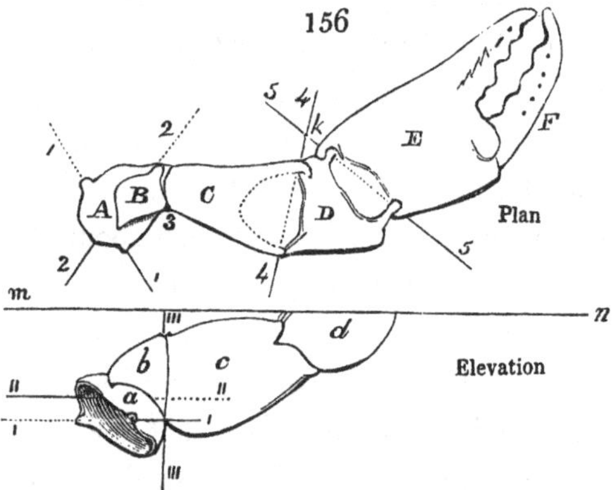

intermediate piece *D*, and the axes of the joints which connect them are shewn by the line 5, 5 between *E* and *D*, and 4, 4 between *D* and *C*. These axes meet in a point *k*, and therefore by what has preceded, it appears that *E* moves with respect to *C* about the point *k*, and that it is at liberty to turn round any axis of flexure passing through that point and in the plane 5, *k*, 4. So that this is in fact a natural Hooke's joint. The compound joint which connects the piece *C* with the body of the animal is more complex; and to exhibit its arrangement, two projections are given, one upon a plane perpendicular to the other, and intersecting it in the line *mn*.

We may suppose the claw to be laid down on the table in the upper figure, in which case this becomes the Plan and and the lower the Elevation, although the figures are drawn without any relation to the position of the claw with respect to the body of the animal, but only so as best to exhibit the joints, as will appear presently.

A ring A or a is jointed to the body of the animal by a joint whose axis is 1, 1, in the Plan, and I, I, in the Elevation. This is jointed to a second ring B, or b, by an axis 2, 2, or II, II; and B is jointed to C by a third axis vertical in the Plan, whose projection is therefore a point 3. It is shewn at III, III, in the Elevation. C is therefore connected to the body of the animal by a compound joint of three axes, whose directions nearly meet, but of which no two are parallel, neither are they in three parallel planes, and therefore, by Art. 310, C is at liberty to move about an axis situated at any angle with respect to the body. The compound joint, in fact, corresponds to the ball and socket joint employed for the shoulder of vertebrate animals. Its motions in different directions are of course limited by the extent of angular motion of which each separate hinge is capable.

The diagram is reduced from a very careful drawing. I found that the axis 2,2 was as nearly as possible in a plane perpendicular to 3, and that when the ring A was placed in its mean position, the axis 1, 1 was also in a plane perpendicular to 3. This determined the choice of the position of the planes of projection.

That of the Plan is parallel to the joints 1,1, 2,2, and therefore perpendicular to the joint 3, which thus becomes a point. The plane of the Elevation is parallel to the point 3.

As to the joints 4,4, 5,5, the joint 4,4 is in the drawing a little overstrained to allow 5,5 to come into parallelism with the plane of the paper; and 4,4, is also not in reality exactly perpendicular to 3. However, it must be understood that my object here is not to shew the relation of the limb to the body of the animal, but merely the principle of arrangement of the joints.

The claw E is shewn in its extreme outward position with respect to C; in its mean position it would be at right angles to the paper; and in the extreme inward position E and C would come into contact, to allow of which the shape of the intermediate piece and position of the hinges are beautifully adapted.

Class B. Division D. COMMUNICATION OF MOTION BY REDUPLICATION.

312. In the examples of Reduplication contained in the corresponding division of Class A, the strings and the motion of the follower are all parallel, and the velocity ratio constant. If the strings and the paths make angles with each other, a varying velocity ratio will ensue; as in the following example. Let the string be fixed at A, fig. 157, and passing over a pin B, let it be attached to a point C; let Bb be the path of the pin, Cc that of the extremity of the string, and when C is moved to c, very near to its first position, let B be carried to b; draw perpendiculars bm, bn, Cp, upon the two directions of the string in its new position.

157

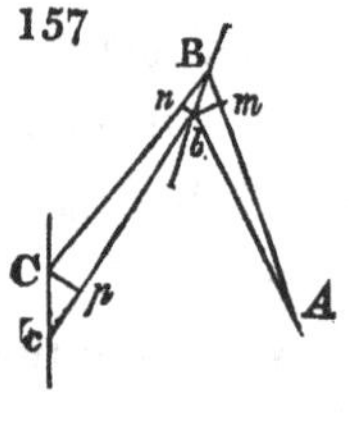

Then since the length of the string is the same in both positions, we have $AB + BC = Ab + bc$, that is,

$$Am + mB + Bn + nC = Ab + bp + pc,$$

But ultimately,

$$Ab = Am, \text{ and } bp = nC; \quad \therefore mB + Bn = pc,$$

$$\text{or} \quad Bb\,(\cos bBA + \cos bBC) = Cc\,.\cos cCB\,;$$

$$\therefore \frac{Bb}{Cc} = \frac{\cos cCB}{\cos bBA + \cos bBC}\,;$$

where the angles are those made by the direction of the string with the respective paths of the pin B and of the extremity C. But by the motion of the system these angles alter, and thus the velocity ratio varies.

If the strings and the path of B become parallel, the cosines become unity, and $\frac{Bb}{cC} = \frac{1}{2}$, as before (Art. 30).

CHAPTER IX.

ELEMENTARY COMBINATIONS.

CLASS C. DIRECTIONAL RELATION CHANGING.

---

313. IN the combinations which have occupied our attention in the preceding Chapters, the directional relation of the pieces has remained constant; but, as I have already explained (Art. 21), there exists a numerous class of combinations, in which the directional relation changes periodically, or in other words, that while one piece pursues its own path with a constant direction of motion, the other piece periodically changes its direction, travelling back and forwards through a constant space. From this it follows that the latter piece must necessarily be limited in the extent of its path by the very nature of the combination, but it will also appear that in the greater number of combinations this reciprocating piece is the follower.

314. The velocity ratio of the pieces may be constant, or may vary; but as the driver may be generally supposed to revolve uniformly, the follower, if the velocity ratio be constant, will in that case travel with a uniform velocity to the end of its path, and instantly reversing the direction, will return with a uniform velocity, and so on. This sudden change, for dynamical reasons, is better avoided; and although, as we shall see, it *may* be effected, yet now that mechanical principles are better understood, those combinations are always preferred in which the reciprocating body is brought gradually to rest, and again gradually set in

motion in the opposite direction, and thus the blows and strains occasioned by the sudden change of direction are got rid of. This is more especially necessary in large and heavy machinery.

CLASS C. DIVISION A. COMMUNICATION OF MOTION BY ROLLING CONTACT.

315. When two spur-wheels act together the axes revolve in opposite directions, but when a spur-wheel acts with an annular wheel the axes revolve in the same direction. By combining a spur-wheel with an annular wheel the *mangle-wheel*, fig. 158, is obtained; in which the directional relation is periodically changed, by causing the driving pinion to act alternately upon the spur-teeth and the annular teeth.

The mangle-wheel in its simplest form is a revolving disk of metal with a center of motion *C*. Upon the face of

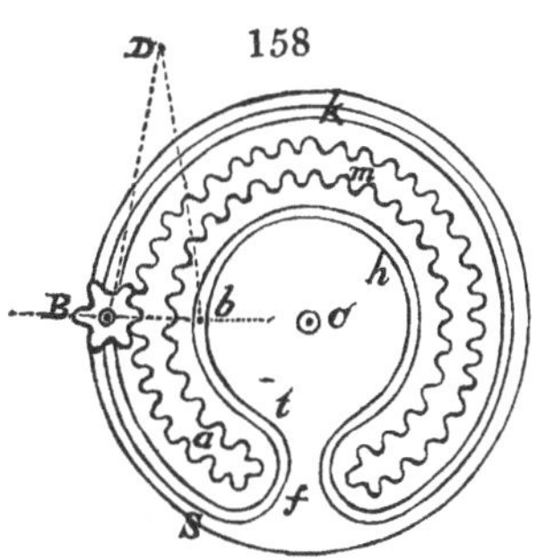

158

the disk is fixed a projecting annulus *am*, the outer and inner edge of which are cut into teeth. This annulus is interrupted at *f*, and the teeth are continued round the edges of the interrupted portion so as to form a continued series passing from the outer to the inner edge and back again.

A pinion *B* whose teeth are of the same pitch as those of the wheel is fixed to the end of an axis, and this axis is mounted so as to allow of a short travelling motion in the direction *BC*. This may be effected by supporting this end of it either in a swing-frame moving upon a center as at *D*, or in a sliding piece, according to the nature of the train with which it is connected. A short pivot projects from the center of the pinion, and this rests in and is guided by a groove *BSftbhk* which is cut in the surface of the disk, and made concentric to the pitch circles of the inner and outer rings of teeth, and at a normal distance from them equal to the pitch radius of the pinion.

Now when the pinion revolves it will, if it be on the outside, as in the figure, act upon the spur-teeth and turn the wheel in the opposite direction to its own; but when the interrupted portion *f* of the teeth is thus brought to the pinion, the groove will guide the pinion from the outside to the inside, and thus bring its teeth into action with the annular teeth. The wheel will now receive motion in the same direction as that of the pinion, and this will continue until the gap *f* is again brought to the pinion, when the latter will be carried outwards, and the motion again reversed.

The velocity ratio in either direction will remain constant, but the ratio when the pinion is inside will differ slightly from the ratio when it is outside, for the pitch radius of the annular teeth is necessarily somewhat less than that of the spur-teeth. However, the change of direction is not instantaneous, for the form of the groove *sft*, which connects the inner and outer grooves, is a semicircle, and when the axis of the pinion reaches *s* the velocity of the mangle-wheel begins to diminish gradually till it is brought to rest at *f*, and is again gradually set in motion from *f* to *t*, when the constant ratio begins; and this retardation will be increased by increasing the difference between the inner and outer pitch circles.

316. The teeth of a mangle-wheel are, however, most commonly formed by pins projecting from the face of the disk, as in fig. 159.

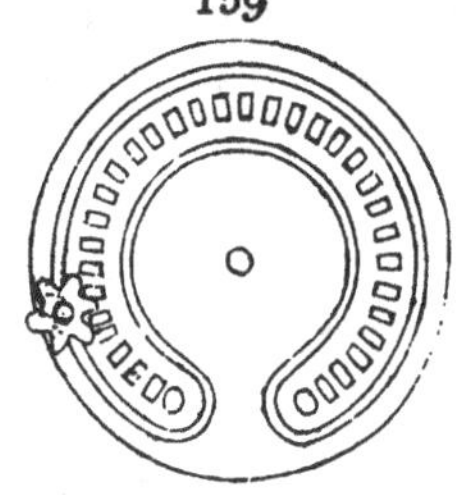
159

In this manner the inner and outer pitch-circles coincide, and therefore the velocity ratio is the same within and without; also the space through which the pinion moves in shifting from the outside to the inside is reduced.

317. This space may be still farther diminished by arranging the teeth as in fig. 160, that is, by placing the

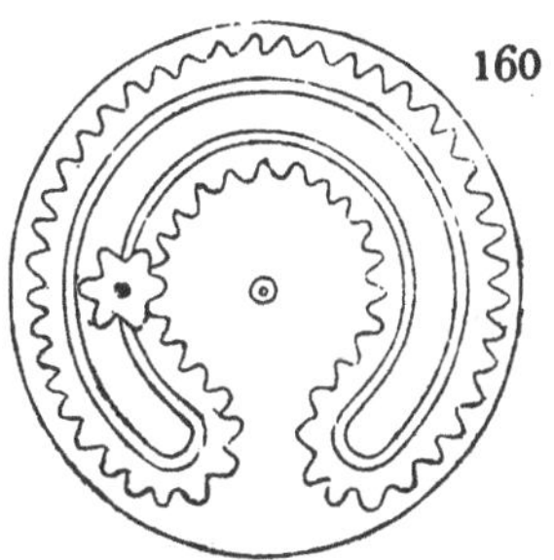
160

spur-wheel within the annular wheel; but at the same time the difference of the two ratios is increased.

318. If it be required that the velocity ratio vary, then the pitch-lines of the mangle-wheel must no longer be

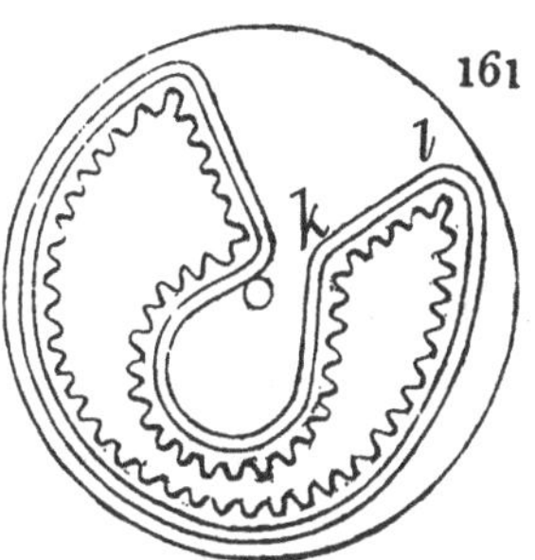

161

concentric. Thus in fig. 161, the groove *kl* is directed to the center of the mangle-wheel, and therefore the pinion will proceed in this portion of its path without giving any motion to the wheel; and in the other lines of teeth the pitch radius varies, and therefore the angular velocity ratio will vary*.

The mangle-wheel under all its forms is a very practical and effective contrivance. It derives its name from the first machine to which it was applied, but has since been very generally employed in manufacturing mechanism.

319. In figs. 158, 160, and 161, the curves of the teeth are readily obtained by employing the same describing circle for the whole of them (Art. 114). But when the form fig. 159 is adopted, the shape of the teeth requires some consideration.

Every tooth of such a mangle-wheel may be considered as formed of two ordinary teeth set back to back, the pitch-line passing through the middle. The outer half, therefore, appropriated to the action of the pinion on the outside of the wheel, resembles that portion of an ordinary spur-wheel tooth that lies beyond its pitch-line, and the inner half which receives the inside action of the pinion resembles the half of an annular wheel tooth that lies within the pitch-circle. But the consequence of this arrangement is, that in both positions the action of the driving pinion must be confined to the approach of its teeth to the line of centers, and consequently these teeth must lie wholly within their pitch-line.

To obtain the forms of the teeth therefore take any convenient describing circle, and employ it to describe the teeth of the pinion by rolling within its pitch-circle, and to describe the teeth of the wheel by rolling within and without

* A mangle-wheel of this kind is employed in Smith's self-acting mule.

its pitch-circle, and the pinion will (Art. 114) then work truly with the teeth of the wheel in both positions. The tooth at each extremity of the series must be a circular one, whose center lies on the pitch-line and whose diameter is equal to half the pitch.

320. If the reciprocating piece move in a right line, as it very often does, then the mangle-*wheel* is transformed into a *mangle-rack*, fig. 162, and its teeth may be simply

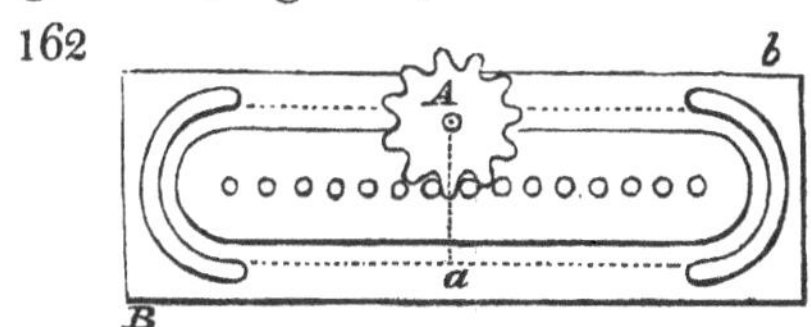

made cylindrical pins, which those of the mangle-wheel do not admit of on correct principle. *Bb* is the sliding piece, and *A* the driving pinion, whose axis must have the power of shifting from *A* to *a* through a space equal to its own diameter, to allow of the change from one side of the rack to the other at each extremity of the motion. The teeth of the mangle-rack may receive any of the forms which are given to common rack-teeth, if the arrangement be derived from either fig. 158 or fig. 160.

321. But the mangle rack admits of an arrangement by which the shifting motion of the driving pinion, which is often inconvenient, may be dispensed with.

Bb, fig. 163, is the piece which receives the reciprocating

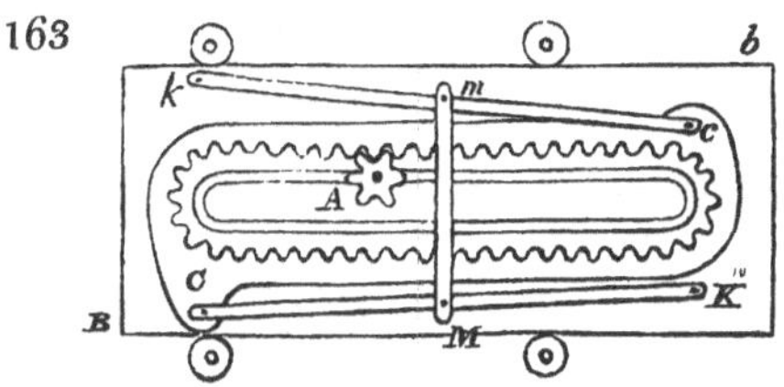

motion, and which may be either guided between rollers, as shewn, or in any other usual way; *A* the driving pinion, whose axis of motion is fixed; the mangle-rack *Cc* is formed upon a separate plate, and in this example has the teeth upon the inside of the projecting ridge which borders it, and the guide-groove formed within the ring of teeth, similar to fig. 160.

This rack is connected with the piece *Bb* in such a manner as to allow of a short transverse motion with respect to that piece, by which the pinion, when it arrives at either end of the course, is enabled by shifting the rack to follow the course of the guide-groove, and thus to reverse the motion by acting upon the opposite row of teeth.

The best mode of connecting the rack and its sliding piece is that represented in the figure, and is the same which is adopted in the well-known cylinder printing-engines of Mr. Cowper. Two guide-rods *KC*, *kc* are jointed at one end *K*, *k* to the reciprocating piece *B b*, and at the other end *C*, *c* to the shifting-rack; these rods are moreover connected by a rod *Mm* which is jointed to each mid-way between their extremities, so that the angular motion of these guide-rods round their centers *K*, *k* will be the same; and as the angular motion is small, and the rods nearly parallel to the path of the slide, their extremities *C*, *c*, may be supposed to move perpendicularly to that path, and consequently the rack which is jointed to those extremities will also move upon *Bb* in a direction perpendicular to its path, which is the thing required, and admits of no other motion with respect to *Bb*.

The earliest shifting rack of this kind is to be found in the work of De Caus*, in which the rack is moved from one side to the other at each end of its trip by a pair of cam-plates, turned by the same pinion which drives the rack.

* De Caus, Les Raisons des forces mouvantes, 1615. L. 1. probs. XVI. and XVII. Copied in Bockler's Theatrum Machinarum, 1662, pl. 94.

322. In the works of the early mechanists a variety of contrivances for reversing motion are to be found, in which the teeth of a driving wheel or pinion are made to quit one set of teeth and engage themselves abruptly with another set, and so on alternately; the two sets being so disposed upon the reciprocating follower as to produce motion respectively in the opposite directions in it.

For example, *Aa*, fig. 164, is an axis which revolves continually in the same direction, *Bb* an axis to which is to be communicated a few rotations to right and left alternately.

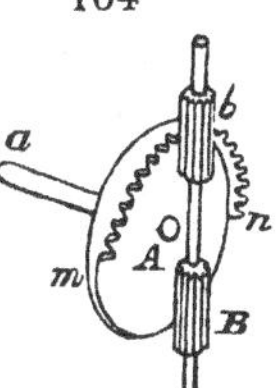

This axis carries two pinions, *B* and *b*, and the first axis has a crown-wheel at its extremity, of which the teeth extend only through half its circumference, as from *m* to *n*.

In the figure the crown-wheel is supposed to revolve in the direction from *n* towards *m*, and its teeth will accordingly act upon those of *b*, and cause the shaft *Bb* to revolve. When the last tooth *n* has quitted *b* this rotation will cease, but at that moment the first tooth *m* of the series will begin to act upon the lower pinion *B*, and turn it in the opposite direction. This contrivance is so manifestly faulty for the two reasons already discussed, of the shock at each change of motion, (Art. 314), and the danger of the first teeth that come together becoming entangled (Art. 271), that I should hardly have thought it worth describing, were it not for the numerous similar forms that present themselves in the early history of machinery, more especially in the work of Ramelli, in which this principle is exhibited in a great variety of forms, and applied not only to wheels but also to racks*.

* Vide Ramelli, I. II. III. IV. et passim. De Caus, pr. III. and IV. Bockler, 109, 110, 111, copied from Ramelli. Bessoni, Theatrum Instrumentorum, 1569, pl. 34.

323. Fig 165 is an application of the same principle to a double rack*, which deserves attention on account of the provision which is made to diminish the shock, and ensure the first engagement of each set of teeth.

Aa is the frame to which the reciprocating motion is to be given, *B* the driving pinion; this is made in the form of a lantern, and the teeth confined to about a quarter of its circumference.

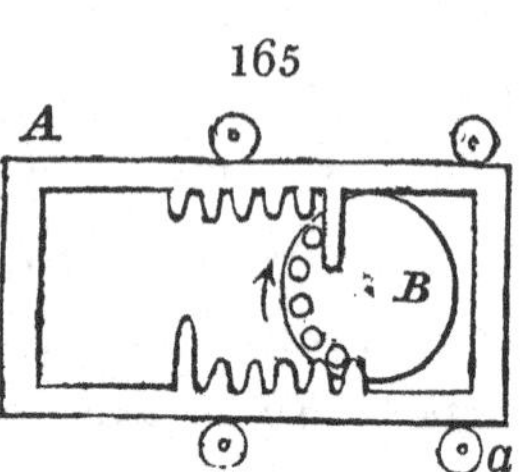

These teeth act alternately upon racks fixed to the opposite sides of the frame, and thus the frame receives a back and forward motion from the continued rotation of the pinion. In the figure the pinion revolving in the direction of the arrow is shewn at the moment of quitting the lower rack to begin its action upon the upper; the tooth of each rack which receives the first action of the pinion is made longer than the others, and straight sided, and is so arranged that the action of the first stave upon it shall be oblique, by which the shock is diminished, while at the same time the stave sliding down the long side is safely conducted into the first space, and thus the proper action of the teeth and staves secured.

324. If the driver be a wheel *A*, fig. 166, and the follower an arm *BC* revolving round a center *B*, and having a wheel of an irregular form *D* turning round a pin at its extremity *C*; its teeth being kept in constant action with those of *A* by means of guide-plates, grooves, or any of the contrivances already described, then the rotation of *A* will produce a reciprocating motion in the arm *BC*, the law of which will vary according to the figure of the wheel

* From Bockler, Theatrum Machinarum, No. 71.

D. For the distance of C from A continually increases or diminishes as A revolves, and therefore C will oscillate to and fro in its path.

CLASS C. DIVISION D. COMMUNICATION BY LINK-WORK.

325. I have thought it necessary to place Link-work in this class, immediately after Rolling Contact, because in some of the combinations by sliding contact I shall have occasion to refer to those which are included under this head. As the order in which these different divisions is taken is otherwise arbitrary, no inconvenience can arise from this change of plan.

The velocity ratio of a pair of arms connected by a link has been already determined (Art. 32); but it is often more convenient to investigate their motion by determining the relative positions of the parts of the system, as follows:

326. Let A, B (fig. 167), be two centers of motion; AP, BQ the arms, PQ the link;

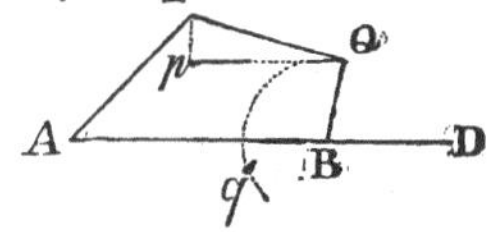

let $AP = R$, $BQ = r$, $AB = d$,

$PQ = l$, $BAP = \theta$, $DBQ = \phi$.

Draw Pp perpendicular and Qp parallel to AB;

$$\text{then } PQ^2 = Pp^2 + pQ^2,$$

$$\text{or } l^2 = (R \sin\theta - r\sin\phi)^2 + (d + r\cos\phi - R\cos\theta)^2$$

$$= R^2 + r^2 + d^2 - 2Rd\cos\theta - 2Rr\sin\theta \,.\, \sin\phi$$

$$+ 2r(d - R\cos\theta)\cos\phi.$$

For convenience assume $m = R^2 + r^2 + d^2 - 2Rd\cos\theta - l^2$,

$$n = 2Rr\sin\theta,$$

$$p = 2r(d - R\cos\theta);$$

$$\therefore n\sin\phi - m = p\,.\cos\phi;$$

whence, squaring and arranging the terms, we have

$$\sin\phi = \frac{mn \pm p\sqrt{p^2 - m^2 + n^2}}{n^2 + p^2}; \quad (1)$$

in which equation m, n, and p, being functions of θ, it appears that for every value of θ, $\sin\phi$ has two values, or in other words, every given position AP of one rod has two possible corresponding positions of the other rod BQ, which is indeed evident; for with center B and radius BQ describe the circle Qq, and take $Pq = PQ$, then will Bq be also a position of this rod corresponding to AP.

If $R = r$, and $l = d$, we have the system of fig. 109 (Art. 196), and when these suppositions are introduced into the equation (1), we obtain

$$\sin\phi = \frac{R^2 \pm d^2 - Rd(\cos\theta \pm \cos\theta)}{R^2 + d^2 - 2Rd\cos\theta} . \sin\theta$$

$$\left\{\begin{array}{l} = \sin\theta \text{ when the upper} \ldots\ldots\ldots\ldots \\ = \dfrac{R^2 - d^2}{R^2 + d^2 - 2Rd\cos\theta} . \sin\theta \text{ when the lower} \end{array}\right\} \text{signs are taken.}$$

The first value corresponds to the system when in the position of a parallelogram, and the second to that in which the link lies across the line of centers.

If, on the other hand, we make $R = l$, and $r = d$, we have

$$m = 2d^2 - 2Rd\cos\theta = p;$$

$$\text{whence } \sin\phi = \frac{mn \pm mn}{n^2 + m^2}$$

$$= \frac{2mn}{n^2 + m^2} \text{ or } 0;$$

and in fact it will be seen that if these proportions are given to the rods, Q will always coincide with A in the second position Bq, since $AP = PQ$ and $BQ = BA$. Consequently, in that position AP will revolve without producing any change in the angular position of BQ, which will coincide with AB in all positions of AP, and therefore $\sin\phi = 0$.

In the first position, however, we have

$$\sin\phi = \frac{R\sin\theta\,(d - R\cos\theta)}{R^2 + d^2 - 2Rd\cos\theta}.$$

327. In a system of this kind the continued rotation of one arm, as AP, may produce either a continued rotation or a reciprocating motion in the other. This is determined by the proportions of the four sides of the figure. For example, if AP and BQ be greater than AB, the arms will both revolve, as in Art. 300. But if AP be less than AB, then the rotations of AP will cause BQ to reciprocate. To enable AP to complete a revolution it is necessary that

$$AB + BQ > PQ + AP, \text{ and } AB - BQ > PQ - AP;$$

for if AP move towards AB, the two rods AP, PQ must first come into one straight line at the moment when Q reaches its greatest distance from A; but this straight line is impossible, unless

$$AP + PQ < AB + BQ;$$

and similarly, when AP has revolved so as to bring Q to its least distance from A, the lines AP, PQ will form a straight line passing through A, which line will be impossible, unless also

$$PQ - AP < AB - BQ.$$

The positions which correspond to these two straight lines are the *dead points* (Art. 196). But in practice the reciprocating point Q generally moves in a straight line directed towards A, the axis of the revolving piece; or else is suspended from an arm BQ at right angles to this line in the mean position, and so long, that the path of Q may be taken as a right line. This simplifies the examination of the motion.

328. Thus let A, fig. 168, be the center of motion of

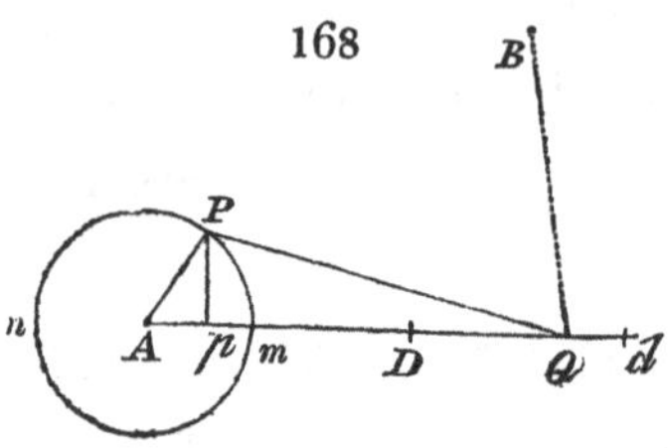

a revolving driver, P a pin carried by a disk or arm fixed to the axis, PQ a link jointed at P to the pin, and at Q to a piece which travels along the line Ad. The pin P may either be carried by an arm, as at P in fig. 111 (p. 187), or by a disk as at p, or it may be a crank as in fig. 112, for the remarks in Art. 199 are completely applicable in the present case. Upon the line Ad of the follower's path set off md and nD each $= PQ$, then as the axis A revolves the point Q will travel back and forwards between d and D, performing one complete excursion or double oscillation for every revolution of A; also $Dd = 2AP$. Draw Pp perpendicular to AD, and let $mAP = \theta$, $AP = R$, $PQ = l$, $AQ = s$, then in the triangle APQ we have $AQ = Qp \pm Ap$, according as p falls upon one side or other of A;

$$\therefore s = \sqrt{l^2 - R^2 \sin^2\theta} \pm R\cos\theta, \quad (1)$$

in which the positive sign is used when mAP is acute, and the negative when it is obtuse.

The extent Dd of the motion of the reciprocating piece is termed the *throw* of the crank or excentric pin, also m and n are the *dead points*.

Generally the length of the link is so great with respect to AP, that its inclination may be neglected, and pQ supposed equal to PQ, in which case $s = l \pm R\cos\theta$; or if the space be measured from d in the opposite direction, we have $s = dQ = mp = R$ versin θ.

329. By Cor. 3, Art. 32, we have

velocity of P : velocity of Q :: cos AQP : sin APQ;

and if the link be long, this becomes $\dfrac{\text{vel. of } P}{\text{vel. of } Q} = \dfrac{1}{\sin\theta}$.

330. To the different forms under which the arm and link appears in Art. 199, may be added *the excentric*, fig. 169.

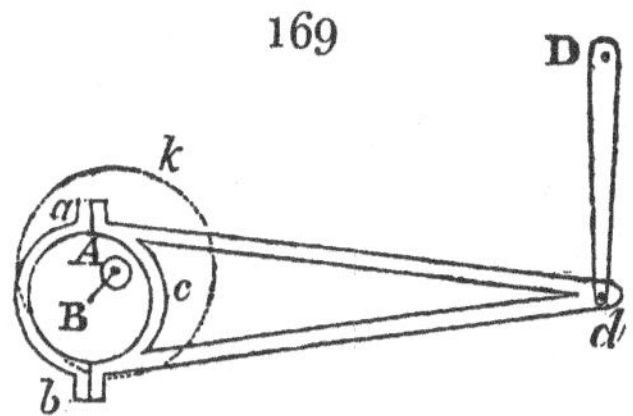

Let A be the axis or center of motion, to which is fixed an excentric circular pully of which B is the center; a hoop abc is made to embrace this pully so as just to allow the pully to turn freely within its circle, for which purpose, as well as to allow the machine to be put together, the hoop is generally made in two halves capable of being separated at a and b; a frame adb connects this hoop with the extremity d of the arm dD, to which it is jointed in the manner of a link. When A revolves the distance Bd from the center of the excentric to the extremity of the arm remains constant, and therefore the motion communicated is precisely the same as that which would be given by an arm AB, and a link Bd. But this contrivance allows the axis to be continued straight through the excentric, whereas when an arm is employed the axis must be cut short, or else bent into a crank, as explained in Art. 199. On the other hand, the magnitude of the hoop and excentric is so great with respect to the radius of motion AB, that this contrivance is necessarily limited to the production of vibrations of small extent. The dotted circle radius Ak includes the space required for the rotation of the excentric, the radius of which is equal to the sum of the radius of the excentric and of AB, and the former must be greater than the latter. A common crank or pin would occupy a circle of about half this radius.

331. The excentric, arm or crank, under the different forms thus described, is by far the most simple mode of converting rotation into reciprocation, and it has the valuable property of beginning the motion in each direction gently, and again gradually retarding it, so as to avoid jerks. Nevertheless the law of variation in the velocities is not always the best adapted to the requirements of the mechanism; but the reciprocation is produced so simply that it is often worth while to retain the crank, and correct the law of velocity by combining other pieces with it in a train. By trains of link-work very complex laws of motion may be derived from a uniformly revolving driver. This will be best illustrated by the examples which follow.

332. Ex. 1. If the crank, instead of being fixed to the uniformly revolving axis, be carried by a second axis, and these two axes connected by one of the combinations in Chapter VIII. for the production of varying velocity ratio, the inequality of velocity in the reciprocating piece may be almost entirely got rid of. Thus, let these two axes be connected by a pair of rolling curve wheels, (Art. 273), let A_1 be the constant angular velocity of the first axis, A_2 the angular velocity of the second axis, upon which is also fixed the crank, let ρ be the radius of the crank, and θ the angle it makes with the path of the reciprocating piece; then if V be the linear velocity of this piece, we have $V = \rho \sin \theta A_2$ (by Art. 329), which is to be constant by hypothesis. Let r_1 and r_2 be the radii of contact of the rolling curves which connect the first and second axis respectively;

$$\therefore \quad \frac{A_2}{A_1} = \frac{r_1}{r_2} = \frac{c - r_2}{r_2},$$

if c be the distance of the axes.

$$\therefore \quad \frac{V}{A_1} = \frac{c - r_2}{r_2} \rho \sin \theta = k;$$

a constant by hypothesis;

$$\text{whence} \quad r_2 = \frac{c\rho \sin \theta}{\rho \sin \theta + k}$$

is the equation to the rolling curve of the second axis, whence that of the first may be found by Arts. 260. or 269.

In practice, however, the figures thus obtained must be altered so as to correct the sudden change of direction.

Any contrivance however that produces two equal periods of variation in the angular velocity in each revolution, will serve to correct the velocity of the crank-follower sufficiently for practice. The rolling curves, as just described, are used in some silk-machinery; but their figure is not so completely formed upon principle.

If the axis of the crank be connected to the uniformly revolving axis of the driver by means of a Hooke's joint, and these axes meet at a sufficient angle, the rotation of the crank will have two maximum and two minimum velocities in each revolution, which, if carefully opposed to those produced by the crank, will nearly correct the unequal motion of the reciprocating piece.

333. Ex. 2. *To equalize the velocity by link-work.* The velocity of the reciprocating piece may be also nearly equalized by a train of link-work only. Thus let A, fig. 170,

170

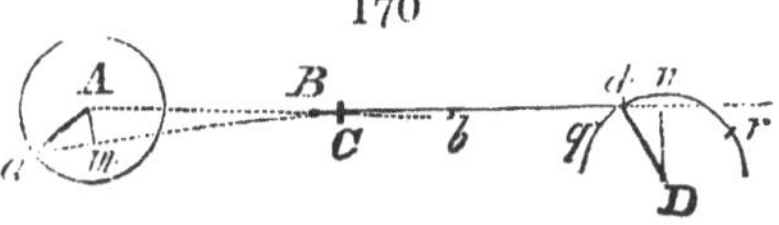

be the axis of the crank Aa, which by means of a link

aC communicates in the usual way a reciprocating motion to a point C, which travels in the line Ab between B and b. A second link Cd connects C with an arm Dd, moving on a center D, and the motion of C between B and b thus moves d between q and r; so that the rotation of the crank Aa causes the arm Dd to reciprocate between the positions Dq and Dr.

In any given position of this system draw perpendiculars Am, Dn from the centers of motion upon the links; then if A_1 A_2 be the angular velocities of Aa, Dd respectively, and V the velocity of C, we have very nearly

$$A_1 \cdot Am = V = A_2 \cdot Dn, \text{ (Art. 329)}; \quad \therefore \frac{A_2}{A_1} = \frac{Am}{Dn}.$$

If Aa and Dd both reach the position perpendicular to the link at the same time, then Am and Dn will reach their maximum values together, and will decrease and increase together, so that the ratio $\frac{Am}{Dn}$ may be made nearly constant; and thus, if Aa revolve uniformly, the reciprocating piece Dd will move in each direction with a velocity much more nearly uniform than that of the piece C.

This latter piece may either slide or may be fixed to a long arm so as to make Bb an arc of large radius; or the intermediate piece C may be even omitted, and ad connected by a single link*; but this is not so good.

334. Ex. 3. *To produce a rapidly retarded velocity.* A, B, D, fig. 171, are centers of motion, Aa an arm revolving round A, bBC an arm revolving round B, and Dd an arm revolving round D; these arms are connected by links ab and Cd, by which the motion of Aa is communicated to Dd. Let Aa move only through an arc of a circle $a1$, 2, 3,

* Hornblower in 1795 applied this latter method to the steam-engine. (Rees Cyc. Steam-engine, Pl. V. fig. 7.)

and let the three points 1, 2, 3 be at equal angular distances from each other, and so placed that the line *bA*, which is a

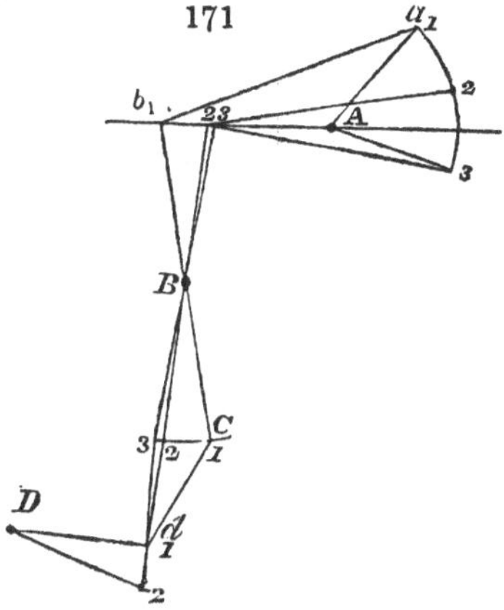

tangent to the small arc described by *b*, shall bisect the angle 2*A*3, described by *a* in its passage from 2 to 3. Now, since the motion given to the arm *Bb* will vary as the versed sine of the angular distance of *A a* from the line *bA*, the motion which *b* receives while *a* moves from 1 to 2 will be very much greater than that which it receives while *a* moves from 2 to 3. The corresponding positions of *a* and *b* are numbered with the same figures. In fact, practically, the second motion is so small that this combination may be employed when the arm *Bb* is required to remain at rest during the second motion of *Aa* from 2 to 3, as well as when the arm *Bb* is required to receive a rapidly retarded velocity from the uniform velocity of *Aa**.

But the third arm *Dd* is so placed with respect to *BC* that the tangent to the arc described by its extremity *d* shall bisect the small angle 2*B*3 described by *C* in its passage between the second and third positions; the motion therefore which *Dd* receives during the second motion of

* This combination was first employed by Watt in the mechanism for opening the valves of the steam-engine.

Aa from 2 to 3 is very much less than the small motion given to *Bb* This third arm is therefore added when a more perfect repose is required*.

335. Ex. 4. *To multiply oscillations by link-work.* If a common crank, *Aa*, fig 172, be jointed by a link *ab*

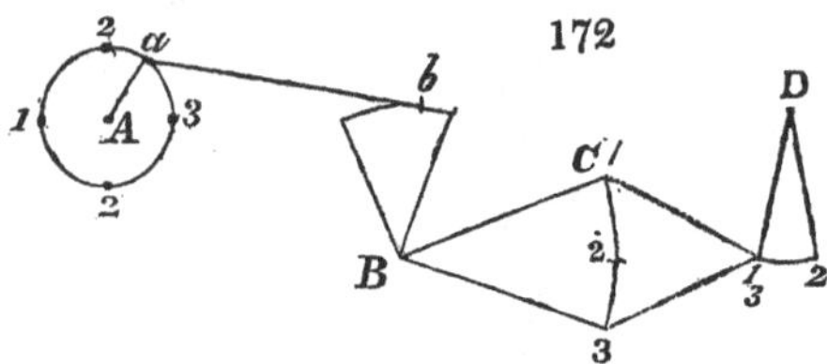

to an arm moving round a center *B*, we have seen that every revolution of the crank will produce one complete double oscillation† of the arm *Bb*, and therefore of an arm *BC* upon the same axis.

Let an arm *D*2 moving round a center *D* be joined by a link to the arm *BC* in such a relative position to it that the tangent to the arc described by the extremity of *D*2 may bisect the angle described by the arm *BC*. The figures 123 upon the circular path of the crank, upon the arc of motion of the arm *BC*, and upon that of the arm *D*2, shew the corresponding positions of these pieces. The motion of *BC* from *B*1 to *B*3 in either direction will produce one complete double oscillation of *D*2 from the position D^1_3 to *D*2 and back again, as shown in the figure; and therefore one double oscillation of *BC*, or one revolution of the crank will produce two complete double oscillations of the arm *D*2. If another arm be connected with *D*2 in the same manner as the latter is connected with *BC*, then one revolution of the crank will produce four double oscillations of the

* This is employed by Erard in the double-action harp.

† In pendulums and other vibrating bodies *one oscillation* includes the motion from one end of the path to the other, in either direction. *A double oscillation*, therefore, is the motion from one end to the other and back again, and thus contains all the phases of the periodic motion.

last arm; and thus with a train of n axes one revolution of a crank may produce 2^{n-2} complete double oscillations of an arm.

336. Ex. 5. *To produce an alternate intermitting motion by link-work.* *A*, fig. 173, is the center of motion of a common crank which by means of the link 2, 2, causes an arm *Bb* to oscillate between the positions *B*1 and *B*3. The extremity *b* of this arm is also jointed to two other links

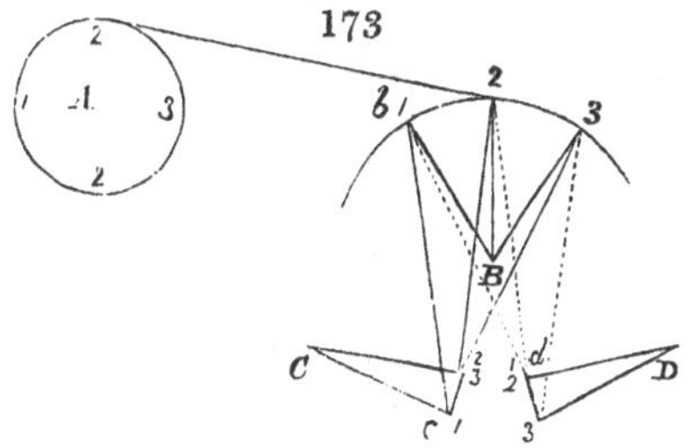

bc and *bd*. The link *bc* connects it with an arm *Cc* whose center of motion is *C*, and the tangent to the path of its extremity passes through *B*, and bisects the angle 2*B*3; therefore by Ex. 3, when *b* moves from 1 to 2, *Cc* will move from *Cc*1 to C^{2}_{3}, but when *b* moves from 2 to 3, *Cc* will remain nearly at rest in the position C^{2}_{3}. On the other hand, the link *bd*, which is shewn by a dotted line, is jointed to an arm *Dd*, the tangent of whose path passes through *B*, and bisects the angle *b*1*B*2; so that while *b* passes from 1 to 2, *Dd* remains nearly at rest in the position Dd^{1}_{2}; but when *b* passes from 2 to 3, *Dd* receives a motion from Dd^{1}_{2} to *D*3. The effect of this arrangement is, that when the crank *A* revolves, the arms *Cc* and *Dd* oscillate with intervals of rest, the one moving when the other rests, and *vice versa*: which may be traced by the corresponding figures, if we follow the motion of the crank at *A* round its circle, as thus:

crank moves from	1 to 2	*Cc* rises and	*Dd* rests
	2 to 3	— rests ...	— falls
	3 to 2	— rests ...	— rises
	2 to 1	— falls ...	— rests.

337. But for shewing the exact nature of the motion produced in this manner, graphic representations are the best (Art. 14). Thus in fig. 174, *Bb* is the vertical axis of a curve which represents the motion of the arm *Bb*; *Cc* and *Dd* the axes of curves which represent the cotemporaneous motions of the arms *Cc*, and *Dd* respectively. The circle described by the crank is divided into twelve equal angles, and the axes of abscissæ are divided into equal parts corresponding to these twelve positions, and numbered accordingly from 0 to 12. The figure represents one revolution and a half, for the better exhibition of the motion; and supposing the crank to revolve uniformly, the vertical abscissæ of the curves will be proportional to the time. The ordinates of these curves are proportional to the spaces or arcs described by the extremities of the arms respectively. Thus the ordinates of the curve *Bb* are proportional to the distance of the extremity *b* of *Bb* from the extreme position *Bb*1. These curves are easily obtained by drawing the figure 173 upon a large scale, and setting out upon it the twelve relative positions of all the arms of the system, in the same way as the three principal positions are there shewn. To return to fig. 174. It appears that the double oscillation of *Bb* from 0 to 12 is converted in *Cc* into two double oscillations, one of which extends from 2 to 10, and is large, while the other from 10 to 2 is so small that it may be considered as a state of rest. The oscillation of *Dd* is similar, but the large wave of the latter is opposed to the small wave of the former, and *vice versa*. Now if these small waves be required to be re-

174

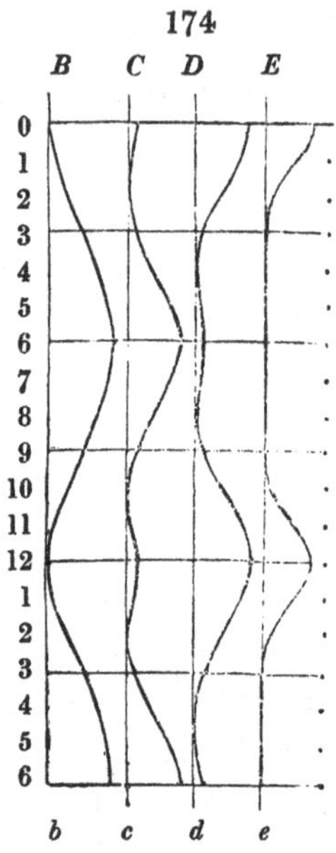

duced, a second arm (as *Dd* fig. 171) must be attached to each of the arms *Cc*, *Dd* of the present system. The curve *Ee* represents the motion of this second arm supposing it to be attached to *Dd*, and from this it appears that while the oscillation of the large wave is rendered more nearly constant in its velocity, the small wave is obliterated and reduced to a line coinciding with the axis of the abscissæ.

338. These examples may serve to shew that very complex motions may be produced by combining link-work in trains, and the mechanism thus obtained is so simple and certain in its action, that it is always desirable, if possible, to employ it. Curves should always be used as a test for the motions, because in these intricate combinations formulæ would not, even to the best mathematicians, give the same clear notion of the cotemporary action of the various pieces of the train that is conveyed in this manner.

339. When a reciprocating and revolving piece are connected by a single crank and link, the revolving piece must be the driver, unless it be heavy; for if the reciprocating piece be made the driver, it is evident that at the dead points (Art. 328) of the system it could communicate no motion to its follower. But if the revolving piece be heavy, it will by its inertia be carried across the dead points, and thus allow the reciprocating piece to continue its action in the reverse direction. This mode of operation belongs to Dynamics, and therefore will not be examined in the present Work. In fact, in Pure Mechanism the only methods by which a reciprocating driver can be made to give continuous rotation to a follower, are by *Escapements*, for which see Sliding Contact in the present Chapter; and by *clicks and ratchet-wheels*, which, as they properly belong to Link-work, I shall proceed to explain.

340. The driver is an arm whose center of motion is *A*

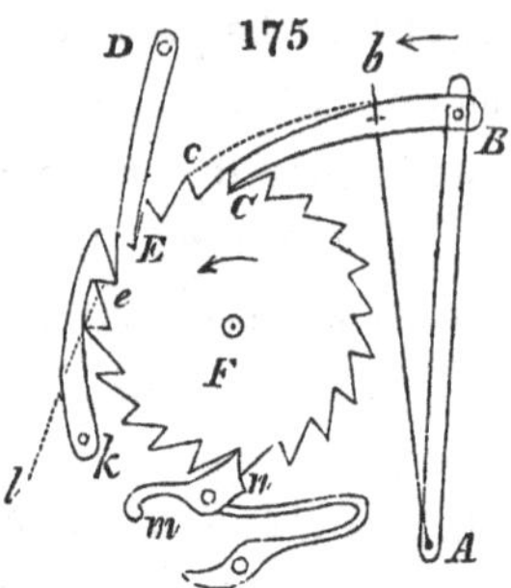

fig. 175. The follower *F* is a wheel termed a *ratchet-wheel*, having teeth formed like those of a saw.

The piece *BC* is freely jointed to the driving arm at *B*, so that it rests by its weight upon the teeth of the wheel. If the arm be moved in the direction of the arrow into the position *Abc*, the extremity *C* will abut against the radial sides of the teeth, and push the wheel as if *BC* were a link jointed to its circumference at *C*. But when the arm is moved backwards towards *AB*, the point *C* will rise over the sloping sides of the teeth, and communicate no motion to the wheel.

If a continuous reciprocation be given to the driver, the follower will advance a few teeth during every motion of the driver in the direction of the arrow, and will remain at rest during its return in the opposite direction.

To ensure the wheel against an accidental motion in the reverse direction, an arm *DE* similar to *BC* is jointed to a fixed center of motion *D*, and by abutting against the teeth in a similar way to *BC*, only allows the wheel to be moved in the one direction required. A detaining arm of this kind is termed a *detent* or latch, and the arm *BC* which communicates motion a *click*, or *ratchet*, or *paul*; but these latter names are frequently used in common for both the moving and detaining pieces *BC* and *DE*.

341. This is a very useful and practical combination*, and admits of great variety of arrangement. Thus the arm *AB* may be made to move concentrically to the ratchet-wheel. This method, when practicable, is to be preferred, for the arm, ratchet and wheel then move together as one piece during the advance of the latter.

Or the crown-wheel form may be given to the ratchet-wheel, as in fig. 176, in which case, the click *B* may be either jointed to an arm *Aa*, which moves concentrically to the wheel, or to an arm *cd*, which is attached to an axis *Cc* at right angles to that of the wheel.

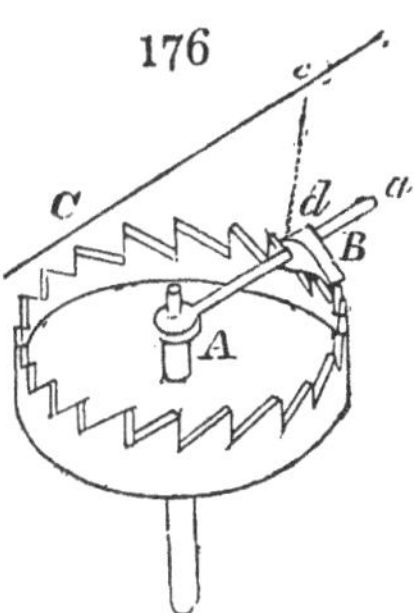

342. The reciprocating arm may also be made to drive the wheel both during its approach and recess. Thus, let *A*, fig. 177, be the center of motion of the arm, *D* that of the ratchet-wheel, and let the arm have two clicks *ab*, *ac*, jointed to its extremity *a*, and engaged with the opposite sides of the wheel.

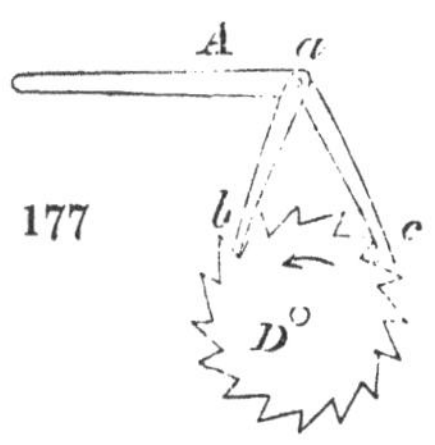

When *a* is depressed the click *b* will push the teeth, but the click *c* will slide over them. On the other hand, when *a* is raised, the click *c* will act upon the teeth, but *b* will now slip over them, so that whether *a* rise or fall the wheel is made to move in the direction of the arrow.

343. A similar contrivance is shewn in fig. 178, where *A* is the center of motion of the arm, and clicks *ab*, *dc* are jointed at equal distances on each side of *A*. When *a* rises,

* It first appears in Ramelli, fig. 136.

the click *ab* slips over the teeth, and *dc* pushes them; but when *a* falls, the click *ab* pushes the teeth and *dc* slips over them. These two latter arrangements are called the levers of Lagarousse, from the name of their inventor*.

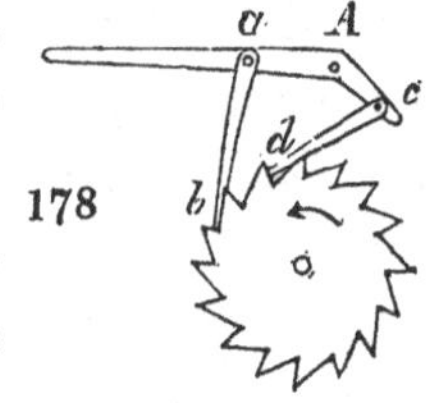

178

344. Levers either of this latter kind with two clicks, or with a single click accompanied by a detent, are also employed to move racks.

345. Instead of jointing the clicks and detents to their levers or centers of motion, they are sometimes made in the form of a slender spring. Thus if *ab* instead of hanging loose from *a*, or being pressed by a spring into contact with the teeth, be itself a slender spring fixed to the lever at *a*, it will act precisely in the same manner as it does in the figure, merely giving way from its elasticity when it is required to slip over the teeth, instead of turning upon the joint for that purpose.

346. The shape of the extremity either of the detent or click, as well as of the teeth against which they act, may be determined as follows:

179

If we examine the action of the detent and wheel, it appears that the two conditions which determine the form are these. If the wheel be urged in one direction, the action of its teeth shall have no effect in raising the detent, but shall rather tend to keep it in its place. If the wheel be urged in the opposite direction, the contrary shall happen.

Now the tooth and detent act upon each other by sliding contact. Let *A*, fig. 179, be the center of motion of

* Machines App. 1702.

the wheel, *B* of its detent, and let *pq* be the normal of contact between the tooth and the end of the detent, and let *Ap*, *Bq* be perpendiculars upon this normal from the centers of motion. Then if the wheel be urged in the direction from *p* to *q*, this normal is the line of action upon the detent, (Art. 33,) which therefore tends to turn the detent round *B* in the direction *pq*, that is, to press it more closely into contact with the teeth.

If, on the contrary, the center of the detent were at *B′*, on the other side of the normal, the action of the teeth would be to turn it in the direction *pq* round *B′*, that is, to raise it out of the teeth. To make the detent hold, therefore, its acting extremity and the teeth must be of such figures that the normal of contact shall pass between its center and that of the wheel. If the wheel be urged in the opposite direction, then it can be shewn in like manner, that to enable the wheel to lift the detent, the normal of contact in this new direction *rs* must also pass between the two centers of motion.

If however the hook form be given to the detent, as at *ke*, fig. 175, then the normals of contact in both directions must pass on the same side of the two centers of motion as *el*.

347. By attending to this principle, which applies equally to the detents and the clicks, we may make them and the teeth of different forms, as in fig. 180, where *B* is a

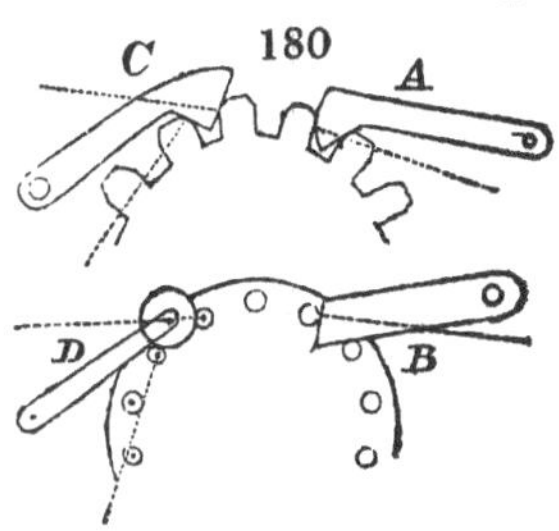

detent adapted to act with a pin-wheel, and *A* with a

common spur-wheel: the dotted lines shew the normals of contact.

A pin projecting from the face of a bar which lies behind the wheel makes an excellent detent.

When the detent requires to be released by hand from the teeth, it may be provided with a tail, as at *m*, fig. 175, the usual form of a detent when it is urged by a spring against the wheel, as in clock and watch-work.

348. But a detent is sometimes required to act in a different manner, that is, to hold the teeth of a wheel in a sort of stable equilibrium, so that they admit of being disturbed either to the right or left of the position of rest, but will still return to it if left to the action of the detent. This is effected by forming the detent as at *C* fig. 180, so that its normals of contact shall pass on the opposite sides of its center of motion, and at the same time providing the detent with a spring or a weight by which it is pressed against the teeth. This pressure will always hold the teeth in such a position that both sides of the detent shall be in contact, but at the same time the teeth of the wheel, whether urged to the right or left, will raise the detent, and pass under it, which is shewn by the direction of the normals.

If the end of the detent carry a roller, and act upon a pin-wheel as at *D*, the same effect will be produced. It is evident that the detention of the wheel in these latter arrangements is entirely effected by the pressure of the spring or weight by which the detent is kept in contact with the teeth, and not by the form of the detent, as in the first examples at *A* and *B* fig. 180, or in fig 175.

349. In fig. 175 the oscillating arm moves the wheel through an arc equal to its own motion. If the arm be

required to move through an indefinite arc, and yet to move the wheel a constant quantity in each of its oscillations, the click must be arranged as in fig. 181. *AD* is the arm, the extremity of which moves in the arc *bc*; the click is mounted on a center *D* at the end of the arm, and urged by a spring *f* against a pin or stop *e*. The ratchet-wheel *G* has a detent *F*, which must also have a spring or weight to keep it in contact. When the arm moves from *b* towards *c*, the click encounters a tooth of the wheel, and having thus carried the wheel through the space of one or more teeth, leaves it and passes onwards towards *c*. The pressure against the end of the click tends to turn it round its center *D*, but the stop *e* prevents this action; on the contrary, when the arm returns from *c* towards *b*, the click *D* again strikes against a tooth of the wheel, but the pressure now being in the opposite direction, the click gives way by turning round its center *D*, and the wheel is held fast by its detent *F*; when the click has passed the wheel the spring *f* restores it to its first position.

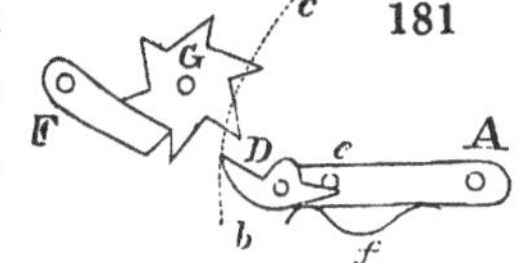

Thus whatever be the extent of the motion of the arm from *b* to *c* and back, the wheel will receive only a constant motion.

350. In all click-work the slipping of the clicks and detents over the teeth occasions a disagreeable noise or *clicking*, whence the former probably derive their name. This moreover tends to wear out the teeth.

To avoid this inconvenience *silent clicks or ratchets* are employed, which are arranged in various ways, one of the simplest of which is shewn in fig. 182. *D* is the ratchet-wheel whose teeth in this method may be made with sides nearly radial, *B* is the ratchet-arm concentric with the wheel, and carrying the ratchet *gh* jointed to it at *g*, *AC* an arm also concentric with the wheel, and moving very freely upon the

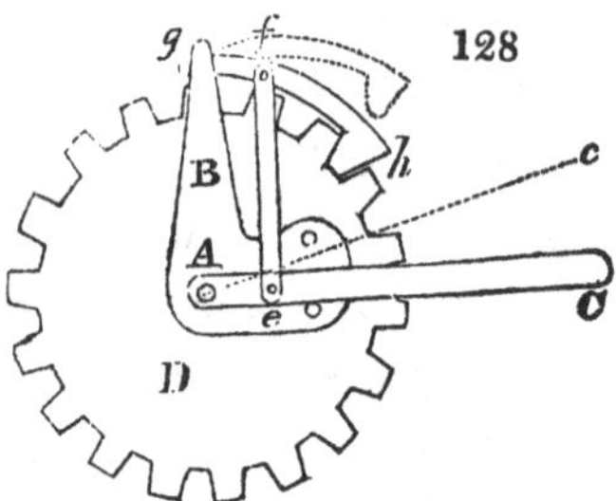

center *A*. This arm is joined by a link *ef* to the ratchet, and lies between two pins which project from the face of the ratchet-arm.

The action of the contrivance is as follows. If the arm *AC* be moved upwards towards *Ac*, it will at the beginning of its motion raise the ratchet *gh* out of the teeth of the wheel by means of the link; proceeding still farther it will then encounter the upper pin of the ratchet-arm, and will therefore carry this latter arm with it, the two arms and ratchet now moving as one piece in the direction from *f* towards *g*, but without disturbing the wheel, because the ratchet is disengaged from its teeth, as shewn by the dotted lines.

On the other hand, when the arm *AC* is moved in the opposite direction, that is from *c* towards *C*, it first passes through the small space *cC* without moving the ratchet-arm *B*, and thus by the link *ef* depresses the ratchet and engages it with the teeth, the arm *AC* then strikes the lower pin of the ratchet-arm, and the two arms, ratchet and wheel now move as if in one piece, so long as the motion of *AC* continues in this direction.

The action of this combination is perfectly silent; the arm *AC* is moved back and forwards just as the ratchet-

* Clicks of this kind are employed under different forms by Mr. Roberts in his self-acting mule, and by Mr. Donkin. Vide also White's Century of Inventions, pl. 6, fig. 18.

arm of fig. 175, but at every change of direction it begins by either engaging or disengaging the ratchet from the teeth, and thus prevents the disagreeable and mischievous noise of the common arrangement.

351. An intermittent motion may be produced from link-work, by making a slit in either end of the link. Let *B*, fig. 183, be the center of motion of a crank, which by

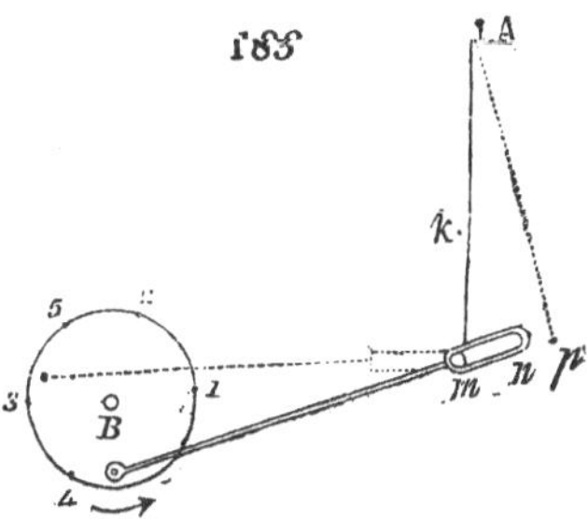

means of a link gives oscillation to a swinging arm *Am*; at the end of the link is a slit *mn*, which nearly fits a pin *m* projecting from the end of the arm *Am*. This arm may either move with friction upon the center *A* so that it will remain where it is left, or it may be urged by a spring or weight in a constant direction, as for example, towards the crank-axis, so as to press it against a stop *k* if left to itself. In the first case, if it remains where it is left, then when the link moves from left to right, the left end *m* of the slit will push the pin and arm from *m* towards *p*; but when the link changes its direction, the arm will receive no motion until the other end *n* of the slit has reached the pin; the arm will then be carried from right to left together with the link, and at the next change of direction will again rest until the end *m* of the slit has reached the pin.

The motion of the arm will thus be intermitted at each end of its course for a time which will be greater or less according to the length of the slit. Thus as 1 and 3 are the points where the changes of direction of the link occur, let 2 and 4 be the points at which the ends of the slit come into action, then the arm *Am* will remain at rest while the crank moves from 1 to 2, and from 3 to 4, and will move during the intermediate motion, thus:

crank moves from {
1 to 2 ... arm rests at *p*
2 to 3 ... — moves from *p* to *m*
3 to 4 .. — rests at *m*
4 to 1 ... — moves from *m* to *p*.

But in the second case, if the arm be pressed by a force towards the center of the crank, the slit will not come into operation unless a stop *k* be provided, then the pin *m* will be always in contact with the extremity *m* of the slit in both directions of its motion; but when the arm *Am* reaches the stop the link will proceed without it by means of the slit to the end of its course, and will take it up on its return. Take 3 5 equal to 3 4 upon the circular path of the crank, then the motion will be as follows,

crank moves from {
1 to 5 ... arm moves from *p* to *m*
5 to 4 ... — rests
4 to 1 ... — moves from *m* to *p*.

Class C. Division B. COMMUNICATION OF MOTION BY SLIDING CONTACT.

352. By means of a properly formed revolving cam-plate a reciprocating motion may be given to a follower which will vary periodically according to any required law.

Thus let *A*, fig. 184, be the center of motion of a cam-plate *nmqp*, *BD* the follower, which in this case is an arm turning on a center *B*, and furnished with a friction-roller *D* which rests upon the edge of the cam. But the follower may also be a sliding bar, as in fig. 71 (p. 153);

Let Am be the least radius of the cam, and Ap the greatest, and let the radii gradually increase along the edge mnp, and decrease along the edge pqm. Then if the cam revolve continually in the direction of the arrow, the roller D will be by the action of the edge pushed away from the center A, during the passage of mnp under it, and will return to the center during the passage of pqm; it being supposed to be kept in contact with the edge by weight or by a spring.

184

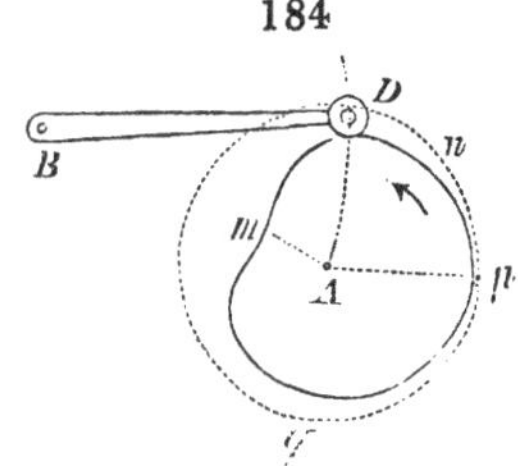

In this manner a series of periodic oscillations are communicated to the bar BD, and the velocity ratio of this bar to that of the cam can be adjusted at pleasure to any required law, by shaping the edge of the plate accordingly (Art. 33).

This may be set out by points in the method of which an example has already been given in Art. 296. If the bar be required to remain at rest during a given angular portion of the revolution of the cam, the edge will be an arc of a circle through that angle. If the follower be a straight bar, as in fig. 71, and this bar be required to perform its motion in both directions with a constant angular velocity ratio to that of the cam, then must a cam-plate be formed of two of the curves given in Art. 161, each occupying half the circumference, and set back to back, so as to produce a heart-shaped figure.

353. If the cam-plate be required to communicate more than one double oscillation in each revolution, its edge must be formed into a corresponding number of waves, as A, fig. 185; and if the follower is to be raised gently and let fall by its own weight, the waves must terminate abruptly, as in

B. If the follower is to receive a series of lifts with intervals

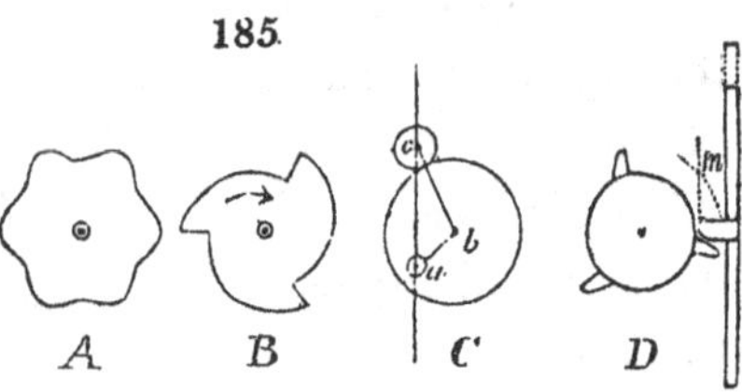

of rest, the cam becomes a set of teeth projecting from the circumference of a wheel, as in *D*. When the cam is employed to lift a vertical bar or stamper, these separate teeth are often termed wipers or tappets.

354. The axis of the follower, if it be a revolving bar, as in fig. 184, is not necessarily parallel to that of the cam; but may be set at any angle to it, if the bar revolve only through a small angle, whose tangent in the mean position is in the plane of rotation of the cam.

355. The simplest form of a cam is that of an excentric circle, as at *C*, fig. 185. Let *a* be the excentric center of motion, *b* the center of the cam, *ac* the direction of motion of the follower, which is a roller whose center is *c*. Then *bc* is plainly constant, and the motion given to the follower the same as if a link *bc* and crank *ab* were employed (Art. 328).

356. If the weight or spring be inconvenient, the cam may be made to press the follower in both directions by means of a double curve. This cannot be made in the form of a slit, as in fig. 71, p. 153, because the motion is now to take place indefinitely in the same direction; but a groove in the face of a plate may be employed, as at *A*, fig. 186.

357. If the cam revolve always in the same direction, the outside curve is only required during that portion of the

motion in which the follower approaches the cam, and it

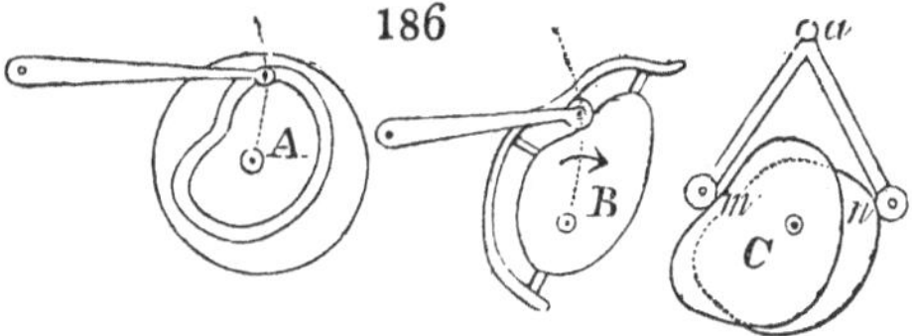

may be supplied by a bar attached to the cam by a few bridge pieces at the back, as at *B*, fig. 186.

358. Or motion may be communicated in the two directions by a double cam, as at *C*, fig. 186, in which the piece that receives the reciprocating motion has two arms, the roller of one of which rests on one cam, and that of the other upon another cam which lies behind the first on the same axis, and the figure of which corresponds to that of the first in such a way that the arc *mn* between the points of contact is constant and equal to the distance between the rollers. Thus when the edge of one cam is retiring from its roller, that of the other is always advancing, and *vice versa*.

359. In fig. 187, *Ee* is a revolving axis, *Gg* a bar capable of sliding in the direction of its own length, and having a friction roller at *g*; a flat circular plate *F* is fixed to the extremity of the axis, but not perpendicular to it; the bar *Gg* may be pressed into contact with the plate by a spring or weight. Now if the plate *F* were perpendicular to the axis, the rotation of the latter would communicate no motion to the bar, but the effect of the inclination is to communicate a reciprocating motion to the bar in the direction of its length, the quantity of which varies with the

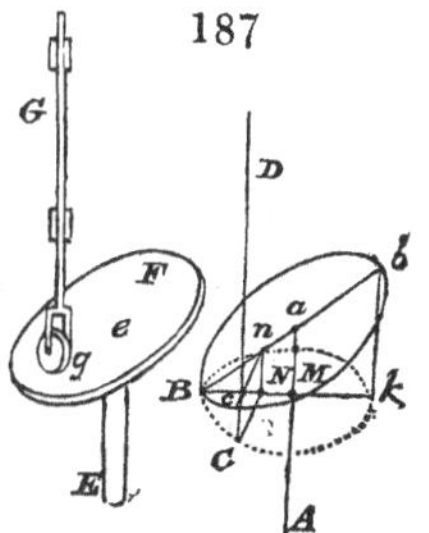

inclination of the plate to the axis; and if the plate be so attached to the axis as to admit of an adjustment of this inclination, a ready mode is obtained of adjusting the length of the excursion of the bar. This plate is termed a *swash-plate;* the law of its motion may be thus found.

Let Aa be the vertical axis of the swash-plate Bb, B its lowest point, and therefore BaA the angle of its inclination to the axis.

Let cD be the sliding bar, BCk the plane of rotation of the point B.

The motion therefore of BM from MC through the angle BMC has moved the extremity c of the bar through the space cC. Draw CN and Nn perpendicular to BM, then will Nn be equal and parallel to Cc;

$$\therefore \quad Cc = \frac{BN}{\tan BaA},$$

$$\text{also } BN = BM \,.\, \text{versin } BMC;$$

$$\therefore \quad Cc = \frac{BM \,.\, \text{versin } BMC}{\tan BaA} = aM \text{ versin } BMC;$$

so that the motion of the bar is the same as that produced by a crank with an infinite link and a radius $= aM$ (Art. 328).

360. If the path of the follower bar of a cam-plate be not parallel to the plane of rotation of the plate, then, as in Arts. 165, 166, a cone, a hyperboloid, or a cylinder, may be employed exactly in the manner there described; but as the velocity ratio of cam and bar is no longer constant, we are no longer confined to the curves there given. Instead of a groove a projecting rib acting between two rollers may be employed, either in these combinations, or in those of the Articles already referred to.

361. If the motion of the bar from one end to the other of its path be required to occupy more than a single revolution of the cam-axis, the double screw of fig. 188 may be employed*. This arrangement has a cylinder and sliding bar exactly corresponding to fig. 76, p. 157, but that on the circumference of the cylinder is traced two complete screws, one a right-hand screw beginning at *a*, and extending from *a* by *mbcdf* to *g*; the other a left-hand screw which begins as a continuation of the right-hand screw at *g*, and extends from *g* by *ohkl* to *a*, where it also joins the other screw; so that the two screws form one continuous path, winding round the cylinder from one end to the other and back again continuously. When the cylinder revolves, the piece *e* which lies in this groove and is attached to the sliding bar, will be carried back and forwards, and each oscillation will correspond to as many revolutions of the cylinder as there are convolutions in the screw.

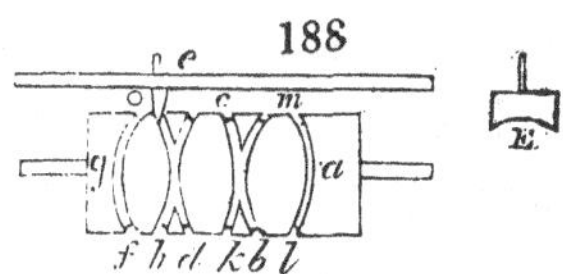

As the screw-grooves necessarily cross each other twice in each revolution, the piece *e* must be made long, so as to occupy a considerable length of the groove, as shewn sideways at *E*; thus it will be impossible for it to quit one screw for the other at the crossing places. Also, as the inclination of the screws to the bar are in opposite directions, it is necessary to attach the piece *e* to the bar by a pivot, as shewn in the figure, so as to allow it to turn through a small arc as the inclination changes. If the bar be required to move more rapidly in one direction than the other, the one screw may be of greater pitch than the other, and similarly, by varying the inclination of the screw at different points, a varying velocity ratio may be obtained.

* Lanz and Betancourt, Analytical Essay on Machines, by whom it is attributed to M. Zureda.

362. In the endless screw, fig. 142, p. 265, if the inclination of the threads be made to vary from right to left in each revolution, the wheel, when the screw revolves uniformly, will revolve with continual change of direction, advancing by long steps, and retreating by short steps alternately.

363. If a single series of changes in velocity and direction be required, and which are too numerous to be included within a single rotation of a cam-plate; then the *spiral-cam* or *solid-cam*, fig 189, may be employed. *A a* is the axis of

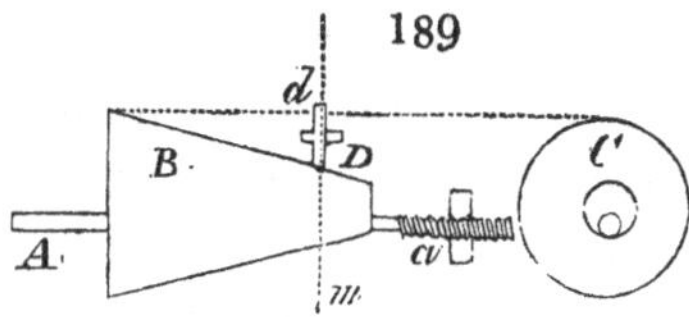

the cam, on one extremity *a* of which a common screw is cut, which works in a nut in the frame of the machine, so that as the axis revolves it also travels endlong. *B* is the solid cam. *Dd* the roller of the follower whose path is *md*, and which is kept in contact with the cam by a weight or spring as usual. As the axis revolves the follower *D* will receive from it a motion in its path, the velocity and direction of which will be governed by the figure of the cam, as in Art. 352. But by means of the screw at *a* the cam will be gradually carried endlong, so that at the completion of each revolution the same point of the cam will be no longer presented to the follower, as in fig. 184, in which the same cycle of changes is repeated in each revolution. On the contrary, the path traced by *D* upon the surface of *B* will be a spiral or screw of the same pitch as that at *a*, and by properly shaping the cam, we can thus provide a series of changes that will extend through as many revolutions of the cam as the length of the cam contains the pitch of the screw *a*.

C is an end-view of the cam. In the figure the transverse sections of the cam are represented as being every where circles of the same excentricity, but of continually increasing diameter. The effect of this would be to communicate to *Dd* a reciprocating motion in its path, of which the trip in one direction would be shorter than that in the opposite direction.

364. In the previous examples the pin or roller has been given to the follower, and the curve to the driver, but either the contrary arrangement may be made, or curves may be given to both pieces, and the pin dispensed with. In fig. 190, *A* is an arrangement by which an excentric re-

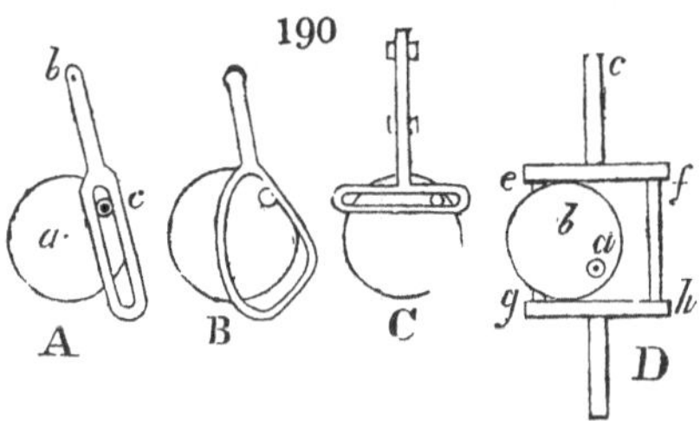

volving pin *c*, working in the slit of an arm whose center of motion is *b*, gives it a reciprocating motion. This is the same combination as that of Art. 290, but that in this case the pin *c*, by revolving always on the same side of the center *b*, produces reciprocation, while in fig. 141 the pin having the center *b* within its path produces a rotation in the follower.

The same formula will therefore apply in the two cases, making *R* less than *E* for reciprocation, and greater than *E* for rotation.

In *B*, fig. 190, it is shewn how by giving a curved outline to the sides of the slit a different velocity ratio may be obtained. In *C* the slit is attached transversely to a bar which slides in the direction of its length; and in this case it is easy to see that the law of motion is the same as in a crank with an infinite link.

Again, by increasing the diameter of the pin of *C*, we obtain an excentric, as at *D*, where *a* is the center of motion, *b* the center of the excentric. The slit now appears in the form of two parallel bars *ef*, *gh*, attached at right angles to the sliding bar; but the combination is exactly equivalent to that of *C*, *ab* being the radial distance of the pin from the centre of motion.

365. Any curve however may be substituted for this excentric circle if it possess this property, that every pair of parallel and opposite tangents are at a constant distance equal to the distance of the bars *ef*, *gh*. For thus the bars will touch the cam in all positions.

For example, fig. 190[b] has such a curve, and is adapted for the production of intermitting motion.

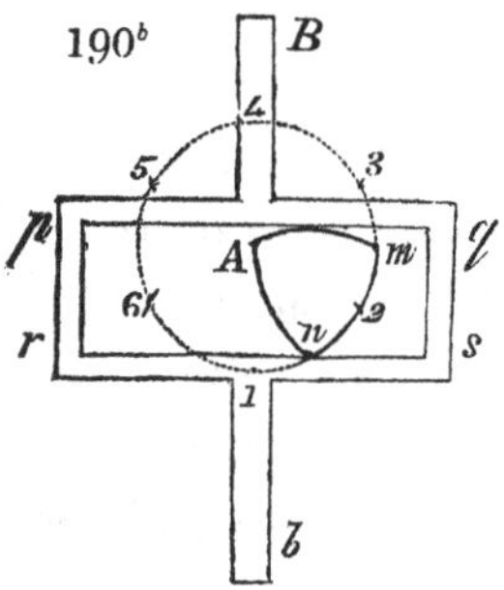

A is the center of motion of the cam, the form of which is a kind of equilateral triangle *Anm*, whose sides are arcs of circles each described from the opposite angle, the center of motion being one angle. The follower is a bar *Bb*, and the cam acts upon two straight edges *pq*, *rs*, fixed at right angles to the bar, and at a distance from each other equal to the radius of the arcs of which the cam consists; consequently the bars will be in contact with an angle and a side of the cam in every position, and the effect of its figure upon the motion is as follows. Let the circle described by its circumference be divided into six equal parts, as in the figure. Then following the point *m* round the circle in the direction of the numbers, it appears that from 1 to 2 no motion is given to the bar; from 2 to 3 the point *n* is in contact with *rs*, and the motion of the bar through that angle will therefore be the same as that by the pin and slit *C*, fig. 190, (*n* replacing the pin,) so that the bar begins to move gently and accelerates; when however *m* reaches 3 this action of *n* terminates abruptly, and *m* begins a similar action upon *pq*, by which the motion of the bar is now retarded, and gradually brought to rest when *m* reaches 4; from 4 to 5 the bar is entirely at rest, from 5 to 6 gradually accelerated, and from 6 to 1 gradually retarded. The motion of the bar is therefore nearly the same as that of the pin and slit of *C*, fig. 190, but with intervals of complete rest*.

ON ESCAPEMENTS.

366. We have now arrived at a class of combinations in which a revolving piece produces the reciprocation of its follower by acting alternately on *two different* pieces attached to it, instead of upon a *single* pin, roller or other piece, as in the combinations we have just been considering. In fig. 191, *abc* is a revolving piece or driver which has three

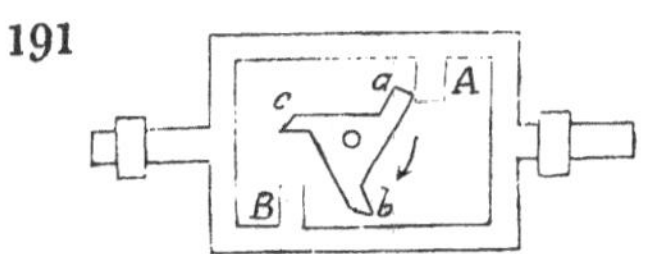

equal wipers or tappets, and the follower is a sliding bar or frame provided with two teeth or *pallets A* and *B* on opposite sides of the center of motion of the driver†. The latter

* This cam was employed by Fenton and Murray to give motion to the valves of their steam-engine.

† This contrivance is taken from De la Hire, Traité de Mécanique, Prop. 111.

revolves in the direction of the arrow, and its wiper *a* is shewn in the act of urging the follower to the right by pressing against the side of the tooth *A*. Revolving a little farther in the same direction, *a* will, by its circular motion, *escape* from *A*, and at the same instant *b* will encounter *B*, and will urge it in the opposite direction, until *b* in like manner escapes from it, when *c* will act upon *A*. In this way the rotation of *abc* will produce the reciprocation of the frame.

367. But the frame may also be made the driver; for if it be moved to the left, *A* will push *a*, and make the wheel revolve in the contrary direction to the arrow, and *c* will pass *B*. When this has happened, let the frame be moved back again; then, after moving a short space, *B* will meet *c*, and move the wheel still farther round, until *b* has passed *A*, when the return of the frame will enable *A* to push *b*. Thus the reciprocation of the frame will cause the wheel to revolve in the opposite direction to that in which itself would produce the reciprocation of the frame. But when the frame is the driver, there will always be a short motion at the beginning of each oscillation, during which no motion will be given to the wheel.

368. Fig. 192 is another method by which a revolving wheel *A* gives a reciprocating motion to a sliding bar *bk**.

192

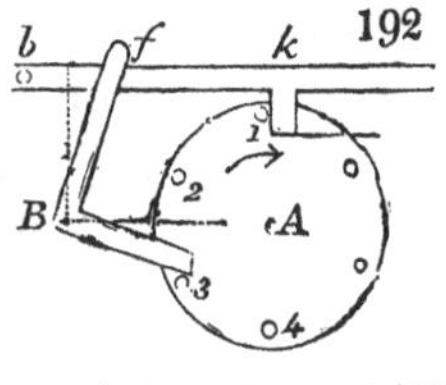

The wheel has six pins projecting from its face. The pin 1 is shewn in the act of driving the bar to the right by acting upon the tooth at *k*. The pin 3 also moves a bell-crank lever, the upper arm *f* of which travels in the contrary direction to the bar. At the moment the first pin 1 escapes from the side of *k* by its circular motion, the pin *b* will have

* From Thiout, Traité d'Horlogerie, t. i. p. 85.

reached the arm *f*, and this will, by acting upon *b*, push the bar in the reverse direction. Again, when the pin 3 escapes from the arm of the bell-crank, the pin 2 will begin to act upon *k*, exactly as the pin 1 had previously done, while the pin 4 will in like manner replace the pin 3, and raise the bell-crank. This action will go on continually, producing a short, alternate, but very abrupt and jerking, motion in the bar.

369. In these two contrivances the teeth of the wheel are made to act upon two distinct pieces attached to the reciprocating piece, and so arranged that as one tooth *escapes* from the reciprocating piece, the other shall begin its action, whence this group of combinations receives the term of *escapements*. Escapements are most largely employed in clock and watch-work (Art. 232), to communicate the action of the moving power to the pendulum or balance; but when so employed they receive many delicate arrangements, which have for their object the distribution of the power in such a manner as will the least interfere with the due action of the pendulum. Such arrangements being governed by dynamical principles, are excluded from our present plan. Escapements are however employed in Pure Mechanism to convert rotation into reciprocation, as for example, in the bell of an alarum-clock. In the two forms already given the reciprocation is communicated to a sliding bar; in those which follow it is given to an axis, which may be either perpendicular or parallel to the revolving wheel.

370. When the axes are at right angles the *crown-wheel escapement*, fig. 193, is commonly employed.

A is the revolving axis, to the extremity of which is fixed a crown-wheel with large saw-shaped teeth; *Cc* the vibrating axis or *verge*. This carries the two pieces or *pallets b* and *a*, which are set in planes making an angle with each other to allow of the escaping action. When the wheel revolves in the direction of the arrow, one of its teeth pressing against the pallet *a* will turn the verge in the same direction, until, by the circular motion of *a*, its extremity is lifted so high that the crown-wheel tooth passes under it, or, in other words, this tooth escapes from the pallet. By the same motion of the verge the pallet *b* is brought into a vertical plane, and the tooth *c* now presses it in the contrary direction, and turns the verge back again until *c* escapes from under *b*, when a new tooth begins to act upon *a*, and so on. Thus the rotation of the crown-wheel produces the vibration of the verge, the crown-wheel being the driver.

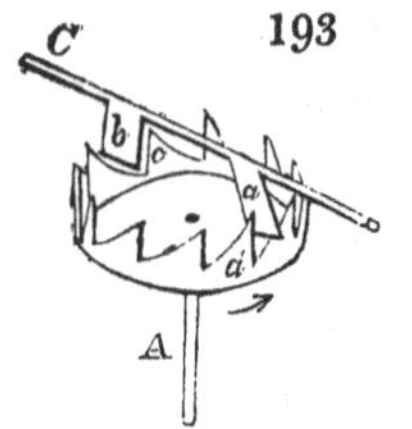

371. The *anchor-escapement*, fig. 194, is adapted to parallel axes.

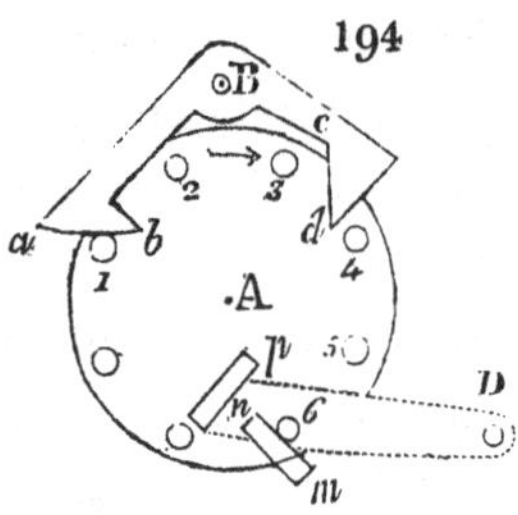

The revolving wheel has pins 1, 2, 3, ... and turns in the direction of the arrow. The vibrating axis *B* has a two-armed piece carrying the pallets at its extremities, and resembling somewhat the form of an anchor; whence the name of the combination. The pin 1 is shewn in the act of pressing against the pallet surface *ab*. Now as the normal of the point of contact passes on the same side of the two axes *A* and *B*, the pin, which acts upon the pallet by sliding contact, will tend to turn the pallet in the same direction as the wheel (Arts. 33, 346). *aB* will therefore revolve upwards, and the pin will slide towards *b*, and there escape from the pallet. At this instant the pin 3 will reach the second pallet-surface *cd*, of which the normal passes between the two axes; the action of this pin will therefore turn the axis *B* in the reverse direction; the second pallet-arm *Bd* will rise, and the pin 3 escape from the pallet at *d*, when a new pin will act upon *ab* as before; and thus the vibration be maintained.

372. This escapement has received a great variety of forms. The teeth of the wheel are more commonly long and slender-pointed spur-teeth, of which many examples may be found in the treatises of Horology.

A very simple arrangement is shewn at the lower part of fig. 194, in which *D* is the verge, *pn*, *nm* the pallets; these are fixed against the face of an arm which lies parallel to the plane of the wheel, and so far from it as to clear the tops of the pins. The pin 6 is shewn in the act of pressing the pallet *mn*, and therefore of depressing the arm; when this pin reaches *n* it escapes from *mn*, and begins to act upon *pn*, by which it raises the arm and escapes at the lower end of the second pallet, when 5 begins to touch and depress the first pallet *mn*, and so on.

373. In all these escapements the verge may be made the driver, and thus a reciprocating motion be made to produce a rotation (Art. 339). The wheel will always revolve the contrary way to that in which it turns when itself drives (Art. 367).

Thus in fig. 194, let the arm *Ba* be depressed, the pallet *ab* will then drive the pin 1 backwards, (that is, contrary to the arrow,) until pin 4 has passed under the point of *d*. If the arm *Bd* be now depressed *dc* will act upon pin 4, and continue the backward rotation until 2 has passed under the point *b*. *Ba* being again depressed will repeat the former action upon 2, and so on. But the rotation of the wheel will be necessarily intermittent, for at each change of direction in the pallet-arm the pallet must pass through a short space before it begins to touch the pin, above which it must have been previously raised to allow the same pin to pass under it. This will also be true of the crown-wheel escapement.

374. In fig. 195 the axes are parallel, but the action is more direct than in the common anchor-escapement.

195

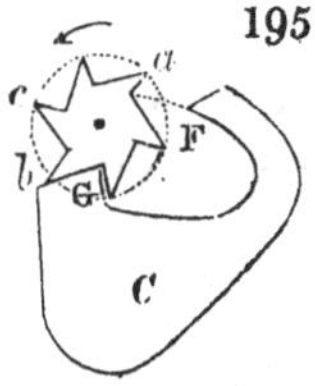

As in the former contrivance, either the wheel or the pallets may drive. I will describe it under the latter action*.

C is the axis of the pallets *G* and *F*. If the pallet-arm be moved to the left, *F* will encounter *a*, and at the same moment *G* will have passed beyond *b*, therefore *F* continuing its motion will turn the wheel in the direction of the arrow, so that when *G* returns it will enter the next space *cb*, and striking the tooth *b* will thus continue the rotation of the wheel, and so on.

Class C. Division C. COMMUNICATION OF MOTION BY WRAPPING CONNECTORS.

375. Let *A*, fig. 196, be the center of the revolving driver which is a pully, as in fig. 146, whose edge is shaped into a curve, and grooved for the reception of a wrapping band; *b* an axis fitted to the reciprocating piece; the path of this axis may be a straight line *Ab*, or it may be carried by an arm *Bb* whose center of motion is *B*. A common circular

* This contrivance, by Meynier, is to be found in the Machines Approuvees, 1724.

pully being fitted to this axis b receives the other end of the wrapping band, and this being kept tight by a weight or spring applied to b, it is evident that when the curve-pully revolves the distance of b from A will change periodically, or, in other words, it will move back and forwards in its path with a velocity the ratio of which to that of A will be governed by the figure of the pully.

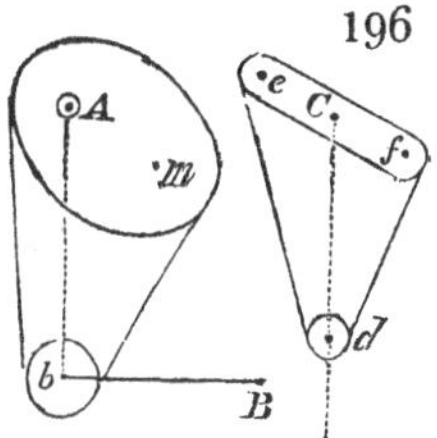

For example, if the pully be an excentric circle whose center is m, mb will be constant, and the motion the same as that produced by a crank with radius Am and link bm.

If the pully have straight parallel sides and be terminated by semicircles whose centers are e and f, and radii the same as that of the small pully d; and if C the center of motion of the large pully be midway between e and f, then Cd will be the the radius of the ellipse whose foci are e and f, and center the center of motion of the pully; so that the motion of d will be determined by the equation of this ellipse round its center.

CHAPTER X.

ON MECHANICAL NOTATION.

---

376. In complex machines of which the parts move according to different laws, and with continually varying relations of velocity and direction, it becomes exceedingly difficult to retain in the mind all the cotemporaneous movements; and a notation is in such cases of almost indispensable service. I have already shewn how in this manner the trains of machines that move with a constant velocity ratio and directional relation may be conveniently represented; and shall now proceed to explain how the more complicated connexions and motions of the last two Chapters may be reduced to notation. The only writer who has endeavoured to form a system for this purpose is Mr. Babbage. His method is not a mere hypothetical device framed to meet an imaginary difficulty; but actually arose from the necessity of the case, during the construction and arrangement of one of the most involved and complicated engines that was ever devised; and having been thus applied to practice, has been found to answer its purpose perfectly. Some parts of this notation belong to mechanical combinations of which we have not yet spoken; I shall therefore, in this place, give an account of the system only so far as it applies to the contrivances hitherto explained*.

377. Every one who has been engaged in the construction and invention of complex machinery, or who attempts

* Vide "A method of expressing by signs the action of machinery," by C. Babbage, Esq., Phil. Tr. 1826, from which paper the following account of the method is derived.

to examine the various motions of an existing machine which is presented to him for the first time, must have experienced great inconvenience from the difficulty of ascertaining from drawings the state of motion or rest of any individual part at any given instant of time; and if it becomes necessary to enquire into the state of several parts at the same moment, the labour is much increased.

In the description of machinery by means of drawings, it is generally only possible to represent an engine in one particular state of its action. If indeed it is very simple in its operation, a succession of drawings may be made of it in each state of its progress, which will represent its whole course; but this rarely happens, and is attended with the inconvenience and expense of numerous drawings.

The difficulty of retaining in the mind all the contemporaneous and successive movements of a complicated machine, and the still greater difficulty of properly timing movements which had already been provided for, led at length to the investigation of a method by which at a glance the eye might select any particular part, and find at any given time its state of motion or rest, its relation to the motion of any other part of the machine, and, if necessary, trace back the sources of its movement through all its successive stages, to the original moving power. The forms of ordinary language being far too diffuse to be employed in this case, and experience having shewn the vast power which analysis derives from the great condensation of meaning in its notation, the language of signs was resorted to for the present purpose.

378. To make the system more easily intelligible, it will be better to apply it as we go on to some machine. The example taken for this purpose in the original paper is a complete eight-day clock with going and striking parts;

but this machine is so complex as to require a large folio plate for its notation, as well as other plates to explain its construction. I shall therefore take a simpler machine, a common saw-mill. Although this machine is so easily understood as not to require the assistance of a notation, it will answer the purpose of exemplifying the method as well, and perhaps better, than a more complicated arrangement.

Fig. 197 is a diagram to explain the connexion of parts

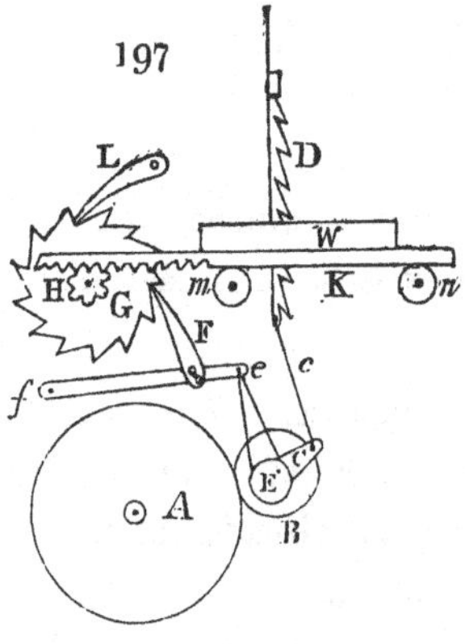

in the saw-mill, but is not drawn with any attention to the exact proportion or arrangement, which may be found in any encyclopædia or elementary book of machinery. *A* is a toothed wheel which may be supposed to be driven either by a water-wheel, or steam-engine, and its teeth are engaged with those of a second and smaller wheel *B*, on whose axis is fixed a crank *C* and an excentric *E*. The crank is connected by a link *c* with the saw-frame *D*; this is fitted between vertical guides, and therefore when the crank revolves receives a vertical oscillating motion.

The timber *W* which is submitted to the action of the saw is clamped to a carriage which moves upon rollers *m*, *n*, in a horizontal direction. While the saw is in motion as above described, the carriage and timber are made to advance in the following manner. The excentric *E* communicates an oscillating motion to a lever *ef*, whose center of motion is *f*; this lever carries a click *F*, which acts upon the teeth of a ratchet-wheel *G*, to which an intermittent rotation is thus given. Upon the axis of *G* is a pinion *H*, which geering with a rack fixed to the wood-carriage, causes the latter to advance towards the saw with the same intermittent motion. This intermission is adjusted to the motion of the saw-frame, so that when the saw rises the wood shall advance, and when the saw descends, and therefore cuts, the wood shall remain at rest. The cut is made by the inclined position of the saw, the toothed edge of which is not vertical but slightly inclined forwards, so as to bring the teeth into successive action during the descent of the frame. The detent *L* serves to hold the ratchet-wheel, and therefore the wood-carriage, firm in its position during the cut. Now all these conditions of motion are very easily represented by the notation which we shall proceed to explain, and which is exhibited in the next page.

	SAW-MILL.									
	Train to Saw.				Train to Wood-carriage.					
Names.	Cog-wheel.	Cog-wheel.	Crank.	Saw-frame.	Eccentric.	Lever and Click.	Ratchet-wheel.	Pinion.	Rack and Wood-carriage.	Detent.
Signs.	A	B	C	D	E	F	G	H	I	K
Number of Teeth.	96	22					60	20		
Linear Velocity per minute.									6^{in}	
Angular Velocity per minute.	11	50	50		50					
Comparative Velocity.										
Origin of Motion.										
Comparison of Motion.										
				up down						holds yields

379. The first thing to be done in reducing any machine to the notation, is to make an accurate enumeration of all the moving parts, and to appropriate, if possible, a name to each; for the multitude of different contrivances in various machinery precludes all idea of substituting signs for these parts. They must therefore be written down in succession, only observing to preserve such an order that those which jointly concur for accomplishing the effect of any separate part of the machine may be found situated near to each other, or in other words, that the succession of parts in each train may be observed as much as possible. Thus in the Saw-mill, against the word "Names" in the first column will be found written in order, first the parts constituting the train from the primary axis to the saw, next those which form the train to the wood-carriage.

Each of these names is attached to a faint line which runs longitudinally down the page, and which may for the sake of reference be called its *indicating line.*

To connect the notation with the drawings of the machine, the letters which in the several drawings refer to the same parts, are placed upon the indicating lines immediately under the names of the things. If there be more drawings than one of the machine, the same letters should always refer to the same parts.

A line immediately succeeding that which contains the references to the drawings, is devoted to the number of teeth on each wheel or sector, or the number of pins or studs on each revolving barrel.

Three lines immediately succeeding this are appropriated to the indication of the velocities of the several parts of the machine. The first must have on the indicating line of all those parts which have a *rectilinear* motion, numbers expressing the velocity with which those parts move, and if this velocity is variable, two numbers may be written, one expressing the greatest, the other the least velocity of the part. The second line must have numbers expressing the angular velocity of all those parts which revolve; the time of revolution of some one of them may be taken as the unit of the measure of angular velocity; or the same may be expressed in the usual method of the number of turns per minute.

If a wheel communicate an intermitting motion to another, the ratios of their angular velocities and comparative velocities will differ; for example, if the two wheels have the same angular velocity when they both move, but one of them remain at rest during half a revolution of the other. In this case their angular velocities are equal, but their comparative velocities as 1 to 2, for the latter wheel makes two revolutions while the other makes only one. A line is devoted to the numbers which thus arise, and is entitled, "Comparative Angular Velocity." No example, however, of this occurs in our Saw-mill.

380. The next compartment of the notation is appropriated to shewing the origin of motion of each part, that is, the course through which the moving power is transmitted, and the particular modes by which each part derives its movement from that immediately preceding it in the order of action. The sign chosen to indicate this transmission of motion (an arrow) is one very generally employed to denote the direction of motion in mechanical drawings; it will therefore readily suggest the *direction* in which the movement is transmitted. As there are various ways by which the motion is communicated, the arrow is modified so as to exhibit them as far as is necessary. Our author reduces them to the following:

One piece may receive its motion from another by being permanently attached to it, as a pin on a wheel, or a wheel and pinion on the same axis.	This may be indicated by an arrow with a bar at the end. +————————>
One piece may be driven by another in such a manner that when the driver moves the other also always moves; as happens when a wheel is driven by a pinion.	An arrow without any bar. ————————>
One thing may be attached to another by stiff friction.	An arrow formed of a line interruped by dots. —·—·—·—>
One piece may be driven by another, and yet not always move when the latter moves; as is the case when a stud or pin lifts a bolt once in the course of its revolution.	By an arrow the first half of which is a full line, and the second half a dotted one. ————·········>
One wheel or lever may be connected with another by a ratchet, as the great wheel of a clock is attached to the fusee.	By a dotted arrow with a ratchet tooth at its end. ...^............>

Each of the vertical indicating lines must now be connected with that representing the part from which it receives its movement, by an arrow of such a kind as the preceding Table indicates. Thus in the Saw-mill Notation, the cog-wheel *A* is connected with the cog-wheel *B* by a plain arrow; the wheel *B*, upon whose axis is fixed the crank *C* and the excentric *E*, is accordingly connected with them both by barred arrows; *F* with *G* by a ratchet-arrow; and *G* with *K* by an interrupted arrow.

381. The last and most essential circumstance to be represented is the succession of the movements which take place in the working of the machine. These movements are generally periodic, for almost all machinery after a certain number of successive operations re-commences the same

course which it had just completed, and the work which it performs usually consists of a multitude of repetitions of the same course of particular motions.

One of the great objects of the notation in question, is to furnish a method by which at any instant of time in this course or *cycle* (Art. 17) of operations of any machine we may know the state of motion or rest of every particular part; to present a picture by which we may on inspection see not only the motion at that moment of time, but the whole history of its movements, as well as that of all the cotemporaneous changes from the beginning of the cycle. In order to accomplish this, the compartment termed *Comparison of Motion* contains adjacent to each of the vertical indicating lines, which represent any part of the machine, other lines drawn in the same direction; these accompanying lines denote the state of motion or rest of the part to which they refer, according to the following rules, and may be called *the motion lines.*

1. Unbroken lines indicate motion.

2. Lines on the right side indicate that the motion is from right to left.

3. Lines on the left side indicate that the direction of the motion is from left to right.

4. If the movements are such as not to admit of this distinction, then when lines are drawn adjacent to an indicating line and on opposite sides of it, they signify motions in opposite directions. (*Thus in the Saw-mill* A *and* B *revolve opposite ways, and their motion lines are accordingly drawn on opposite sides of their indicating lines*).

5. Parallel straight lines denote uniform motion.

6. Curved lines denote a variable velocity. It is convenient as far as possible to make the ordinates of the curve proportional to the different velocities (Art. 13). (*The motion of the saw-frame* D, *and of the lever and click* F, *are examples of this rule*).

7. If the motion may be greater or less within certain limits; then if the motion begin at a fixed moment of time, and it is uncertain when it will terminate, the line denoting motion must extend from one limit to the other, and must be connected by a small cross line at its commencement with the indicating line. If the beginning of its motion is uncertain, but its end determined, then the cross line must be at its termination. If the commencement and the termination of any motion are both uncertain, the line representing motion must be connected with the indicating line in the middle by a cross line.

8. Dotted lines imply rest. It is also convenient sometimes to denote a state of rest by the absence of any line whatever. (*This rule, combined with No. 6, is employed in exhibiting the intermittent motion of the ratchet-wheel* G, *pinion* H, *and rack* I).

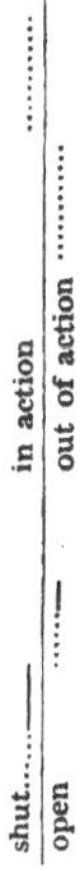

9. The thing indicated may be of such a nature that instead of motion it may be required to exhibit rather the periods of its being in action or out of action, open or closed, bolted or unbolted, and so on; as in the case of clicks, bolts, or valves; in which cases lines may be used in the above manner, but words must be added in explanation of this new employment of the signs. The line should be on the right side when the piece is out of action, unbolted, or open, and on the left side when in the reverse state. Dotted lines will be employed if the piece rests in both states; and if it be necessary to exhibit the time occupied by the motion of transition from one state to the other this can be done by a short continuous line at the beginning of each; thus if a valve fly open suddenly and close gently, it will be represented as in the margin. (*The detent* K *is an example of this rule*).

If any other modifications of movement should present themselves, it will not be difficult for any one who has rendered himself familiar with the symbols and method just explained, to contrive others adapted to the new combinations which may present themselves.

382. As an example of the way in which very minute circumstances of motion are shewn in this manner, it may be remarked, that the motion of the saw-frame, excentric, and click-lever, is necessarily continuous; but that the motion

given to the ratchet-wheel by the click does not begin at the instant the change of motion in the click takes place. The click must first move through a small space until it abuts against the tooth of the ratchet-wheel which is ready to receive it. On the other hand, it is evident that the ratchet-wheel and the click will both *cease* their motion in that direction together. When the click moves backwards the ratchet-wheel with the pinion and wood-carriage will remain at rest until the saw begins its cut, when they will be driven slightly backwards until the ratchet-tooth abuts against the end of the detent. All these accidents of motion in the ratchet-wheel and its connected pieces are exhibited by the notation, as will appear by comparing the motion lines of *G* with those of *F*. It is true, that in the actual machine these small motions are reduced exceedingly by giving a great number of teeth to the ratchet-wheel; but I have exaggerated them to shew the susceptibility of the notation, which when applied to complex machinery is of the very greatest service; more especially in assisting in the invention or improvement of machines.

383. The system of motion lines is not intended to exhibit accurately the law of motion of the pieces, as in the graphic representation of Art. 13, although it is founded upon the same principle; but merely its general phases.

When the simultaneous motions are required to be precisely exhibited, their motion curves may be, however, exactly laid down and compared, by placing them side by side; their parallel axes of abscissæ then become the indicating lines of Babbage's system. In this case, however, I am inclined to think the second method (Art. 14) is preferable, in which the ordinates are proportional not to the velocities but to the spaces; of the use of which I have already given an example in Art. 337.

384. I have found some advantages in the amalgamation of the system of Babbage with that of which an explanation has been given in Art. 233.

For in defining trains of mechanism in the present work, I have shewn that they consist of principal pieces moving each according to a given path, and connected one with the other in succession by means of drivers and followers, which are attached to these moving pieces. Now the drivers and followers carried by any one of these pieces must all move according to the same law, since they move as one piece; and a single indicating line with its velocity numbers and motion curves is quite sufficient for every such piece: whereas, as we have seen, in the notation just exhibited every part of the machine has such an indicating line and figure attached to it, and consequently all the parts that are united together merely repeat the same indication as *B*, *C* and *E*; or *G* and *H*, in page 336. In the next page I have shewn the Saw-mill under the form of Notation which I have been in the habit of employing, and which it will be seen at once differs only from that of page 336 by being united with the old clockmakers' form already explained; by which means the *genealogy*, so to speak, of the motion is perhaps more clearly perceived, and the number of indicating lines reduced.

385. To represent a machine in this form, rule as many parallel lines as there are principal moving pieces in the train, writing the name or nature of each in the first column. Upon each line write all the followers and the driver which are carried by the piece to which it belongs; taking care to place every follower vertically under its own driver, if possible. Every follower may be connected with its driver by an arrow formed according to the rules in Art. 380, or by a simple line. The arrow is only

SAW-MILL.

Names of Pieces.	Velocity per minute.	Origin of Motion.	Comparison of Motion.	
Main Shaft.	11	Spur-wheel A. (96)..		
Crank Shaft.	50	.....Pinion B. (22)——Crank C. (30)——Eccentric E. (4)		
Saw-frame.	....	.........................Saw-frame D................................		up down
Lever on Stud.	....	............Click F.——————Lever F.......		up down
Spindle.	....	Ratchet-wheel G. (60)——Pinion H. (20)....................... }		
Wood-carriage.	6	....Rack I.......................... }		
Detent on Stud.	....	..........Detent K...		holds yields

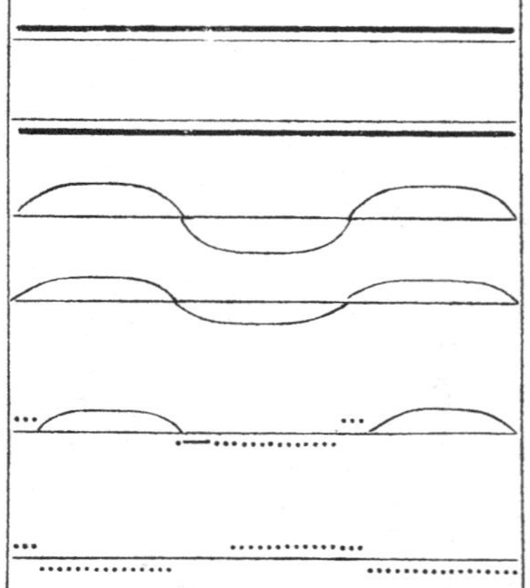

necessary if the nature of the machine renders it necessary to place some of the followers above their drivers. The connecting lines might also receive additions, by which the nature of the connexion, as by sliding, wrapping, link-work, &c. might be shewn; but the names of the parts are generally sufficient for this purpose; and there is a great mischief in unnecessarily multiplying symbols. Numbers attached to toothed wheels are their numbers of teeth, to pullies their diameters in inches, to cranks and excentrics their throw in inches, unless otherwise stated. In the column of Velocity the numbers attached to revolving pieces shew their angular velocity in turns per minute, and to sliding pieces their linear velocity in inches per minute, unless otherwise stated in words. In the column of Comparison of Motion, the rules in Art. 381 are followed, but that when two or more pieces move together in a system, one indicating line is made to serve for them all by connecting those to which it applies by a bracket. Thus the variation of motion in the ratchet-wheel spindle and the wood-carriage being the same, one line is used for them both. Columns may be added for the pitch of the wheels, or any other particulars that may be required.

It rarely however happens that the whole notation is necessary. For some machines the table of the origin of motion is required, for others that of the comparison of the motion; and of the system of the latter, and of its utility when properly applied, it is impossible to speak too highly.

PART THE SECOND.

ON AGGREGATE COMBINATIONS.

CHAPTER I.

GENERAL PRINCIPLES OF AGGREGATE MOTION.

386. THE motion of a point with respect either to its path or velocity may be considered as the resultant of two or more component motions. If it happen that the latter taken separately are more simple and more easily communicated than the resultant motion, it is evident that this may be advantageously obtained by communicating simultaneously to the given point the component motions. For an example of an aggregate path, let it be required to make a point describe an epicycloid. Every epicycloidal path may be resolved into two circular paths, one of which represents the base of the epicycloid, and the other the describing circle. And if the point be attached to a disk or arm which revolves uniformly round its own center, while at the same time that center revolves uniformly round the center of the base in a plane parallel to that of the first revolution, the point will describe an epicycloid, the nature and proportions of which will depend upon the proportion of the radii of the two circular component paths, and upon the relative time and directions of their revolutions. In this example a very complex path is referred to two paths of the simplest nature, and the question is one case of a general problem that may be thus enunciated:—*To cause a point to move*

in a required path by communicating to it simultaneously two or more motions in space.

387. As an example of motion complex in velocity, but simple with respect to its path, let a body be required to travel in a right line by a reciprocating motion, but always making its forward trip through a space greater than its backward trip, and thereby gradually advancing from one end of the path to the other. This motion may be resolved into a reciprocating motion of equal advance and retreat, combined with a simple slow forward motion.

If therefore the body be mounted on a carriage or frame which advances slowly in the required direction, and if at the same time a common reciprocating motion be given to the body with respect to the carriage; the question will be answered by referring the given compound motion to two of a simple and practicable nature.

388. Again, let a body be required to move so very slowly in a right line, that in the ordinary methods a long train of wheel-work or of other combinations would be required to reduce sufficiently the velocity of the original driver. But if this small velocity be considered as the difference of two velocities in opposite directions, then it may be obtained by mounting as before the body on a carriage which proceeds with any convenient velocity in one direction, while the body moves with respect to the carriage with a nearly equal velocity in the opposite direction.

These examples belong to a second problem which may be thus stated:—*To produce the motion of a piece in a given path by communicating to it simultaneously two or more motions in that path, either in the same or in opposite directions.*

389. In these examples, however, it appears that the frame or part of the machine which determines the path of

one of the component motions is itself in motion. In the first example, the center of motion of the revolving piece which carries the describing point itself travels in a circle; and in the second example, the slide upon which the point that receives the aggregate motion is made to move, is itself also in motion. And this, from the nature of Aggregate Combinations, will always be the case; and as these bodies which travel in moving paths have to derive their motion from a driver whose path is in the usual manner stationary, it appears that to carry this aggregate principle into effect, requires that we should have the means of communicating motion from a driver to a follower, when the respective position of their paths is variable.

I shall therefore begin by giving examples of the methods by which this may be effected.

To connect a Driver and Follower, the relative position of whose paths is variable.

390. If the center of motion of a toothed wheel itself travel in a circle parallel to the plane of rotation, then a second wheel concentric with the circular path, and in geer with the travelling wheel will remain in geer with it in all positions of its center; or if the center of the wheel travel in a right line parallel to the plane of rotation, a rack parallel to its path will always remain in geer with the wheel, and communicate a motion to it; as will also an endless screw, as in fig. 198, where *Aa* is a long endless screw, *B* the travelling wheel whose center of motion moves in the path *Bb*, parallel to the axis of the screw. The screw

198

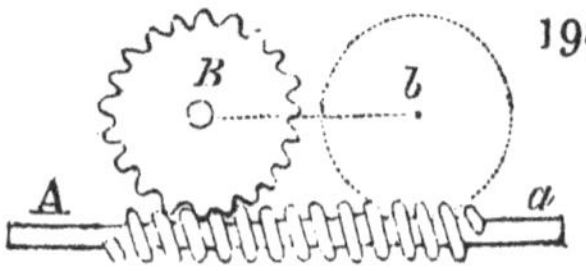

B b A a

will therefore act upon the wheel whatever be the position of its center upon this line, and will also allow the center to be moved into any position upon the surface of the cylinder that would be generated by the motion of *Bb* round *Aa*, the plane of the wheel of course always passing through the axis *Aa*.

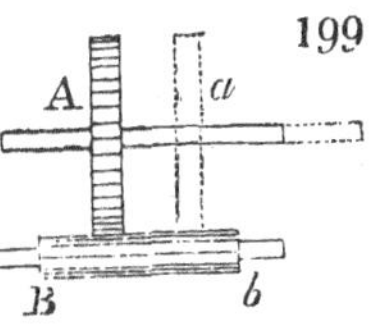

Again, if the wheel be required to travel in the direction of its own axis, as from *A* to *a*, fig. 199, a *long pinion Bb* will retain its action upon it in all its positions.

But if the center of the wheel is to travel in any other curve in a plane perpendicular to its axis, let *A*, fig. 200, be a fixed center of motion, *B* the travelling center of motion, and let *AC*, *CB* be a frame jointed at *C*; then if *B* be moved into any position within the circle whose radius is *AC* + *CB*, the frame will follow it, the angle *ACB* becoming greater or less according to the radial distance of *B* from *A*. Let a center of motion be placed at *C*, then will three wheels whose centers are *A*, *C*, and *B*, remain in geer in all these positions of the frame, and thus allow *B* to travel in any curve without losing its connexion with the central wheel at *A*.

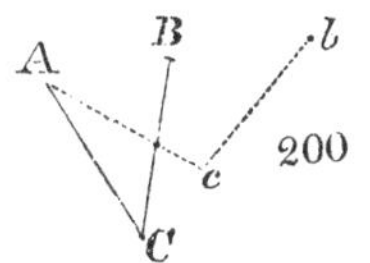

391. The same principles also apply to centers of motion connected by sliding contact or wrapping connectors; for generally, it is evident, that if two parallel axes be connected by any of the contrivances for communicating unlimited rotation, one axis may travel round the other in the circle whose radius is the perpendicular distance of the axes, without disturbing their connexion. Other expedients are also employed, which belong rather to constructive mechanism. Thus, instead of the long pinion *Bb*, fig. 199, a

short pinion may be used which can slide along its axis, but not turn with respect to it, and this pinion may be made to follow the wheel A in its motions. But, in fact, as we advance in our subject, the combinations necessarily increase in number and complexity under each head to such a degree, that it becomes impossible to include them all in the limited space of such a treatise as this. I shall therefore merely give examples of one or two of the least obvious arrangements; others will occur during the calculations of Aggregate Motion in the succeeding Chapters.

392. A travelling pully which derives its rotation from another pully with a fixed axis of motion, may have its own axis carried about to any relative position with the first, provided the wrapping band have a suspended stretching pully to keep it tight in all these changes of distance, and that the pully travel only in its own plane, and consequently its axis always remains parallel to that of the other pully. For if it move out of that plane the wrapping band will be thrown off the pully (Art. 184). Fig 201 is one arrangement by which the pully may be also allowed to move in the direction of its axis*.

201

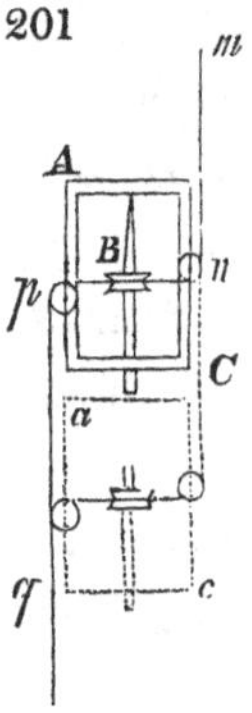

B is the pully whose axis is mounted in a frame AC, to whose sides are fixed the axes of guide-pullies n, p; the wrapping band is passed over these pullies as at $mnpq$, making one turn round the pully B in its passage; the ends mn, pq of the band are carried parallel to the axis of B, and passed over proper guide-pullies to the driving wheel. The frame AC may evidently be moved into any other position ac, in the plane mq, without disturbing either the tension of the band or its connexion with B.

* Lanz and Betancourt (Anal. Essay, D. 20.) have a somewhat similar arrangement.

393. Two arms AP, CD (fig. 114, p. 190), being connected by a link PD, the center of motion C of one of them may be shifted into various positions with respect to A, without breaking the connexion of the system; but the velocity ratio of the arms will necessarily be different in every new position. If the arms have only a small angular motion, as in the Article referred to, the center C may receive a small travelling motion in a direction perpendicular to PD, without materially altering the velocity ratio.

Fig. 202 is an expedient by which this communication can be maintained between shifting centers without affecting the velocity ratio.

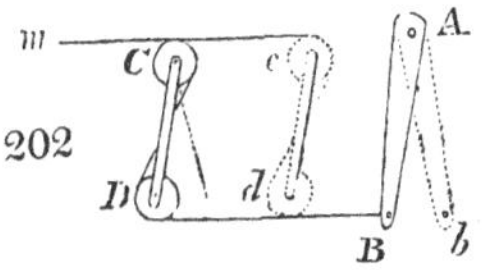

AB is the arm whose center of motion A is fixed, CD the arm whose center of motion travels in the line Cc; guide-pullies C, D are mounted, one concentric to C, and the other at the extremity D of the arm. A line is fixed at m, passed over the pullies C and D, and attached to B. If B be moved to b it will, by means of this line, communicate the same motion to CD round C as if it were a link jointed in the usual way at D and B. But the peculiar arrangement of the line allows the center of the arm to be removed to any other point in Cc, as to c, without interrupting the connexion of B with its extremity. The arm is supposed to be returned by a spring or weight.

CHAPTER II.

ON COMBINATIONS FOR PRODUCING AGGREGATE VELOCITY.

---

394. I SHALL in this Chapter proceed to shew the principal methods of obtaining the complex motion of a body in a given path by the simultaneous communication to it of two or more simple motions in that path; arranging the solutions under the same divisions as in the first part of this Work, but taking them in a somewhat different order, for the sake of convenience.

BY LINK-WORK.

395. Let a bar ABC, fig. 203, be bisected in B, and let a small motion Aa perpendicular to the bar be communi-

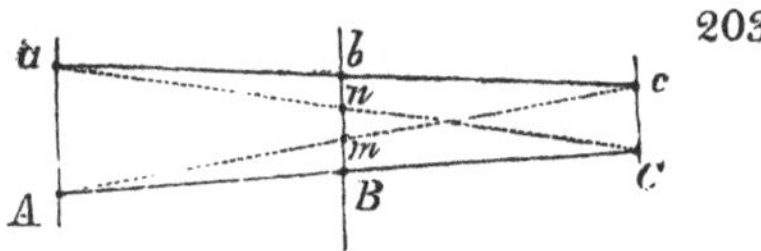

cated to the extremity A, C remaining at rest; then will the central point B move through a space $Bn = \frac{Aa}{2}$. On the other hand, had A remained at rest, and a small transverse motion Cc been given to the other extremity C, the central point B would have moved through a space $Bm = \frac{Cc}{2}$. If these two motions are communicated either simultaneously or successively to the two extremities, the center B will be carried through a space $Bb = \frac{Aa + Cc}{2}$. Or, if starting from the position Ac, the two motions had been communicated in the opposite directions, so as to carry the bar into the position aC, then the center of the bar would receive a motion $mn = \frac{Aa - Cc}{2}$. The length of the bar being always supposed so great, compared with the motions, that its inclination in the different positions may be neglected, and therefore the lines Cc, Bb, Aa, be all considered perpendicular to AC. Hence *two small independent motions being communicated to the extremities of a bar; its center receives half their sum or difference, according as the motions are in the same or in opposite directions.*

If the motions be communicated to A and B, then C will receive the whole motion of A in the opposite direction, and twice the motion of B in the same direction. The bar AC has been divided in half at B for simplicity only, for it is evident that by dividing it in any other ratio we can communicate the component motions in any desired proportions. But in general it is the law of motion which is to be communicated, and the quantity is of less consequence, especially if reduced for both motions in the same proportion.

396. Let FG, fig. 204, be a bar whose center is E, and to whose extremities are fixed pins F and G, upon which the centers of other bars AB, CD, turn. Then if four independent motions be communicated to the points A, B, C, D, the motions of A and B will be concentrated upon F, and those of C and D upon G, and the motions of F and G being concentrated in like manner upon E, this point will receive the four motions. By jointing other levers to the extremities of these, and so

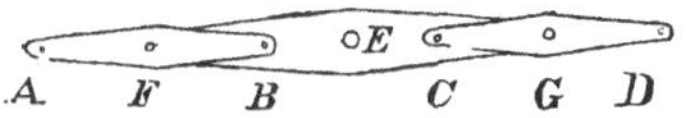

204 — E; A F B C G D

on, any number of independent motions may be concentrated upon the point *E**.

BY WRAPPING CONNECTORS.

397. If a bar *Bb*, fig. 205, be capable of sliding in the direction of its length and carry a pully *A* round which is passed a cord *DE*, then it can be shewn in the same manner, that the bar will receive half the sum of independent motions communicated to the extremities *D*, *E*, the bar being supposed to be urged in the direction *bB*, by a weight or spring. This is a more compendious contrivance than the former, as the motions may be of considerable extent. If the component motions be communicated to one extremity of the string *D* and to the bar, then will the other extremity *E* receive the entire motion of *D* in the reverse direction, and also twice the motion of *Bb* in the same direction†.

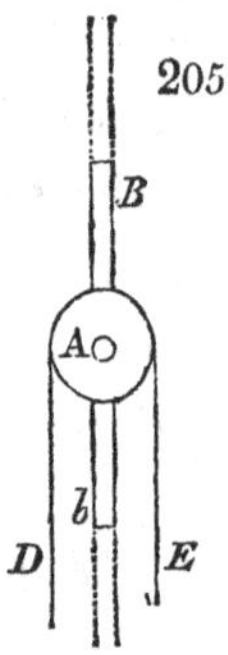

398. If a second similar combination be placed at the side of this, with its bar parallel to that of the first, and if a cord whose ends are tied to the upper extremities of each bar be passed over a third intermediate pully, the center of this latter pully will receive the aggregate motion of the cords of the two systems, as shewn for the lever in Art. 396.

399. As an example of the employment of these combinations, let *C*, fig. 206, be an axis of motion upon which is fixed a small barrel round which the cord *e* is rolled, and also a disk with an excentric pin *c*, which by means of a

* Another example of aggregate velocity by Link-work is the well-known reticulated frame termed *Lazy tongs*, which resembles a row of X's, thus XXXXX. It is too weak from its numerous joints to be of much practical service.

† The first application of this principle appears to be the *Rouet de Lyon*, for winding silk. Vide Enc. Meth. Manufactures, t. II. p. 44.

link *cb* communicates a reciprocating motion to an arm *Aa*, whose center of motion is *A*. The extremity of this arm carries a revolving pully *D*, and the cord which is coiled round the band is laid over this pully and fixed to a heavy piece *E*, which moves in the vertical path *Ef*. Now when *C* revolves, the center *a* of the pully *D* moves up and down through a small arc which is nearly a right line parallel to *fE*, and by virtue of this motion the string *f* and the body *E* will receive a reciprocating motion of double its extent. But the string *e* will be also slowly coiled upon the barrel by which it, as well as *E*, will receive a slow travelling motion in a constant direction upwards. By what has preceded, therefore, the body *E* receiving these motions simultaneously, will, as in the example of Art. 387, move vertically with a reciprocating motion, of which the downward trip is shorter than the upward one.

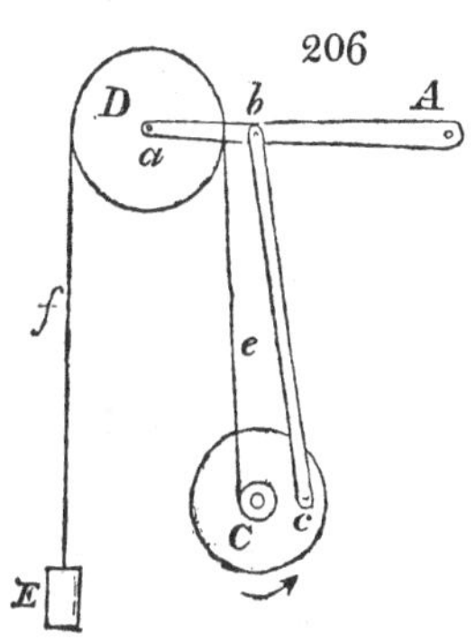

400. Let *Aa*, fig. 207, be an axis to which are fixed two cylinders *B* and *C*, nearly of the same diameter, and let a cord be coiled round *B*, passed over a pully *D*, and then brought back and coiled in the opposite direction round *C*. When *Aa* revolves, one end of the cord will be coiled and the other uncoiled, and if *R* be the radius of *B*, and *r* of *C*, *A* the angular velocity of the axis, the velocities of the two extremities of the cord will be *AR* and *Ar*; and by Art. 397, the center of the pully *D* will travel with a velocity equal to half the difference of these velocities, since they are in opposite

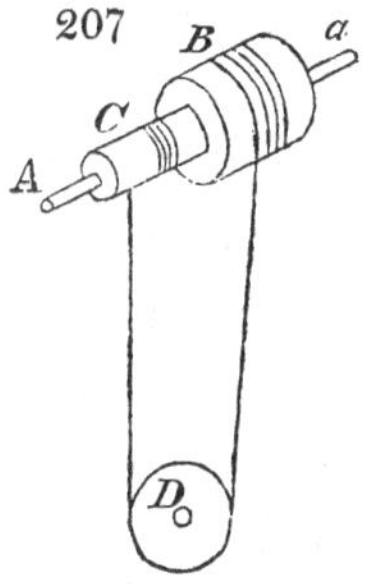

directions, or to $\frac{A(R-r)}{2}$. This velocity is the same as would be obtained if the center of the pully D were suspended from the axis Aa by a cord wrapped round a single barrel whose radius $= \frac{R-r}{2}$.

401. This combination belongs to a class which has received the name of *differential motions*, their object being to communicate a very slow motion to a body, or rather to produce by a single combination such a velocity ratio between two bodies that under the usual arrangement a considerable train of combinations would be required *practically* to reduce the velocity, for, *theoretically*, a simple combination will always answer the same purpose. Thus in the above machine, although theoretically a barrel with a radius $\frac{R-r}{2}$ would do as well as the double barrel, yet its diameter in practice would be so small as to make it useless from weakness. Whereas each barrel of the differential combination may be made as large and as strong as we please.

If a considerable extent of motion however be required, this contrivance becomes very troublesome, on account of the great quantity of rope which must be wound upon the barrels. For by one turn of the differential barrel the space through which the pully is raised $= \pi(R-r)$, but the quantity of rope employed is the sum of that which is coiled upon one barrel, and of that which is uncoiled from the other $= 2\pi(R+r)$. Now in the equivalent simple barrel the quantity of rope coiled is exactly equal to the space through which the body is moved, and therefore in this case $= \pi(R-r)$, so that for a given extent of motion

$$\frac{\text{rope for differential barrel}}{\text{rope for common barrel}} = 2\,\frac{R+r}{R-r},$$

when $R - r$ is by hypothesis very small. This inconvenience has been sufficient to banish the contrivance from practice, for although it is represented in all mechanical books under the name of the Chinese windlass, it is never actually employed.

BY SLIDING CONTACT.

402. *Aa*, fig. 208, is an axis upon which are formed two screws *B* and *D*, whose pitches are *C* and *c* respectively. *B* passes through a nut *b* fixed to the frame, and *D* through a nut *d*, which is capable of sliding parallel to the axis of the screw*.

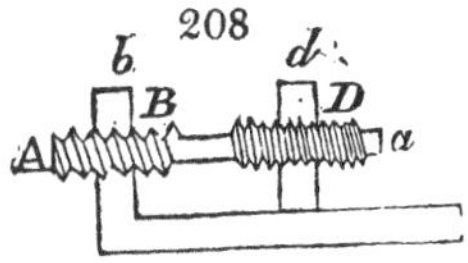

Now when a screw is turned round it travels with respect to its nut through a space equal to one pitch for each revolution, consequently one turn of *Aa* will cause it to move with respect to *b* through the space *C*. But the same motion will cause the nut *d* to move with respect to its screw through a space *c*. The nut *d*, therefore, receives two simultaneous motions, for by the advance of the screw *Aa* through the fixed nut *b*, the nut *d* is carried forwards through the space *C*, but by the revolving action of the screw *Aa* it will be at the same time carried backwards through the space *c*; its motion during one rotation of the screw *Aa* is therefore equal to the difference of the two pitches $= C - c$. If *C* be greater than *c* this will be positive, and the nut will advance slowly when the screw *Aa* advances; but if *c* be greater than *C*, the nut will move slowly in the opposite direction to the endlong motion of the screw. If $C = c$ then $C - c = 0$, and the nut *d* receives no motion, which is indeed obvious. All this supposes that the threads of the two screws are both right-handed or both left-handed. If

* This contrivance is claimed by White, (Century of Inventions, p. 81.) and also for M. Prony, by Lanz and Betancourt, (Essay, D. 3).

one be right-handed and the other left-handed, each revolution of the screw Aa will cause the nut d to advance through a space $= C + c$.

403. In fig. 209*, Ff is a screw which passes through a nut g, this nut is mounted in a frame so as to be capable of revolving but not of travelling endlong in the direction of the axis of the screw. So that if the nut were turned round, and the screw itself prevented from revolving, this screw would receive an endlong motion in the usual manner, at the rate of one pitch for each revolution of the nut. A toothed wheel E is fixed to the nut, and engaged with a pinion C, which is fixed to the axis Aa, parallel to the screw. To the screw is also fixed a toothed wheel D, which engages with a long pinion B upon the same axis Aa which carries the pinion C. When Aa revolves therefore, it communicates rotation both to the screw and to the nut. If B and C, D and E were respectively equal, it is plain that the nut and screw would revolve as one piece, and consequently no relative motion take place between them; but as these wheels are purposely made to differ, the nut and screw revolve with different velocities, and thus a motion arises between the nut and its screw, which causes the latter to travel in the direction of its length, with a velocity ratio that may be thus calculated.

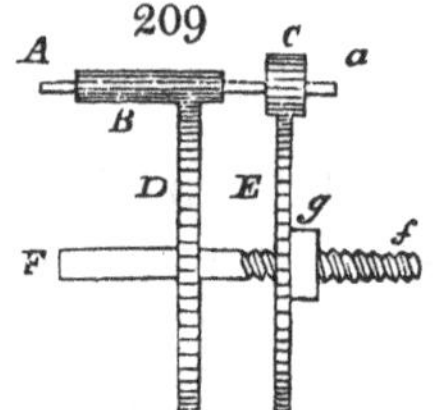

Let the letters $B\,C\,D\,E$ applied to the wheels, represent their respective numbers of teeth, and let P be the pitch of the screw. Also, let the synchronal rotations of the axis Aa, the nut and the screw, be $L\,L_n$, and L_s respectively,

$$\therefore\ L_n = \frac{LC}{E} \text{ and } L_s = \frac{LB}{D}$$

* This combination occurs in White Century of Inventions.

But the endlong motion of the screw depends upon the *relative* rotations of the screw and nut, and not upon their absolute rotations. Now it is obvious, that if the screw make L_s rotations, and the nut L_n rotations in the same direction, that the screw and nut will have made $L_s - L_n$ rotations with respect to each other, and therefore that the screw will have advanced endlong through a space

$$= (L_s - L_n) . P = L . P\left(\frac{B}{D} - \frac{C}{E}\right),$$

which may be made very small with respect to L.

This combination is applied to machinery for boring, for the motion of a boring instrument consists of a quick rotation combined with a slow advance in the direction of its axis, which is precisely the motion given to the screw Ff. Nothing more is therefore required than to fix the boring tool to one end of this screw.

The long pinion B (Art. 390) is employed for the obvious purpose of maintaining the action of B upon D during the endlong motion of the screw, and this endlong motion is in fact the difference of two motions that are simultaneously given to the screw. For Aa revolving, if B and D were removed the rotation of the nut would cause the screw to travel endlong with one velocity, and if C and E were removed instead of B and D, then the rotation of the screw in its fixed nut would cause it to travel endlong with another velocity; but these two causes operating simultaneously, the screw travels with the difference of these velocities.

404. A slow relative rotative motion of two concentric pieces may be produced, as in fig. 210, in which Dd is a fixed stud, B an endless screw-wheel revolving upon the stud, and C a second endless screw-wheel revolving upon the tube which carries the preceding wheel B. A is an

endless screw so placed as to act at once upon both wheels*. Now if these wheels had the same number of teeth they would move as one piece, but if one of them has one or two teeth more or less than the other, this will not disturb the pitch of the teeth sufficiently to interfere with the action of the endless screw. And as the revolutions of this screw will pass the same number of teeth in each wheel across the plane of centers, it follows that when one wheel has thus made a complete revolution, the other will have made more or less than a complete revolution by exactly the number of deficient or excessive teeth.

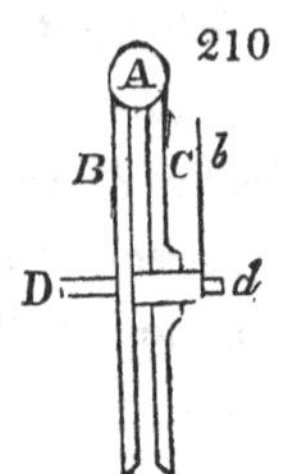

Let B have N teeth, and C, $\overline{N+m}$ teeth, then since the same number of teeth in each wheel will simultaneously pass the plane of centers, $N \times \overline{N+m}$ teeth of each will pass during N rotations of C, and $\overline{N+m}$ of B, which are therefore their synchronal rotations, and their relative rotations in the same time are $\overline{N+m} - N = m$.

This contrivance is used in counting the revolutions of machinery, for by attaching an index to the tube which carries B, and graduating the face of C into a proper dial-plate, b revolves so slowly with respect to C, that it may be made to record a great number of rotations of A before it returns again to the beginning of the course. Thus if B have 100 teeth and C 101, the hand will make one rotation round the dial during the passage of 100×101 teeth of either wheel across the plane of centers, that is, during 10100 rotations of the screw. Also the same hand b may read off sub-divisions upon a small dial attached to the extremity of the fixed axis d.

405. This contrivance does not strictly belong to the problem we are at present considering, but it has a kind of

* From Wollaston's Odometer, for registering the number of turns made by a carriage-wheel.

natural affinity with it that induced me to give it a place here. Similarly, a thick pinion upon an axis parallel to Dd, may be employed to drive the two wheels in lieu of an endless screw, but the relative motion will not be so slow*. But by employing two pinions of different numbers of teeth to drive the two wheels a very slow relative motion may be obtained; thus, if in fig. 209, the screw and nut be suppressed, and the wheel E be the dial-plate, and the wheel D carry the index, as in fig. 210, then we have found

$$\frac{L_s - L_n}{L} = \frac{C}{E} - \frac{B}{D},$$ which may be made very small.

BY EPICYCLIC TRAINS.

406. A train of mechanism the axes of which are carried by an arm or frame which revolves round a center, as in figs. 211, 212, 213, is termed in this work an *Epicyclic train.*

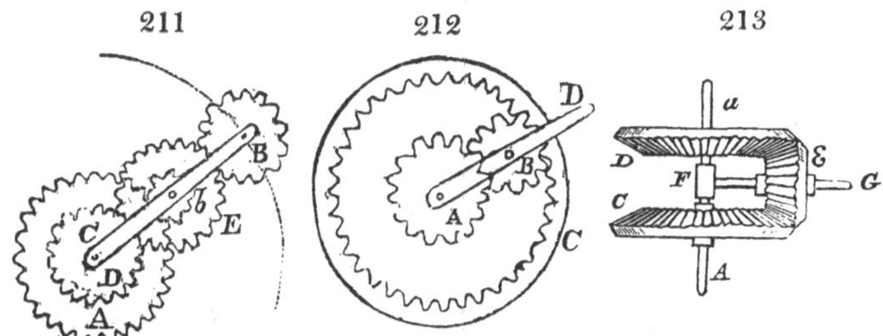

The two wheels which are at each end of such a train, or at least one of them, will be always concentric to the revolving frame.

Thus in fig. 211, CB is the frame or *train-bearing arm*, a wheel A concentric to this frame geers with a pinion b, upon whose axis is fixed a wheel E that geers with a wheel B. And thus we have an epicyclic train A (Art. 233)

$$\begin{array}{l} A \\ b \text{———} E \\ \quad\quad\; B, \end{array}$$

* This combination occurs in a clepsydra, by Marcolini, described in the notes to the ninth book of Vitruvius, by Dan. Barbaro, 1556. Vide also Art. 256.

of which if the first wheel *A* be fixed, and a motion be given to the arm, the train will then revolve round the fixed wheel, and the relative motion of the arm to the fixed wheel will communicate rotation through the train to the last wheel *B*; or the first wheel as well as the arm may be made to revolve with different velocities, in which case the last wheel *B* will revolve with a motion that will be presently calculated.

If the wheel *E*, instead of geering with *B*, be engaged with a wheel *D*, which, like the wheel *A*, is concentric to the arm, then we have an epicyclic train *A*

b——*E*

D,

of which both the extremities are concentric to the arm. In such a train we may either communicate motion to the arm and one extreme wheel in order to produce an aggregate rotation in the other extreme wheel, or motion may be given to the two extreme wheels *A* and *B* of the train, with the view of communicating the aggregate motion to the arm.

Fig. 212 is a simple form of the epicyclic train, in which the arm *AD* carries a pinion *B*, which geers at once with a spur-wheel *A* and an annular wheel *C*, both concentric with the train-bearing arm.

Fig. 213 is another simple form in which *FG* is the arm, *Aa* the common axis; *D*, *C*, two bevil-wheels moving freely upon it, and *E* a pinion carried by the arm, and geering at once with the two bevil-wheels. These two arrangements contain the least number of wheels to which an epicyclic train can be reduced, if its two extreme wheels are to be concentric to the arm; and, as in fig. 211, motion may either be given to the two wheels in order to produce aggregate motion in the arm, or else to the arm and one wheel, in order to produce aggregate motion in the other. Or very commonly, one of the concentric wheels is fixed, and motion being then given to the arm, will be communicated to the other wheel, or *vice versa*, according to a law which we shall proceed to investigate. In these examples, toothed wheels only are employed, but the subsequent formulæ will apply as well to epicyclic trains in which any of the combinations of Class A are used.

407. *To find the velocity ratios of Epicyclic trains.*

Let AB, fig. 214, be the train-bearing arm revolving round A, and carrying a train of which the first wheel A is concentric to the arm, and the last wheel B may either be concentric with A or not. These two wheels are connected by a train of any number of axes carried by the arm or frame AB Now the revolutions of the wheels of the train may be estimated in two ways; First, with respect to the *fixed* frame of the machine, that is, by measuring the angular distance of a given point on the wheel from the fixed line Af; or, if the wheel be excentric as B, from a line Bk parallel to Af. Secondly, they may be measured with respect to the arm which carries them. The first may be termed the absolute revolutions, and the second the relative revolutions, or motions relative to the train-bearing arm.

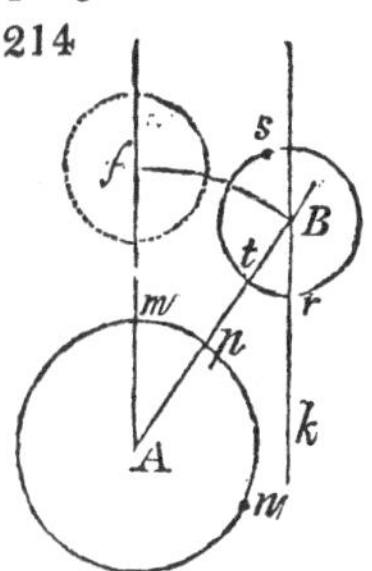

Let the arm with its train move from the position Af to AB, and during the same time let a point m in the wheel A move to n from any external cause, and the point r in the wheel B move to s by virtue of its connexion with the wheel A, all being supposed for simplicity to revolve in the same direction as the arm. Then mAn, rBs are the *absolute* motions of the wheels A and B, and pAn, tBs their *relative* motions to the arm,

$$\text{but } mAn = mAp + pAn, \text{ and } rBs = rBt + tBS$$
$$= mAp + tBS;$$

where mAp is the motion of the arm.

If, on the other hand, the wheels had moved in the opposite direction to the arm, then

$$mAn = pAn - mAp, \text{ and } rBs = tBs - mAp,$$

and these are true whatever be the magnitude of the angles described, and are therefore true for entire revolutions, for the angular velocity ratios in these trains are constant. Hence it appears that the absolute revolutions of the wheels of epicyclic trains are equal to the sum of their relative revolutions to the arm, and of the revolutions of the arm itself, when they take place in the same direction, and equal to the difference of these revolutions when in the opposite direction.

408. Let a, m, n, be the synchronal absolute revolutions of the train-bearing arm, of the first wheel of the train, and of the last wheel respectively; and let ϵ be the epicyclic train, that is, let it represent the quotient of the relative revolutions of the last wheel divided by those of the first; ϵ is therefore the quantity which is represented by $\frac{L_m}{L_1}$, or by $\frac{D}{F}$ in Chapter VII, the motions of the wheel-work being estimated with respect to the train-bearing arm alone. Also, the first and last wheel of the epicyclic train are included in the expression ϵ, although one or both of them may be concentric to the arm.

Then the relative revolutions of the first wheel with respect to the arm $= m - a$, and of the last wheel $= n - a$, and as the motions of the train, considered with respect to the arm, will be the same as those of an ordinary train, we have $n - a = \epsilon \,.\, \overline{m - a}$.

$$\epsilon = \frac{n-a}{m-a};$$

$$\text{whence } a = \frac{m\epsilon - n}{\epsilon - 1}, \quad n = a + \overline{m - a} \,.\, \epsilon,$$

$$\text{and } m = a + \frac{n - a}{\epsilon}.$$

If the first wheel of the train be fixed, which is a common case, its absolute revolutions = 0; $\therefore$ $m = 0$, and we have

$$a = \frac{n}{1 - \epsilon}, \text{ and } n = \overline{1 - \epsilon} \,.\, a.$$

If the last wheel of the train be fixed, then $n = 0$, and

$$\text{we have } a = \frac{m\epsilon}{\epsilon - 1}, \text{ and } m = \left(1 - \frac{1}{\epsilon}\right) a.$$

But when these wheels are not fixed,

$$a = \frac{m\epsilon - n}{\epsilon - 1} = \frac{m\epsilon}{\epsilon - 1} + \frac{n}{1 - \epsilon},$$

that is, the revolutions of the arm are equal to the sum of the separate revolutions which it would have received from the train, supposing its extreme wheels to have been fixed in turn.

In the formulæ of this Article the rotations of the first and last wheel and of the arm are all supposed to be in the same direction; if either of them revolve in the opposite, the sign of m, n, or a must be changed accordingly. With respect to the sign of ϵ, see Art. 412.

409. But in trains of this kind it often happens that if neither the first nor last wheel of the epicyclic train be fixed, then either motion is communicated from some original driver to the two extreme wheels of the epicyclic train with a view to produce an aggregate motion of the arm, or else the original driver communicates motion to one of these extreme wheels and to the arm, for the purpose of producing the aggregate motion of the other extreme wheel.

Fig. 215 is an example of the first case. mn is an axis to which is fixed the train-bearing arm kl, which carries the two wheels d and e united together and revolving upon the arm itself. The wheels b and c are united and revolve together upon the axis mn, but are not attached to it. Likewise the wheels f and g are fixed together, and revolve freely round the axis mn. The wheels c, d, e, and f constitute an epicyclic train, of which c is the first, and f the last wheel. An axis A is employed as a driver, and carries two wheels a and h, the first of which geers with the wheel b, and thus communicates motion to the first wheel c of the epicyclic train, and the wheel h drives the wheel g, which thus gives motion to the last wheel f of the epicyclic train. When the axis A is turned round it thus communicates motion to the two ends of the epicyclic train, through which the train-bearing arm kl receives an aggregate rotation, which we shall presently calculate.

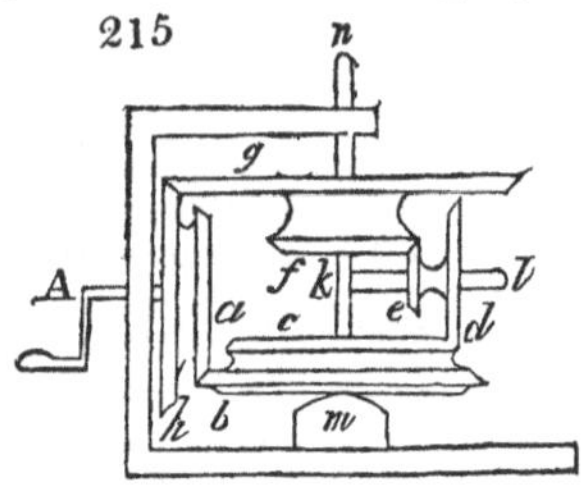

As an example of the second case, we must suppose the wheels g and f to be disunited, g being now *fixed* to the axis mn, and f only running loose upon it. The driving axis A will thus communicate, as before, rotation to the first wheel of the epicyclic train c by means of the wheels a and b, and will also by h cause the wheel g, the axis mn, and the train-bearing arm kl to revolve, by which the compound rotation will be given to the loose wheel f. In this second combination however, the last wheel f of the train is not necessarily concentric to the train-bearing arm, which it must be in the first case.

410. To obtain a formula adapted to this first case. Let the driving axis be connected with the first wheel of the train by a train μ, and with the last wheel by a train ν; and

let the synchronal rotations of this driver with these wheels be p;

$$\therefore \ m = \mu \, . \, p, \text{ and } n = \nu \, . \, p;$$

$$\therefore \ \frac{a}{p} = \frac{\mu\epsilon - \nu}{\epsilon - 1} = \frac{\mu}{1 - \frac{1}{\epsilon}} + \frac{\nu}{1 - \epsilon}$$

The first part of which is due to the action of the train μ, and the second to that of the train ν.

For suppose the train μ removed, then would the firs wheel of the epicyclic train remain fixed, and $m = \mu p = 0$;

$$\therefore \ \frac{a}{p} = \frac{\nu}{1 - \epsilon},$$

and in like manner, if the train ν were removed,

$$\frac{a}{p} = \frac{\mu}{1 - \frac{1}{\epsilon}}.$$

The arm moves, therefore, with the sum or difference of the separate actions of the two trains from the original driving axis.

411. In the second case, let the driving axis be connected with the first wheel of the epicyclic train by a train μ, and with the arm by a train α,

$$\text{then } m = \mu p, \text{ and } a = \alpha p;$$

$$\therefore \ n = \alpha p \, \overline{1 - \epsilon} + \mu p \epsilon,$$

$$\frac{n}{p} = \alpha \, . \, \overline{1 - \epsilon} + \mu \epsilon.$$

The revolutions, therefore, of the last wheel of the epicyclic train are the aggregate of those due to the train α, which produces the motion of the arm, and of those due to the train μ, which produces the motion of the first wheel of the epicyclic train.

412. The only difficulty in the application of these formulæ lies in the signs which must be given to the symbols of the trains. But these it must be remembered, are each of them the representatives of a fraction, whose numerator and denominator are respectively equal to the synchronal rotations of the last follower and first driver of the train.

One direction of rotation being assumed positive, the opposite one will be negative, and therefore if the extreme wheels revolve in the same direction, whether that be back or forwards, the symbol of the train will be positive; and if they revolve in the opposite direction it will be negative. The rotations of the train μ, ν are *absolute;* and those of ϵ *relative to the arm.* To find the sign of ϵ, we must suppose the arm to be for the moment fixed, and then analyse the train in the usual manner to find whether the motions of its extreme wheels are in the same or in opposite directions, and the directions of rotation must be estimated accordingly. In a similar way, the signs of μ and ν are easily determined by considering them separately, and observing whether their extreme wheels move in the same or in opposite directions. If in the same, then μ and ν have the same signs; and if in opposite, then different signs. In the formulæ the symbols are all supposed positive, and therefore in every particular case positive trains retain the signs which are already given to them in these formulæ, but negative trains take the opposite signs. And although the term epicyclic train strictly implies that all the axes of the train are carried excentrically round the centre of the arm, yet I must repeat that the first and last wheel must be included in it, although one or both may happen to be concentric with the arm.

413. Let, for example, these principles and formulæ be applied to the simple epicyclic trains in figs. 211, 212, 213,

and suppose the letters to represent the numbers of teeth. The epicyclic train formed by the wheels A, B, C, in fig. 212, is of such a nature that the extreme wheels A and C revolve in opposite directions, therefore ϵ is negative, and so also in the train C, E, D, in fig. 213, but in the train $\begin{matrix} A & \\ b\text{——}E & \\ B & \end{matrix}$ or $\begin{matrix} A & \\ b\text{——}E & \\ D & \end{matrix}$ of fig. 211, the extreme wheels revolve the same way, and therefore ϵ is positive. Also in fig. 211,

$$\epsilon = + \frac{AE}{bB}, \text{ in fig. } 212 \ \epsilon = - \frac{A}{C},$$

$$\text{and in fig. } 213 \ \epsilon = - \frac{C}{D} = -1.$$

Let the first wheels of these trains be fixed, then when the arm revolves we have

$$\text{for } 211. \quad n = \left(1 - \frac{AE}{bB}\right) a,$$

$$212. \quad n = \left(1 + \frac{A}{C}\right) a,$$

$$213. \quad n = 2\,a,$$

where n and a are the synchronal rotations of the last wheel of the train and of the arm respectively.

In fig. 213, therefore, it appears that when one wheel C is fixed, the other revolves twice as fast as the arm in the same direction.

In fig. 215, in its first case $\epsilon = \frac{ce}{df}$, and if the arm were fixed, c and f would revolve opposite ways, therefore ϵ is negative; $\mu = \frac{a}{b}$ and $\nu = \frac{h}{g}$, also g and b revolve opposite ways, and therefore μ and ν must have different signs, and thus the formula becomes

$$\frac{a}{p} = \frac{\mu\epsilon - \nu}{1 + \epsilon} = \frac{\dfrac{ace}{bdf} - \dfrac{h}{g}}{1 + \dfrac{ce}{df}} = \frac{aceg - hbdf}{bg\,(df + ce)}.$$

But under the second case, ϵ is negative, as before;

$$\mu = \frac{a}{b} \quad a = \frac{h}{g},$$

and these have different signs;

$$\therefore \ \frac{n}{p} = a(1+\epsilon) - \mu\epsilon = \frac{h}{g}\left(1 + \frac{ce}{df}\right) + \frac{ace}{bdf}.$$

414. Epicyclic trains are employed for several different purposes, each of which will be exemplified in turn.

(1.) For the representation of planetary motion, and for all machinery in which epicyclic motion is a part of the effect to be produced, as in the geometric pen and epicycloidal chuck, where real epicycloids are to be traced, or in the machinery for laying ropes. Some of these effects more properly belong to the next chapter.

In all these cases a frame containing mechanism is carried, by the action of machinery, round other fixed frames, and the motion can only be communicated to the machinery in this travelling frame upon the principle of epicyclic trains.

(2.) When a velocity ratio is required to be accurately established between two axes whose centers are fixed in position, and this ratio is composed of unmanageable terms when applied to the formation of a simple train, the epicyclic principle will generally effect the decomposition required, as we shall presently see.

(3.) For producing a small motion by what is termed the Differential principle, of which examples by other aggregate combinations have been already given.

(4.) To concentrate the effect of two or more different and independent trains upon one wheel or revolving piece, when one or both of them are variable in their action.

This was first applied to what are termed Equation clocks, in which the minute-hand points to true time, and its motion therefore consists of the equable motion of an ordinary minute-hand, plus or minus the equation, or difference between true and mean time.

The same principle has been applied with the greatest success to the bobbin and fly-frame.

415. The train which is carried on the arm, and the arm itself, receive various forms; the train should be as light as possible, and consist of few wheels, especially when it revolves in a vertical plane; because being excentric its weight interferes with the equable rotation of the arm or wheel which carries it, unless it be balanced very carefully. When the excentric train is necessarily heavy, this difficulty is in some degree got over by making the train-bearing axis vertical, as in planetary machinery and in rope-laying machinery.

EXAMPLES OF THE FIRST USE OF EPICYCLIC TRAINS.

416. Ex. 1. *Ferguson's Mechanical Paradox.*

This was contrived to shew the properties of a simple epicyclic train, of which the first wheel is fixed to the frame of the machine.

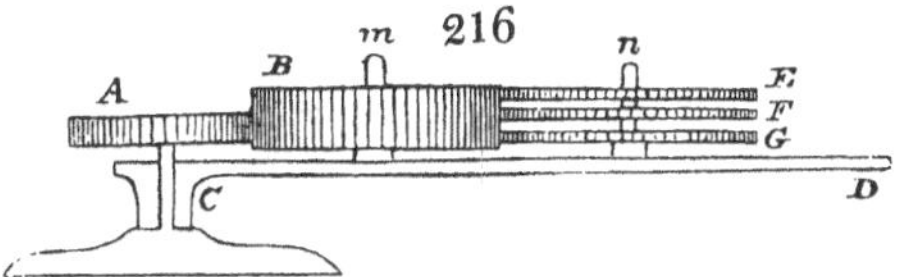

It consists of a wheel A, fig. 216, of 20 teeth, fixed to the top of a stud which is planted in a stand that serves to support the apparatus. An arm CD can be made to revolve round this stud, and has two pins m and n fixed into it, upon one of which is a thick idle wheel B of any number of teeth, which wheel geers with A and also with three loose wheels E, F, and G, which lie one on the other about the pin n.

When the arm CD is turned round, motion is given to these three wheels which form respectively with the intermediate wheel B and the wheel A three epicyclic trains.

Now in this machine the extreme wheels of each epicyclic train revolve in the same direction, and therefore ϵ is positive, and the formula applicable to this case is $\frac{n}{a} = 1 - \epsilon$, where n and a are the absolute synchronal rotations of the last wheel and of the arm. But the object of this machine is only to shew the directions of rotation.

If $\epsilon = 1$ $\frac{n}{a} = 0$, and the last wheel of the train will have no absolute rotation. If ϵ be less than unity $\frac{n}{a}$ will be positive, and the last wheel will revolve absolutely in the same direction as the arm. But if ϵ be greater than unity $\frac{n}{a}$ will be negative, and the absolute rotations of the arm and wheel will be in opposite directions.

Let E, F, G have respectively 21, 20, and 19 teeth, then

$$\text{in the upper train } \epsilon = \frac{A}{E} = \frac{20}{21}$$

is less than unity, and E will revolve the same way as the arm.

$$\text{in the middle train } \epsilon = \frac{A}{F} = \frac{20}{20}$$

equals unity, $\frac{n}{a} = 0$ and F will have no absolute revolution.

and in the lower train $\epsilon = \frac{A}{G} = \frac{20}{19}$

is greater than unity, and G will revolve backwards.

It follows from this that when the arm is turned round, E will revolve one way, G the other, and F will stand still, or rather continually point in the same direction. Which being an apparent paradox, gave rise to the name of the apparatus, which is well adapted to shew the more obvious properties of trains of this kind. But Ferguson was not the first who studied the motions of epicyclic trains; Graham's orrery in 1715, appears to be the original of this curious class of machinery, but for which no general formula appears to have been hitherto given*.

417. Ex. 2. The contrivance termed *sun and planet-wheels* was invented by Watt as a substitute for the common crank in converting the reciprocating motion of the beam of the steam engine into the circular motion of the fly-wheel. The rod DB, fig. 217, has a toothed wheel B fixed to it, and the fly-wheel has a toothed wheel A also attached

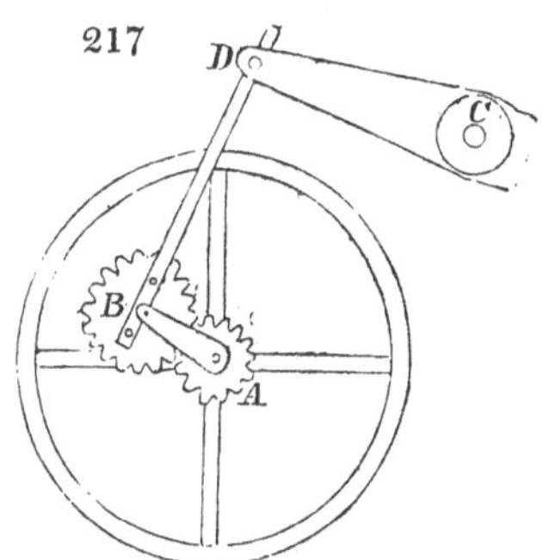

to it, a link BA serves to keep these wheels in geer. Now when the beam is in action the link or arm BA will be made to revolve round the center A, just as a common crank

* In Rees' Cyclopædia, Art. Planetary Numbers, are a few arithmetical rules for the calculation of planetary trains, given without demonstration.

would, but as the wheel B is attached to the rod DB so as to prevent it from revolving absolutely on its own center B, every part of its circumference is in turn presented to the wheel A, which thus receives a rotatory motion, the proportionate value of which is easily ascertained by the formula already given.

The wheels AB with the arm constitute an epicyclic train $\frac{A}{B} = \epsilon$, in which ϵ is negative, since the wheels revolve in opposite directions considered with respect to the arm, and in which the last wheel B has no absolute rotation, being pinned to the arm D; the formula

$$m = a + \frac{n - a}{\epsilon}$$

becomes

$$\left.\begin{array}{l}\text{making } n = 0 \\ \text{and } \epsilon = -\frac{A}{B}\end{array}\right\} \frac{m}{a} = 1 + \frac{B}{A}.$$

In Watt's Engine the wheels were equal and therefore $m = 2a$, and the fly-wheel revolved twice as fast as the crank-arm.

418. Ex. 3. *Planetary Mechanism.* mn is a fixed central axis, upon which a train-bearing arm fg turns, carrying two separate epicyclic trains ϵ_1 and ϵ_2.

218

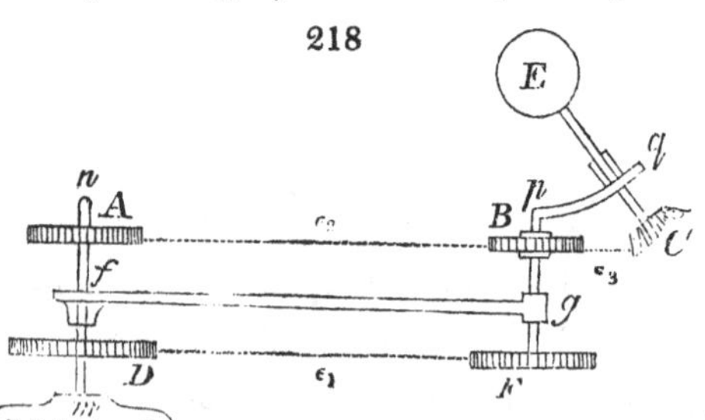

One of these, ϵ_1, has a first wheel D, and a last wheel F, connected by any train of wheel-work, and the axis of this last wheel passes through the end of the arm fg, and carries a second arm pq.

The other train ϵ_2 has a first wheel A connected to its last wheel B, by any train of wheel-work, but this last wheel is united to the first wheel of an epicyclic train ϵ_3 borne by the arm pq, of which train the last wheel is C. The question is, to find the absolute rotations of this last axis. The arrangement is one that occurs in some shape or other in most orreries, for the purpose of representing the Diurnal rotation of the Earth's axis, in which case fg is the annual bar, and E a ball representing the Earth.

Let the absolute synchronal rotations of the bar $fg = a$, those of $D = m_1$; of F (and therefore of the arm pq) $= n_1$; of $A = m_2$; of B (and therefore of the first wheel of the train ϵ_3) $= n_2$; and of C (and therefore of the Earth) $= n_3$.

$$\text{Then } n_1 = a \,.\, \overline{1 - \epsilon_1} + m_1\epsilon_1$$

$$n_2 = a \,.\, \overline{1 - \epsilon_2} + m_2\epsilon_2$$

$$n_3 = n_1 \,.\, \overline{1 - \epsilon_3} + n_2\epsilon_3.$$

In an orrery by Mr. Pearson for equated motions, described in Rees' Cyclopædia, the arm or annual bar fg, is carried round by hand, and the wheels A and D are fixed to the central axis. In this case m_1 and m_2 vanish, and we obtain the formula

$$\frac{n_3}{a} = 1 - \epsilon_1 + \epsilon_1\epsilon_3 - \epsilon_2\epsilon_3.$$

But the arm pq which carries the Earth's axis must preserve its parallelism, and therefore having no absolute rotation $n_1=0$. The train ϵ_1 will therefore $= +1$;

$$(1.) \qquad \text{and } \frac{n_3}{a} = \epsilon_3 - \epsilon_2\epsilon_3 = \epsilon_3 \,.\, \overline{1 - \epsilon_2},$$

which must be positive, since the Earth performs its daily

and annual revolutions in the same direction. The train ϵ_3 in Mr. Pearson's orrery consists of three wheels of 40 each *en suite*; $\therefore \epsilon_3 = +1$,

$$\text{also his train } \epsilon_2 = \frac{269 \times 26 \times 94}{10 \times 10 \times 18},$$

in which the extreme wheels revolve in opposite directions, therefore ϵ_2 is negative;

$$\therefore \frac{n_3}{a} = 1 + \frac{269 \times 26 \times 94}{10 \times 10 \times 18} = \frac{164809}{450}.$$

In making these calculations it must be remembered that the absolute period of E is a sidereal day and its period relative to the arm fg is a solar day, also the period of fg is a year. Now from Art. 407 it appears that the absolute revolutions of any wheel or piece of an epicyclic train are equal to the sum of its relative revolutions and of the revolutions of the arm when they revolve in the same direction, and the same reasoning shews that the number of sidereal days in a year is equal to the number of solar days + 1.

Also n_3 and a are the synchronal absolute rotations of the arm or annual bar fg, and Earth's axis CE; therefore $\frac{n_3}{a}$ = number of sidereal days in a year; but the fractions in Art. 247 represent the number of solar days in a year, and we may therefore employ them for $\frac{n_3}{a}$ by adding unity as above. We may thus obtain other and simpler trains than that already given. The train ϵ_3 being carried by a small arm should be as simple and light as possible. But it may be reduced to only two wheels by making ϵ_3 negative, and at the same time ϵ_2 positive, since $\frac{n_3}{a}$ *must* be positive.

For example, employing the fraction $\frac{94963}{260}$ (vide p. 233) and remembering that the rotations n_3 are sidereal days, we have

$$\frac{n_3}{a} = 1 + \frac{94963}{260} = \frac{95223}{260} = \frac{3}{2} \times \left(\frac{7 \times 29 \times 157}{2 \times 5 \times 13} - 1\right),$$

which, compared with (1), gives

$$\epsilon_3 = -\frac{3}{2}, \text{ and } \epsilon_2 = \frac{7 \times 29 \times 157}{2 \times 5 \times 13} = \frac{203 \times 157}{10 \times 13}$$

Otherwise,

$$\frac{10 \times 164809 - 27 \times 58965}{10 \times 450 - 27 \times 161}$$

$$= \frac{56035}{153} = \frac{5 \times 7 \times 1601}{3^2 \times 17}$$

$$= \frac{7}{3} \times \frac{8005}{51} = \frac{7}{3} \times \left(\frac{8056}{51} - 1\right)$$

$$= \frac{7}{3} \times \left(\frac{2^3 19 . 53}{3 \times 17} - 1\right)$$

with an error of $33''.9$ in defect.

$$\text{Again } \frac{7 \times 164809 - 18 \times 58965}{7 \times 450 - 18 \times 161} = \frac{92293}{252} = \frac{17 \times 61 \times 89}{2^2 \times 3^2 \times 7}$$

$$= \frac{61}{9} \times \left(\frac{23 \times 67}{4 \times 7} - 1\right)$$

with an error of $13''.7$ in defect.

419. Ex. 4. In the ordinary construction of a planetarium, difficulty arises on account of the number of concentric tubes which are required to communicate the motion of the wheels to the arms which carry the planets. This is avoided in a planetarium by Mr. Pearson. By interposing an epicyclic train between each pair of planetary arms he makes them each derive their motion from the next one in the series, so that the tubes are entirely dispensed with. Referring to Rees' Cyclopædia, Art. Planetary Machines, for

an elaborate description and drawings of this machine, I shall quote one portion as an example of the use of our formulæ.

A fixed stud *mn*, fig. 219, carries the whole of the arms

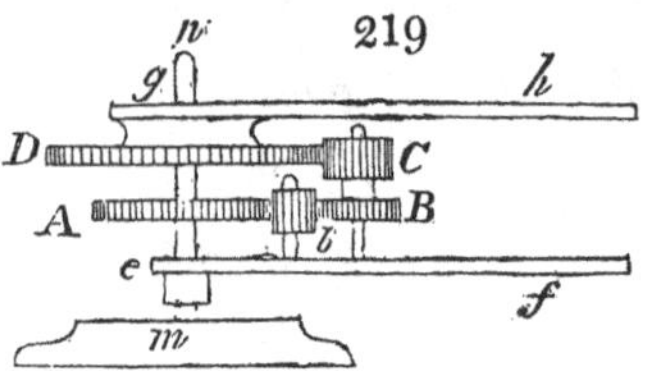

in order, of which the arms of Mercury and of Venus are only shewn in this diagram, the others being disposed in the same manner. Between these arms a wheel A is fixed to the stud, and the arm of Venus carries an epicyclic train, of which A is the first wheel, and the last wheel D is fixed to the arm of Mercury. If, then, the period of Venus = ♀ and of Mercury = ☿ , we have

$$\frac{n}{a} = 1 + \epsilon,$$

since ϵ by virtue of the intermediate idle wheel b is negative,

$$\text{where } \frac{n}{a} = \frac{\text{♀}}{\text{☿}} = \frac{1553}{608}, \text{ nearly;}$$

$$\therefore \quad \epsilon = \frac{AC}{BD} = \frac{945}{608} = \frac{63 \times 30}{16 \times 76},$$

which are Mr. Pearson's numbers.

If on the other hand *ef* were the Earth's arm, and *gh* that of Venus, we should have

$$\frac{\oplus}{\text{♀}} = \frac{3277}{016} = 1 + \frac{AC}{BD}; \therefore \frac{AC}{BD} = \frac{\oplus - \text{♀}}{\text{♀}} = \frac{1261}{2016} = \frac{13 \times 97}{2^5 . 3^2 . 7}.$$

To examine whether the idle wheel b cannot be dispensed with, it must be observed that it is introduced to make ϵ negative, and that if it were removed ϵ would be positive, and $\frac{n}{a} = 1 - \epsilon$. Now because the two arms must revolve in the same direction, $\frac{n}{a}$ is positive, therefore ϵ if positive must be less than unity, which makes n less than a, and the train-bearing arm revolve quicker than the other. If, then, the arm of Mercury were to carry the train instead of the arm of Venus, the idle wheel would be got rid of.

Supposing, therefore, in the figure, that Mercury is changed for Venus, the whole being inverted, we have

$$\epsilon = +\frac{AC}{BD}, \quad \text{and} \quad \frac{☿}{♀} = 1 - \frac{AC}{BD} = \frac{608}{1553},$$

$$\text{whence } \frac{AC}{BD} = 1 - \frac{☿}{♀} = \frac{945}{1553} = \frac{2 \times 5 \times 53}{13 \times 67} \text{ nearly},$$

$$\text{or on the second supposition } \frac{♀}{⊕} = \frac{2016}{3277} = 1 - \frac{AC}{BD};$$

$$\therefore \frac{AC}{BD} = \frac{⊕ - ♀}{⊕} = \frac{1261}{3277} = \frac{13 \times 97}{29 \times 113}.$$

EXAMPLES OF THE SECOND USE OF EPICYCLIC TRAINS.

420. The second use which I have mentioned of epicyclic trains is for the establishment of an exact ratio of angular velocity between two axes when the terms of the ratio are unmanageable if applied to the arrangement of the ordinary trains of wheel-work, and when an approximation (Art. 243) is not admissible.

In Art. 410 we have shewn that if ϵ be an epicyclic train, and if a driving axis be connected with the first wheel of the train ϵ by a train μ, and with the last wheel of the train ϵ by a train ν, we have

$$\frac{a}{p} = \frac{\mu}{1 - \frac{1}{\epsilon}} + \frac{\nu}{1 - \epsilon},$$

when a and p are the synchronal rotations of the train-bearing arm and of the driving axis respectively.

As the epicyclic train is in this case employed merely to concentrate the effect of the two trains μ and ν upon the axis of the train-bearing arm, the epicyclic train itself may be employed in the simplest form, as in fig. 220, which shews one form of the mechanism which results.

220

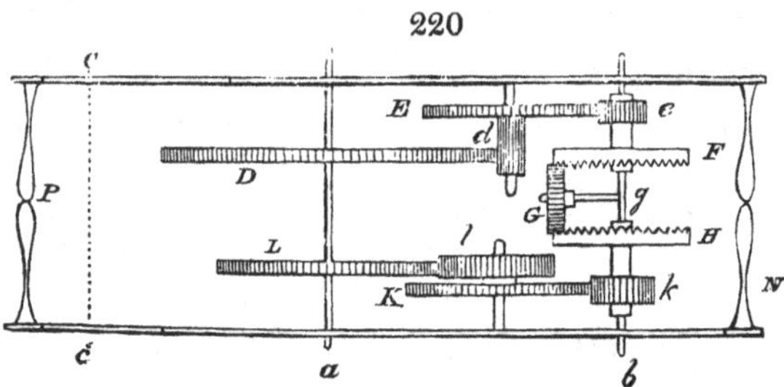

Bb is the axis of the train-bearing arm *Gg*, this arm carries a wheel *G* which geers with two equal crown-wheels *F* and *H*, which are concentric to the axis *B b*, but are each fixed to tubes or cannons which run freely upon it.

The epicyclic train consists therefore of these three wheels, *F*, *G* and *H*, of which *F* may be considered to be the first wheel, and *H* the last wheel.

Aa is the driving axis, and this carries two wheels *D* and *L*; *D* serves to connect the axis with the first wheel *F* of the epicyclic train by means of the train of wheel-work *d*, *E* and *e*; and *L*, together with *l*, *K* and *k*, constitute a train of wheel-work which connects the axis *Aa* with the last wheel *H* of the epicyclic train. We have therefore

$$\mu = \frac{DE}{de}, \text{ and } \nu = \frac{LK}{lk}.$$

If the motion of the epicyclic train be considered with respect to the arm, it is clear that its extreme wheels *F*, *H* move in opposite directions, therefore ϵ is negative and equal to $-\frac{FG}{GH} = -1$;

$$\therefore \quad \frac{a}{p} = \frac{1}{2}(\mu + \nu) = \frac{1}{2}\left(\frac{DE}{de} + \frac{LK}{lk}\right).$$

If therefore a ratio of angular velocity $\frac{a}{p}$ be given, of which the numerator or denominator, or both, are not decomposable, we must endeavour to find two manageable fractions whose sum shall be equal to the proposed fraction, and employ them to form a train of wheel-work similar to that shewn in fig. 220.

This employment of epicyclic trains is given by Francœur*, from whom I have derived the calculations in the following articles. He attributes the mechanism to Messrs. Péqueur and Perrelet, about 1823, but the first idea of this method appears due to Mudge, who obtained an exact lunar train by epicyclic wheels before 1767†.

421. *First case.* Let $\frac{a}{p}$ be a fraction of which the denominator is decomposable into factors, but not the numerator.

Let the denominator $p = fgh$, therefore the fraction which represents the ratio of the velocities will be $\frac{a}{fgh}$. The denominator may often be susceptible of a division into three factors in various manners, each of which will furnish a distinct solution of the problem, subject to a condition which will presently appear. To decompose $\frac{a}{fgh}$ into two reducible fractions, assume

$$\frac{a}{fgh} = \frac{fx}{fgh} + \frac{gy}{fgh},$$

that is to say, $a = fx + gy$. It is easy to resolve this equation in prime numbers for x and y, and obtain an infinity of values for x and y that will satisfy the problem, and give

* Dict. Technologique, t. XIV. p. 431.

† Vide Mudge on the Timekeeper, or Reid's Horology, p. 70.

$$\frac{a}{fgh} = \frac{x}{gh} + \frac{y}{fh};$$

f and g must however be prime to each other, since a is prime, which is the condition already alluded to.

For example, let $\frac{271}{216}$ be the fraction proposed. Since

$216 = 4 \times 9 \times 6$ we may assume $271 = 9x + 4y$, $f = 9$, $g = 4$. The ordinary methods employed in equations of this kind will give $x = 31 - 4t$, $y = 9t - 2$, where t is any whole positive or negative number, $gh = 24$, $fh = 54$. Hence we have

$$\begin{array}{lllll} & x = 27, 23, 19 \ldots & 31, & 35, & 39, \\ & y = 7, 16, 25 \ldots & -\ 2, & -11, & -20, \\ \text{corresponding to} & t = 1, 2, 3 \ldots & -\ 0, & -\ \ 1, & -\ \ 2, \end{array}$$

The fraction $\frac{271}{216}$ is therefore equal to

$$\frac{27}{24} + \frac{7}{54}, \quad \frac{23}{24} + \frac{16}{54}, \quad \frac{19}{24} + \frac{25}{54},$$

or to $\frac{31}{24} - \frac{2}{54}$, $\frac{35}{24} - \frac{11}{54}$, $\frac{39}{24} - \frac{20}{54}$, and so on.

The first set referring to the case in which the crown-wheels turn in the same direction, the second to that in which they turn different ways.

But since 8 and 3 have no common factor, the denominator 216 might have been decomposed into $8 \times 3 \times 9$, whence assuming $271 = 8x + 3y$, we should have had

$$x = 3t - 1, \quad y = 93 - 8t, \quad \text{and}$$

$$x = 2, 5, 8, \ldots\ldots -1, \ -4, \ -7 \ldots$$

$$y = 85, 77, 69, \ldots\ldots \ 93, \ 101, \ 109 \ldots$$

whence the new decompositions

$$\frac{2}{27} + \frac{85}{72}, \quad \frac{5}{27} + \frac{77}{72}, \quad \frac{8}{27} + \frac{69}{72}, \quad \frac{93}{72} - \frac{1}{27},$$

and so on, all of which are solutions of the question.

Generally the proposed denominator must be resolved into prime factors under the form $m^{\alpha}.n^{\beta}.p^{\gamma}$...... and any two of the divisors of this quantity may be assumed for f and g, provided they be prime to each other. Thus if the equation $a = fx + gy$ be resolved in whole numbers, the component fractions will be $\frac{x}{gh} + \frac{y}{fh}$, where h is the product of all the remaining factors of the denominator, after f and g have been removed.

422. Ex. 1. A mean lunation $= 29^{d}.\ 12^{h}.\ 44'.\ 3''$ $= 2551443''$, therefore the ratio of a lunation to twelve hours $= \frac{850481}{14400}$, of which the numerator is a prime. But this fraction may be by the above method resolved into two:

$$\text{thus } \frac{850481}{14400} = \frac{40 \times 50}{6 \times 6} + \frac{71 \times 79}{50 \times 32}.$$

And if these fractions be employed for the trains μ and ν, the axes Aa, Bb will revolve with the required ratio,

$$\text{for } \frac{a}{p} = \tfrac{1}{2}(\mu + \nu) = \tfrac{1}{2}\left(\frac{80 \times 50}{6 \times 6} + \frac{71 \times 79}{25 \times 32}\right) = \tfrac{1}{2}\left(\frac{DE}{de} + \frac{LK}{lk}\right).$$

And the periods are inversely as the synchronal rotations. If therefore a period of twelve hours be given by a clock to the axis Bb, Aa will receive a period accurately equal to a lunation.

The mechanism may be thus represented in the notation already explained.

Axes.	Trains.	Periods.
First Axis	79———80......................	Lunation.
Upper Stud	. ... 6—50	
Upper Cannon	. ... 6—Crown Wheel F.	
Lower Stud............	32—71	
Lower Cannon	......25———— Crown Wheel H	
Train-bearing Axis.	—————————Epicyclic Wheel G..	12 hours.

If the fraction be resolved into a difference instead of a sum, as in the example $\frac{271}{216} = \frac{35}{24} - \frac{11}{54}$, this may be translated into mechanism, by making the trains μ and ν of different signs, that is, by making their extreme wheels revolve different ways.

423. Ex. 2. Mean time is to sidereal time nearly as 8424 : 8401.

$$\text{Now } \frac{8401}{8424} = \frac{31 \times 271}{39 \times 216} = \frac{31}{39} \times \left\{\frac{19}{24} + \frac{25}{54}\right\};$$

$$\therefore \frac{a}{p} = \tfrac{1}{2}(\mu + \nu) = \left(\frac{19}{24} + \frac{25}{54}\right); \quad \therefore \mu = \frac{19}{12} \quad \nu = \frac{25}{27},$$

and we obtain the following train, which differs from fig. 220 only in fixing the wheels E and K upon a single axis, which also carries a wheel of 39, geering with a wheel of 31 upon Aa, as appears in the following notation.

Axes.	Trains.	Periods.
First Axis	31	Sidereal Day.
Second Axis	39—19—25	
Upper Cannon	.......|....27—Crown Wheel F.	
Lower Cannon	......12———Crown Wheel H	
Train-bearing Axis.	............Epicyclic Wheel G...	Solar Day.

424. *Second case.* The fraction in the first case has been supposed to have a decomposable denominator. Let now both denominator and numerator be prime. Form two fractions $\frac{a}{A}$ and $\frac{a_{\prime}}{A}$, in which A is an arbitrary quantity and commodiously decomposable into factors, and proceed to obtain from each of these fractions the sums or differences of two decomposable fractions as before, which may be employed in wheel-work as follows.

Let an axis Aa, fig. 220, be connected to one axis Bb, by two trains and an epicyclic train, as in the figure, and also to another axis Cc by a precisely similar arrangement. Then if the synchronal rotations of the axes Aa, Bb, Cc be A, a and $a_{,}$, μ, ν the trains which connect Aa with Bb, and $\mu_{,} \nu_{,}$ the trains that connect Aa with Cc, we shall have

$$\frac{a}{A} = \frac{\mu + \nu}{2} \text{ and } \frac{a_{,}}{A} = \frac{\mu_{,} + \nu_{,}}{2}; \quad \therefore \frac{a}{a_{,}} = \frac{\mu + \nu}{\mu_{,} + \nu_{,}},$$

will be the ratio of the synchronal rotations of Bb and Cc.

Suppose for example that it be required to make one axis perform 17321 turns, while another makes 11743; both being prime numbers, the fraction $\frac{17321}{11743}$ is irreducible, and indecomposable into factors.

Assume a divisor $5040 = 7 \times 8 \times 9 \times 10$, and form separately two trains whose velocities are represented by

$$\frac{17321}{5040} \text{ and } \frac{11743}{5040}.$$

For the first we have

$$\frac{17321}{5040} = \frac{1480}{630} + \frac{783}{720} = \frac{148}{63} + \frac{87}{80},$$

whence the trains $\frac{74}{63}$ and $\frac{87}{40}$, as in the first method. (Art. 421).

For the second train,

$$\frac{11743}{5040} = \frac{830}{633} + \frac{729}{720} = \frac{83}{63} + \frac{81}{80},$$

whence the trains $\frac{166}{63}$ and $\frac{81}{40}$.

If we represent the wheels which in the left-hand train correspond to F, G and H, by f, g and h, we have the following notation of the resulting machine.

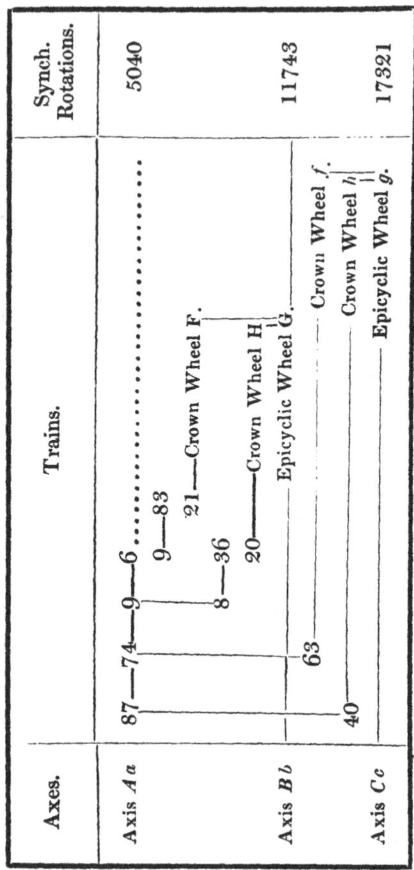

EXAMPLES OF THE THIRD USE OF EPICYCLIC TRAINS.

425. The third employment of epicyclic trains, is to produce a very slow motion. In the formula $\frac{a}{p} = \frac{\mu\epsilon - \nu}{\epsilon - 1}$ Art. 410, all the trains are at present taken positive.

Let ϵ be made negative, and let μ and ν have different signs,

$$\therefore \frac{a}{p} = \frac{\mu\epsilon - \nu}{\epsilon + 1},$$

in which, by properly assuming the numbers of the trains, a may be made very small with respect to p, and therefore the arm to revolve very slowly. This leads to such an arrangement as that of fig. 215, (Art. 409.)

$$\text{for } \frac{a}{p} = \frac{aceg - hbdf}{bg(ce + df)}, \quad \text{(Art. 413.)}$$

and in this expression the two terms of the numerator having no common divisor, may be so assumed as to differ by unity, by which an enormous ratio may be produced.

For example, put a, c, e, g each equal 83,

$$b = 106,\ d = 84,\ f = 65,\ h = 82,$$

and we get

$$\frac{a}{p} = \frac{83^4 - 82 \times 106 \times 84 \times 65}{106 \times 83\,(83^2 + 84 \times 65)} = \frac{1}{108646502}.$$

If in this machine we suppress the wheels h and e by making a turn both b and g, and d turn both f and c, we have*

$$\frac{a}{p} = \frac{a}{bg} \times \frac{cg - bf}{c + f} = \frac{20}{100 \times 99} \times \frac{101 \times 99 - 100^2}{101 + 100} = \frac{1}{99495}.$$

426. If on this contrary we wish to make the shaft, whose revolutions are p, revolve slowly with respect to the arm; then the numerator of the fraction $\frac{a}{p}$ must be a sum, and the denominator a difference; therefore ϵ must in the expression $\frac{a}{p} = \frac{\mu\epsilon - \nu}{\epsilon - 1}$ be positive, and nearly equal to unity, and μ and ν must have different signs.

* Putting $a = 20$, $b = 100$, $c = 101$, $g = 99$, and $f = 100$. This latter combination is given with these numbers by White (Century of Inventions).

Fig. 221 is a combination that will answer the present purpose: mp is a fixed axis upon which turns a long tube, to the lower end of which is fixed a wheel D, and to the upper a wheel E; a shorter tube turns upon this, which carries at its extremities the wheels A and H. A wheel C is engaged both with D and A, and a train-bearing arm mn, which revolves freely upon mp, carries upon a stud at n the united wheels F and G. The epicyclic train therefore is formed of the wheels EFG and H, and is plainly positive, the extreme wheels EH revolving in the same direction.

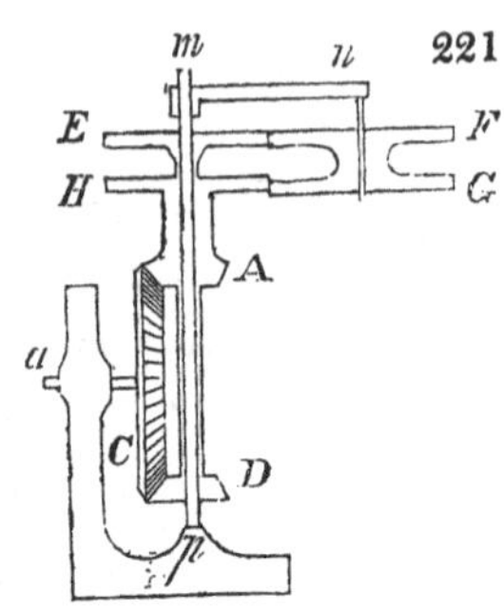

Let H be the first wheel; $\therefore\ \epsilon = \dfrac{HF}{GE}$,

also $\mu = \dfrac{C}{A}$ and $\nu = \dfrac{C}{D}$ with different signs, since A and D revolve different ways;

$$\therefore\ \frac{a}{p} = \frac{\dfrac{C}{A}\cdot\dfrac{HF}{GE} + \dfrac{C}{D}}{\dfrac{HF}{GE} - 1},$$

put $A = 10$, $C = 100$, $D = 10$, $E = 61$, $F = 49$, $G = 41$, $H = 51$, and we shall obtain $\dfrac{a}{p} = 25000$, that is, 25000 rotations of the train-bearing arm mn will produce one of the wheel C.

427. Generally, however, the first wheel of the epicyclic train is fixed, in which case the formula becomes $\dfrac{n}{a} = 1 - \epsilon$. If ϵ be positive and very near unity, this will be very small, or n small with respect to a, that is, the motion of the last wheel of the train slow with respect to that of the arm. In

the simple forms of epicyclic trains, figs. 211, 212, and 213, the two latter are excluded, because ϵ is negative, but the former with the train $\begin{array}{ccc} A & & \\ b & \text{——} & E \\ & & D \end{array}$ is usually selected, A being a fixed wheel, and $\frac{n}{a} = 1 - \frac{AE}{bD}$ is made as small as possible; which is effected by making $AE - bD = \pm 1$.

Thus if $\epsilon = \frac{101 \times 99}{100 \times 100}$ be the numbers of the wheels,

$$\text{we have } \frac{n}{a} = \frac{1}{10000},$$

but as these large numbers are inconvenient for the wheels that are carried upon the arm,

$$\text{let } \epsilon = \frac{111 \times 9}{100 \times 10}; \quad \therefore \frac{n}{a} = \frac{1}{1000},$$

$$\text{or let } \epsilon = \frac{31 \times 129}{32 \times 125}, \quad \therefore \frac{n}{a} = \frac{1}{4000}$$

428. This combination is used for registering machinery for the same purpose as the contrivances in Arts. 404 and 405; and since the concentric wheels A and D (fig. 211) are very nearly of the same size, the pinions b and E carried by the arm may be made of the same number of teeth, or in other words, a thick pinion substituted for them which geers at once with the fixed wheel A and the slow-moving wheel D*.

Let M, $M - 1$, and K be the numbers of teeth of D, A, and the thick pinion respectively, then

$$\frac{n}{a} = 1 - \frac{K \times (M - 1)}{K \times M} = \frac{1}{M},$$

where M is the number of teeth of the slow-moving wheel.

* In Roberts self-acting mule.

EXAMPLES OF THE FOURTH USE OF EPICYCLIC TRAINS.

429. The fourth employment of epicyclic trains consists in concentrating the effects of two or more different trains upon one revolving body when these trains move with respect to each other with a variable velocity ratio. I have already shewn how this may be effected when the extent of motion is small, as in Arts. 395, 397, but by epicyclic trains an indefinite number of rotations may be produced.

As an example of this application I shall take the equation clock, as it is the earliest problem of this class which presents itself for solution in the history of mechanism, and actually occupied the attention of mechanists for a long period*. The object of this machine is to cause the hand of a clock to point on the usual dial, not to mean solar time, but to *true solar time.* For this purpose we may resolve its motion as astronomers resolve the motion of the sun; namely, into two, one of which is the uniform motion which belongs to the mean time, and the other the difference between mean and true time or the equation. If, then, two trains of mechanism be provided, one of them an ordinary clock, and the other contrived so as to communicate a slow motion corresponding to the equation of time, and if we then concentrate the effects of these separate trains upon the hand of our equation clock by means of an epicyclic train, we shall obtain the desired result. There are three possible arrangements, as in Art. 406, (1) the equation may be communicated to one end of the train, and the mean motion to the other, the arm receiving the solar motion†; (2) the equation may be given to one end of the train, and the mean motion to the arm, the other end of the train will then receive the solar motion; (3) the equation may be communicated to the

* Vide the Machines Approuvées of the Acad. des Sciences.

† Employed in the equation clock of Le Bon, 1722.

arm, and the mean time to one end of the train, when the other end of the train will receive the solar motion*. I shall describe the mechanism of the latter arrangement.

430. Fig. 222 is a diagram which will serve to shew the wheel-work of that part of an equation clock by which the motion is given to the hands. This wheel-work is commonly called the dial-work. *G* is the centre of motion of the epicyclic train, *GDe* the train-bearing arm. The wheels *f*

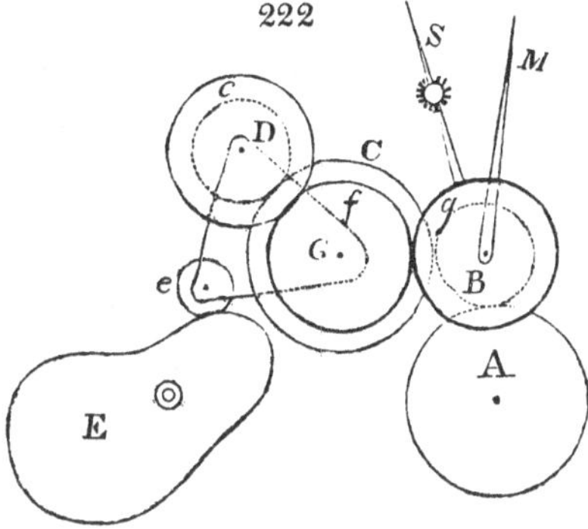

and *C* turn freely upon the axis *G*, and the axis *D* carried by the arm has two wheels *D* and *c* fixed to it, which geer with *f* and *C* respectively.

The epicyclic train consists, therefore, of the four wheels *C*, *c*, *D* and *f*, of which let *C* be the first wheel. In this arrangement the equation is to be communicated to the train-bearing arm, and the mean motion to the first wheel *C* of the epicyclic train. Now for this purpose *C* is driven by the wheel *B*, dotted in the figure, which derives its motion from a wheel *A* connected with an ordinary clock, and as the minute-hand *M* of the clock is fastened to the axis of *B*, this minute-hand will shew *mean time* upon the dial in the usual manner.

The equation is communicated to the train-bearing arm *GDe*, as follows. *E* is a cam-plate, which by its connexion

* In the clocks of Du Tertre, 1712, and Enderlin.

with the clock is made to revolve in a year (Art. 247). A friction roller e upon the train-bearing arm rests upon the edge of the cam-plate, and is kept in contact with it by means of a spring or weight. The cam-plate is shaped so as to communicate the proper quantity of angular motion to the arm. We have seen how one end of the epicyclic train receives the mean motion, and f, which is the other extremity of the train, geers with a wheel g concentric to the minute-wheel B, and turning freely upon it; the solar hand S is fixed to the tube or cannon of g, and thus receiving the aggregate of the mean motion and the equation, will point upon the dial to the true time which corresponds to the mean time indicated by M.

The formula which belongs to this case is, (Art. 411)

$$n = a \cdot \overline{1 - \epsilon} + m\epsilon,$$

in which ϵ is positive and $= \dfrac{Cc}{Df}$. Now if the synchronal rotations of the minute-hand M and of C be M and m respectively, we have $m = M \cdot \dfrac{B}{C}$, and if those of f and g be n and s, we have $n = s \cdot \dfrac{g}{f}$; substituting these values in the formula, we obtain

$$s = a \; \frac{Df - Cc}{Dg} + M \cdot \frac{Bc}{Dg},$$

of which the first part belongs to the equation, and the second to the mean motion.

Now the mean motion of S must be the same as that of M; $\therefore \dfrac{Bc}{Dg} = 1$. And for that part of the motion of S which is due to the equation, the expression $a \cdot \dfrac{Df - Cc}{Dg}$ shews the proportion between the angular motion of the train-bearing arm and of the hand s, synchronal rotations being directly proportional to angular velocity (Art. 20.). If the arm is to move with the same angular velocity as the hand,

$$\text{then } \frac{Df - Cc}{Dg} = 1,$$

and this is readily effected by making $f = c = g$ and $C = 2D$; also, since $Bc = Dg$ where $c = g$, we must have $B = D$, and these are the actual proportions employed by Enderlin. But if it be required that the arm move through a less angle than the hand, through half the angle, for example, then $C = 3D$, and so on.

431. In the treatises on Horology, and in the machines of the French Academy, may be found a great number of contrivances for equation clocks, which was a favourite subject with the mechanists of the last century. The machine itself is merely curious, and the desired purpose may be effected in a much more simple manner, if indeed it be worth doing at all, by placing concentrically to the common fixed dial a smaller moveable dial, and communicating to the latter the equation, by which the ordinary minute-hand of the clock will simultaneously shew mean time on the fixed, and true time on the moveable dial, without the intervention of the epicyclic train*.

Nevertheless, I have selected this machine as the best for the purpose of explanation, as being easily intelligible. The most successful machine of this class is undoubtedly the Bobbin and Fly-frame, in which, by means of an epicyclic train, the motions of the spindles are beautifully adjusted to the increasing diameter of the bobbins and consequent varying velocity of the bobbins and flyers. But this machine involves so many other considerations, that the complete explanation of it cannot be given in the present stage of our subject.

* This is done in the early equation clocks of Le Bon, 1711, Le Roy, &c.

CHAPTER III.

ON COMBINATIONS FOR PRODUCING AGGREGATE PATHS.

---

432. I HAVE already stated in the beginning of this work (Art. 39), that pieces in a train may be required to describe elliptical, epicycloidal, or sinuous lines, and that such motions are produced by combining circular and rectilinear motions by aggregation. The process being, in fact, derived from the well-known geometrical principle by which motion in any curve is resolved into two simultaneous motions in co-ordinate lines or circles.

If the curve in which the piece or point is required to move be referred to rectangular co-ordinates, let the piece be mounted upon a slide attached to a second piece, and let this second piece be again mounted upon a slide attached to the frame of the machine at right angles to the first slide. Then if we assume the direction of one slide for the axis of abscissæ, the direction of the other will be parallel to the ordinates of the required curve. And if we communicate simultaneously such motions to the two sliding pieces as will cause them to describe spaces respectively equal to the corresponding abscissæ and ordinates, the point or piece which is mounted upon the first slide will always be found in the required curve.

This first slide, being itself carried by a transverse slide, falls under the cases described in the first Chapter of this Part, and the motion may be given to it by any contrivance for maintaining the communication of motion between pieces the position of whose paths is variable, as, for example, by a rack attached to the slide and driven by a long pinion. For the purpose of communicating the velocities to the two slides, any appropriate contrivance from the first part of the work may be chosen.

433. If the curve in which the point is to move be referred to polar co-ordinates, these may be as easily translated into mechanism, by mounting the point upon a slide and causing this slide to revolve round a center, which will be the pole. Then connecting these pieces by mechanism, so that while the slide revolves round its pole the point shall travel along the slide with the proper velocity, this point will always be found in the given curve.

434. Fig. 223 is a very simple arrangement, by which a short curve may be described upon the above principles.

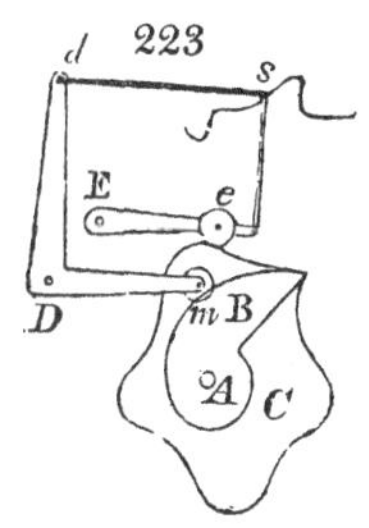

E is the center of motion of an arm *Ee* which is connected by a link with the describing point *s*; *D* is the center of motion of a second arm *Dd* which is connected by a link *ds*, with the same describing point *s*. If now *Ee* be made to move through a small arc, it will communicate to *s* a motion round *d* which will be nearly vertical, and if *Dd* be made to move through a small arc, it will communicate to *s* a motion round *e*, which will be nearly horizontal; and as the motion of the describing point *s* is solely governed by its connexion with these two links, these motions may be separately or simultaneously communicated to it. *A* is an axis, upon which are fixed two cam-plates, the lower of which, *C*, is in contact with a roller *e* at the end of the arm *Ee*, and the upper, *B*, in contact with a roller *m* at the end of an arm *Dm*, fixed at right angles to the arm *Dd*.

When the axis *A* revolves the cams communicate simultaneously motions to the two arms, which motions are given to the describing point, one in a direction nearly perpendicular to the other, the point will thus describe a curve of which the horizontal co-ordinates are determined by the cam *B*, and the vertical by the cam *C*.

In practice the shape of the cams may be obtained by trial: the machine must be previously constructed, and plain disks of a sufficient diameter substituted for the cams, then if the required path of *s* be traced upon paper, and it be placed in succession upon a sufficient number of positions upon this path, the cam-axis being also shifted, the corresponding positions of the rollers *e* and *m* may be marked upon the disks, and the shape of the cams thus ascertained.

435. If the object of the machine be merely to trace a few curves upon paper or other material, the principle of relative motion* will enable us to dispense with the difficulties that are introduced by the necessity of maintaining motion with a piece whose path itself travels. For since every complex path is resolvable into two simple paths, let the describing point move in one component path, and the surface upon which it traces the curve move in the other component path with the proper relative velocity, then will the curve be described by the relative motion of the point and surface.

Thus to describe polar curves, the surface upon which the curve is to be described may be made to revolve while the describing point travels with the proper velocity along a fixed slide, in a path the direction of which passes through the axis of motion of the surface. And as in this arrangement the axis of motion of the surface and the path of the describing point are both fixed in position, the simultaneous

* Already employed in Arts. 256, 404, 405.

motions may be communicated to them by any of the contrivances in our first Part, without having recourse to the principle of Aggregate Motion. And thus, in general, a firmer and simpler machine will be obtained.

Also the tracing of curves upon a surface is sometimes accomplished under the Aggregate principle by causing the *surface* to move with the double motion, while the describing point is at rest*.

436. Screw-cutting and boring machines are reducible to this head. For the cutting of a screw is in fact the tracing of a spiral upon the surface of a cylinder, and the motion of boring is also the tracing of a spiral upon the surface of a hollow cylinder; the tool being in both cases the describing point, and the plain cylinder the surface. Now as the tracing of this spiral is resolvable into two simultaneous motions, one of revolution with respect to the axis of the cylinder, and the other of transition parallel to that axis, we have in the construction of machines for boring and screw cutting the choice of four arrangements.

(1) The cylinder may be fixed and the tool revolve and travel. This is the case in all simple instruments for boring and tapping screws, in machines for boring the cylinders of steam engines, and in engineers' boring machines.

(2) The tool may be fixed and the cylinder revolve and travel. Screws are cut upon this principle, in small lathes with a traversing mandrel, as it is called.

(3) The tool may revolve and the cylinder travel. The boring of the cylinders of pumps is often effected upon this principle.

(4) The cylinder may revolve and the tool travel. Guns are thus bored, and engineers' screws cut in the lathe.

437. But motion in curves may be often more simply obtained by means of some geometrical property that may

* The motion which must be communicated to a plane to enable it to receive a given curve from a fixed describing point, is not the same as that which would cause a point, carried by the moving plane, to trace the same curve upon a fixed plane. Vide Clairaut, Mem. de l'Acad. des Sciences, 1740.

admit of being employed in mechanism, as the ellipse is described by the trammel fig. 224. This consists of a fixed cross *abcd*, in which are formed two straight grooves meeting in *C*, and perpendicular to each other; a bar *PGH*

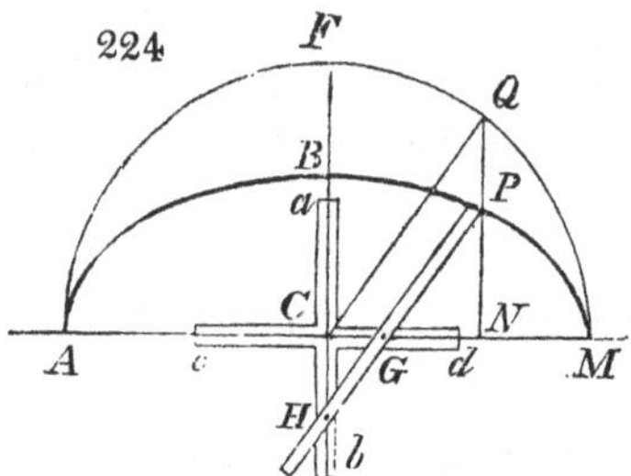

has pins attached to it at *G* and *H*, which fit and slide in these grooves, and a describing point is fixed at *P*. When the bar moves it receives simultaneously the rectilinear motion of the pin *H* in the groove *ab*, and that of the pin *G* in the groove *cd*, by which the describing point *P* traces a curve *MPB*, which can be shewn as follows to be the ellipse.

When *HP* coincides with *ab*, *G* comes to *C*, and therefore $GP = BC$, and when *HP* coincides with *Cd*, *H* comes to *C* and therefore $HP = CM$.

With center *C* and radius *CQ* equal to *HP*, describe a semicircle *AFM*, and through *P* draw *QPN* perpendicular to *cd* produced, join *CQ*, then *QP* is parallel to *CH*, also $HP = CM = CQ$, $\therefore$ *CHPQ* is a parallelogram.

$$\therefore \frac{CQ}{GP} = \frac{QN}{PN}$$

But $CQ = CF$ and $GP = BC$,

$$\therefore \frac{QN}{PN} = \frac{CF}{BC},$$

and the curve is an ellipse.

438. Thus also epicycloids or hypocycloids are described mechanically in Suardi's pen*, by fixing the describing point

* Adams' Geometrical and Graphical Essays.

at the end of a proper arm upon the extreme axis *B*, fig. 211, of an epicyclic train in the manner already explained in the first Chapter (Art. 386.) And in this instance we may also avail ourselves of the principles of Art. 435, and describe these curves by causing the plane and the arm which carries the describing point to revolve simultaneously with the proper angular velocity ratio, round parallel axes fixed in position.

439. But the most extensively useful contrivance of this class is that which is termed a *parallel motion*, by which a point is made to describe a right line by the joint action of two circular motions, and as this is a contrivance of great practical importance, it is necessary to examine it in detail.

ON PARALLEL MOTIONS.

440. A parallel motion is a term somewhat aukwardly applied to a combination of jointed rods, the purpose of which is to cause a point to describe a straight line by communicating to it simultaneously two or more motions in circular arcs, the deviations of these motions from rectilinearity being made as nearly as possible to counteract each other.

The rectilinear motion so produced is not strictly accurate, but by properly proportioning the parts of the contrivance, the errors are rendered so slight that they may be neglected.

441. Let *Aa*, *Bb*, fig. 225, be rods capable of moving round fixed centers *A* and *B*, and let them be connected by a third rod or link *ab* jointed to the extremities of the first rods respectively, as in Art. 326. The rods *Aa*, *Bb* are termed radius rods. This system may be moved in succession through a series of positions, the principal ones of

which are indicated by the figures 1, 1, 2, 2, 3, 3, 4, 4, *a*, *b*,

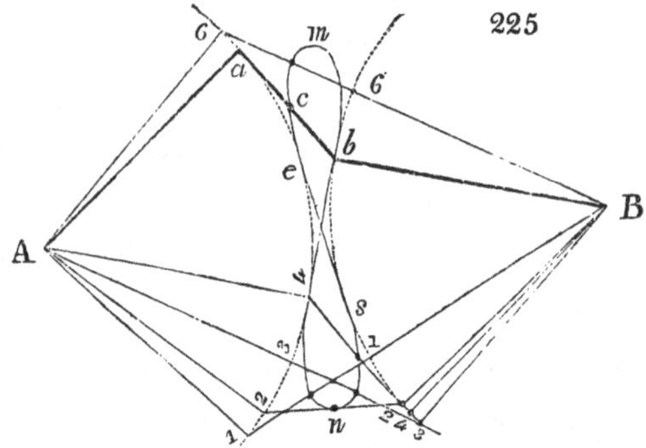

6, 6, 1, 1, and so on repeatedly. If a tracing point *c* be attached to some part of the link near its center, it will describe a curve *mce8n4bm*, somewhat resembling the figure 8. If the position of the tracing point be properly assumed, a very considerable length of the intersecting portion of this curve will be found to approximate so nearly to a right line, that it may, for all practical purposes, be considered and employed as such.

442. For example, let *Ee*, fig. 226, be a crank or excentric, which, by its revolution is intended to communicate a reciprocating motion to the piston *P* through a link *ec*, jointed to the top of the piston rod *Pc*. In the common mode the upper end *c* of the piston rod would be guided in a vertical line, either by sliding through a collar or in a groove. If, however, the end *c* be jointed to the center of a link *ab* connecting two equal radius rods *Aa*, *Bb*, whose centers of motion *B*, *A* are attached to the frame of the machine; then

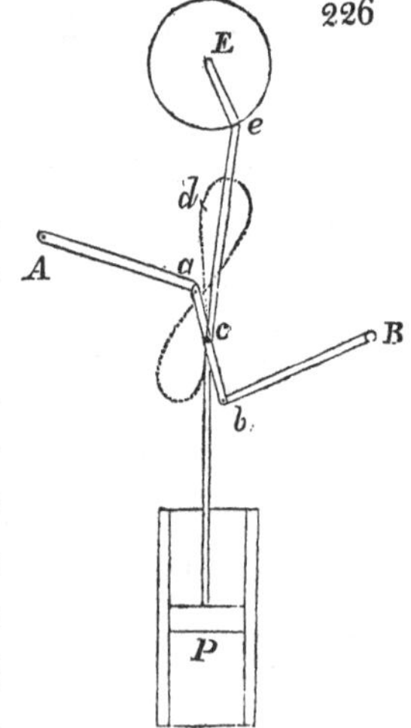

the path of c will be a certain segment cd of the curve described in Art. 441; and if the motion of c be not too great with respect to the length of the radius rods, this curve will vary so slightly from a right line that it may be safely employed instead of a sliding guide. An algebraical equation may be found for the entire curve*, but it is exceedingly involved and complex, and of no use in obtaining the required practical results, which are readily deduced by simple approximate methods, as follows.

443. Let A, C, fig. 227, be the centers of motion, AB, CD the radius rods, BD the link, and let the link be perpendicular to the two radius rods in the mean position of the system $ABDC$.

Let AB be moved into the position Ab, and Cc, bc be the corresponding positions of the other rod and the link. Draw bf parallel to BD. Now in the first position the link BD is perpendicular, and in the second position this link is thrown into the oblique position bc, by which the upper end is carried to the left, and the lower to the right of the vertical line BMd, through spaces be, dc, which are respectively equal to the versed sines of the angles described by

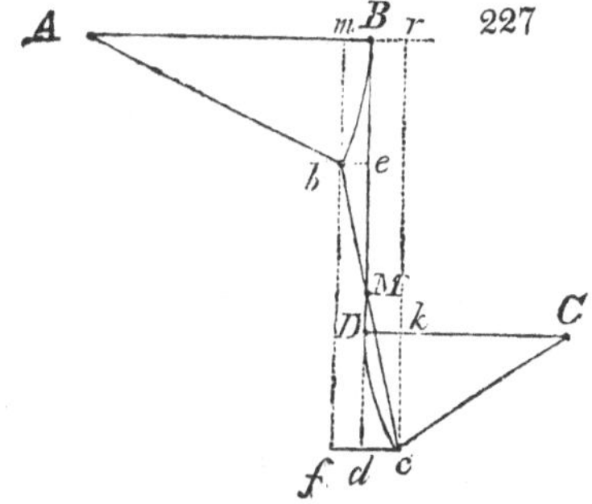

the radius rods AB, DC in moving to their second positions Ab, Cc. But as the ends of the link move different ways,

* This is completely worked out by Prony, Architecture Hydraulique, Art. 1478.

there will be one point M between them that will be found in the vertical line BMd, and its place is determined by the proportion (Art. 395).

$$bM : Mc :: be : dc.$$

$$\text{Let } AB = R, \quad CD = r, \quad BD = l,$$

$$BAb = \theta, \quad DCc = \phi, \quad \text{and } bM = x;$$

$$\therefore \quad \frac{x}{l-x} = \frac{R \text{ versin } \theta}{r \,.\, \text{versin } \phi} = \frac{R \,.\, \sin^2 \frac{\theta}{2}}{r \,.\, \sin^2 \frac{\phi}{2}} = \frac{r}{R} \times \frac{R^2 \,.\, \sin^2 \frac{\theta}{2}}{r^2 \,.\, \sin^2 \frac{\phi}{2}}.$$

Now as the angle BAb never exceeds about 20^0 in practice the inclination cbf of the link is small, and $Bb\,(= R\theta)$ very nearly equal to $Dc\,(= r\phi)$; and as these angles are small we may assume without sensible error $R \sin \frac{\theta}{2} = r \sin \frac{\phi}{2}$;

$$\therefore \quad \frac{x}{l-x} = \frac{r}{R}, \quad \text{and } x = \frac{lr}{R+r},$$

which is the usual practical rule.

This rule may be simply stated in words, by saying that the segments of the link are inversely proportional to their nearest radius rods.

Ex. Let $R = 7$ feet, $r = 4$ feet, $l = 2$ feet.

$$\therefore \quad x = \frac{2 \times 4}{7 + 4} = \frac{8}{11} = .727 \text{ feet} = 8.72 \text{ inches}.$$

444. The deviation of the point M from the line BD may be measured with sufficient accuracy as follows, and it is necessary to know it in order to ascertain how great a value of the angle θ may be safely employed. For simplicity I shall confine myself to the case in which the radius rods AB, CD are equal in length, and taking their length equal to unity, let the link $BD = l$, draw bf, rc parallel to BD, and let the inclination fbc of the link to the vertical $= \gamma$;

$$\therefore \gamma = \frac{fc}{l}, \text{ since } fbc \text{ is small,}$$

$$= \frac{\text{versin } \theta + \text{versin } \phi}{l} = \frac{2 \text{ versin } \theta}{l}, \text{ very nearly.}$$

Now $mf = rc$, that is, $\sin \theta + l \cos \gamma = l + \sin \phi$;

$$\therefore \sin \phi = \sin \theta - l \text{ versin } \gamma = \sin \theta - \frac{l\gamma^2}{2}$$

$$= \sin \theta - \frac{2}{l} \times (\text{versin } \theta)^2.$$

From this expression the value of ϕ corresponding to any given value of θ may be calculated.

When the radius rods are inclined upwards, we have (fig. 228) retaining the same notation,

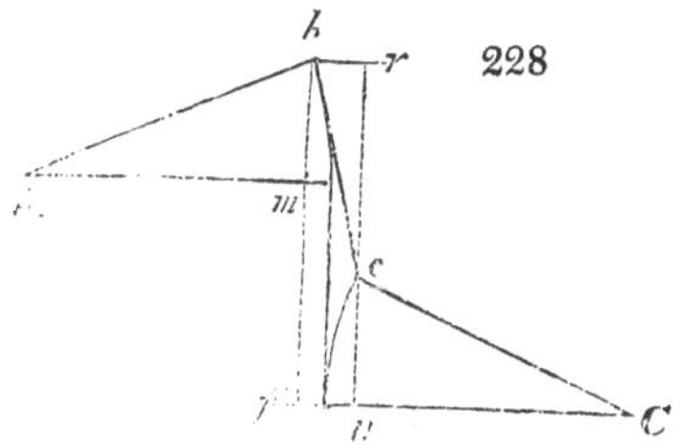

$$bm + mf = rc + cn, \text{ that is, } \sin \theta + l = l \cdot \cos \gamma + \sin \phi;$$

$$\therefore \sin \phi = \sin \theta + \frac{2}{l} \times (\text{versin } \theta)^2.$$

Let the link be half of the radius rods; therefore

$$\sin \phi = \sin \theta \pm 4 (\text{versin } \theta)^2,$$

where the upper sign is taken when the radius rods are above the horizontal line, and the lower sign when below.

Also the deviation of the central point M of the link from a vertical right line is equal to

$$\frac{\text{versin } \theta}{2} - \frac{\text{versin } \phi}{2} = \frac{\cos \phi - \cos \theta}{2}. \quad (\text{Art. } 395.)$$

and the actual values of ϕ, and of the deviation which correspond to the principal values of θ, are given in the following table.

Values of θ.	ABOVE HORIZONTAL LINE.		BELOW HORIZONTAL LINE.	
	Values of ϕ.	Deviation.	Values of ϕ.	Deviation.
25^0	$27^0\ 15'$	.00864	$22^0\ 48'$	.00777
20^0	$20^0\ 54'$	.00274	$19^0\ 7'$	.00258
15^0	$15^0\ 17'$	.00064	$14^0\ 44'$	.00060
10^0	$10^0\ 3'$	.00007	$9^0\ 57'$	.00007

Thus if the radius rod or *beam* AB have 3 ft. radius, the deviation at 25^0 amounts to $.0086 \times 36$ inches $= .31$ inches, and at 20^0 to .097 inch.; generally the entire beam is made equal to three times the length of the stroke, and therefore describes an angle of about 19 degrees on each side of the horizontal line.

445. Even this error may be greatly reduced by a different mode of arranging the rods. Supposing the rods to be of equal length and equal to unity, let Ab, fig. 229, be the extreme angular position of the rod AB, let $BAb = \theta$, and let the horizontal distance AK of the centers of motion A, C, be made equal to $AB + CD -$ versin θ,

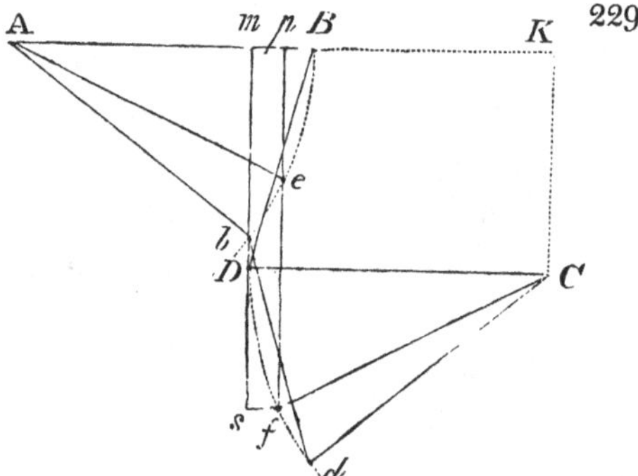

instead of being equal to the sum of the radii AB, CD, as in the former case. In this arrangement the radii being supposed parallel in the first position AB, CD, it is clear from the mere inspection of the figure, that the link is inclined to the left in one position as far as it is inclined to the right in the other very nearly; and therefore the central point of the link in the lowest position *bd* will be in the vertical line which passes through the place of its central point in the position *BD*.

But as the link is continually changing its inclination in the intermediate positions between these two, there will be in these intermediate positions a deviation of the central point from this vertical line, which it is easy to see will be at a maximum when the link is vertical. Let this happen when the radius rod is at an angle $BAe = \theta_{,}$,

and let $DCf = \phi_{,}$, and $mDB = dbs = \gamma$;

then we have $mD + Ds = pe + ef$,

that is, $l \cdot \cos\gamma + \sin\phi_{,} = \sin\theta_{,} + l$;

$$\therefore \sin\phi_{,} = \sin\theta_{,} + l \cdot \text{versin}\,\gamma = \sin\theta_{,} + \frac{l \cdot \gamma^2}{2}.$$

$$\text{But } \gamma = \frac{Bm}{l} = \frac{\text{versin}\,\theta}{l};$$

$$\therefore \sin\phi_{,} = \sin\theta_{,} + \frac{(\text{versin}\,\theta)^2}{2l},$$

also the deviation of the middle point $= \dfrac{\cos\phi_{,} - \cos\theta_{,}}{2}$.

The following table exhibits the corresponding values of the angles and deviation, supposing as before that $l = \frac{1}{2}$; and also that $\text{versin}\,\theta = 2\ \text{versin}\,\theta_{,}$ which is very nearly true.

θ	$\theta_{,}$	$\phi_{,}$	Deviation.
20^0	$14^0\ 7'$	$14^0\ 20'$	.00046
25^0	$17^0\ 36'$	$18^0\ 8'$	.00143
30^0	$21^0\ 1'$	$22^0\ 6'$	.00347
35^0	$24^0\ 33'$	$26^0\ 38'$	.00785

In practice the angle θ never exceeds 20^0. Let the radius rods be 3 feet in length, then the deviation in inches is $36 \times .0005 = .018$ instead of .097, as in Art. 444.

446. If the radius rods AB, DC are arranged on the same side of the link CB, and the link be produced downwards, as in fig. 230, then the upper rod being made shorter

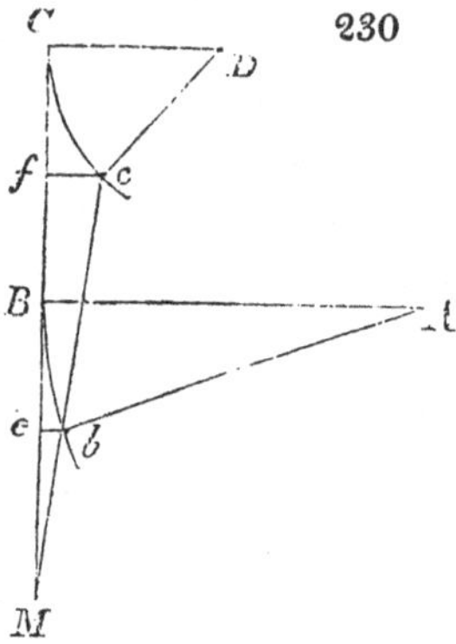

than the lower will move through a greater angle, and carry the upper end c of the link through a deviation cf greater than be, which is that produced by the longer rod. There will therefore be a point M in the link below the lower rod, which will remain in the line CB produced; and this point will be found by the proportion

$$\frac{cM}{bM} = \frac{fc}{be} = \frac{r \text{ versin } \phi}{R \text{ versin } \theta} = \frac{R}{r} \times \frac{r^2 \sin^2 \frac{\phi}{2}}{R^2 \sin^2 \frac{\theta}{2}},$$

when $AB = R$, $CD = r$, $BAb = \theta$, $CDc = \phi$.

But $r \sin \frac{\phi}{2} = R \sin \frac{\theta}{2}$ very nearly, whence $\frac{cM}{bM} = \frac{R}{r}$ gives the position of the point M.

447. The complete parallel motion which is most universally adopted in large steam-engines is shewn in fig. 231.

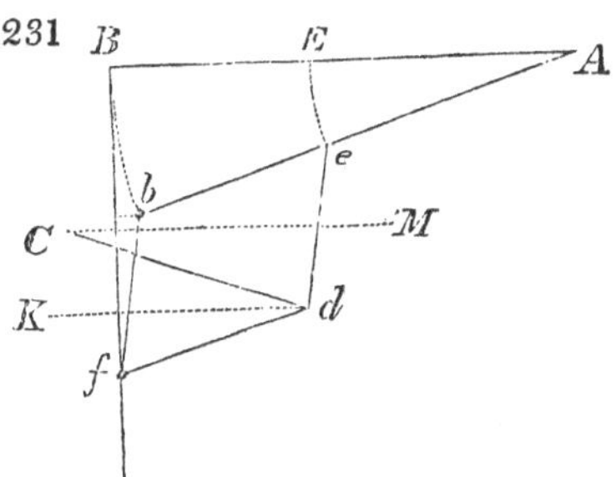

When so employed the beam of the engine becomes one of the radius rods of the system. Ab is half this beam whose center of motion is A. It has two equal links ed, bf jointed to it, of which bf is termed the *main link*, and ed the *back link*, and these are connected below by a third link df, termed the *parallel rod*, and equal to be. The radius rod or *bridle-rod* Cd is jointed to the extremity d of the back link ed, and its center C is fixed at a vertical distance below A equal to ed or bf. The length of the rods are so proportioned that f shall be the point to which the rectilinear motion is communicated, or *parallel point* as it is termed. To find the proportions let

$$Ae = R, \quad be\ (= fd) = R_{,}\,, \quad Cd = r,$$

draw Kd parallel to AB;

$$\therefore\ Kdf = BAb\,(=\theta), \text{ and let } MCd = \phi,$$

then, as before, the point d is carried towards K through a space equal to Cd versin $\phi = r$ versin ϕ, and the point f receives simultaneously this motion towards K, and a motion in the opposite direction arising from the inclination of the parallel rod df, which motion is equal to df versin fdK $= R_{,}$ versin θ. If these two motions be equal the point f will remain in the vertical line Bf, as required;

$$\therefore\ r \,.\, \text{versin}\,\phi = R_{,}\,\text{versin}\,\theta, \text{ or } \frac{r}{R_{,}} = \frac{\sin^2\dfrac{\theta}{2}}{\sin^2\dfrac{\phi}{2}}.$$

But the rods Ae, Cd, connected by the link ed, form a system similar to that of Art. 443, and, as before, we may assume

$$Ae.\sin\frac{\theta}{2} = Cd\sin\frac{\phi}{2}, \text{ or } R\sin\frac{\theta}{2} = r.\sin\frac{\phi}{2} \text{ very nearly};$$

$$\therefore \frac{\sin^2\frac{\theta}{2}}{\sin^2\frac{\phi}{2}} = \frac{r^2}{R^2} = \frac{r}{R_{,}}; \quad \therefore r = \frac{R^2}{R_{,}},$$

that is, Ae is a mean proportional between cd and df.

448. Since the joints Ae, ed, Cd, considered separately, form a system similar to the first simple arrangement, it follows that if the proper point be taken between d and e an additional parallel motion is obtained; so that this form combines two parallel motions in one, and is commonly so employed in steam engines, by suspending the great piston rod from f and the lesser air-pump rod from the link ed. The three parallel motions described (figs. 227, 230, and 231) are all due to Mr. Watt, and are to be found in his patent of 1784.

449. Let $AbfedC$, fig. 232, be an arrangement similar to the last; produce bf, and make $bp = ed\,\frac{Ab}{Ae}$.*

232

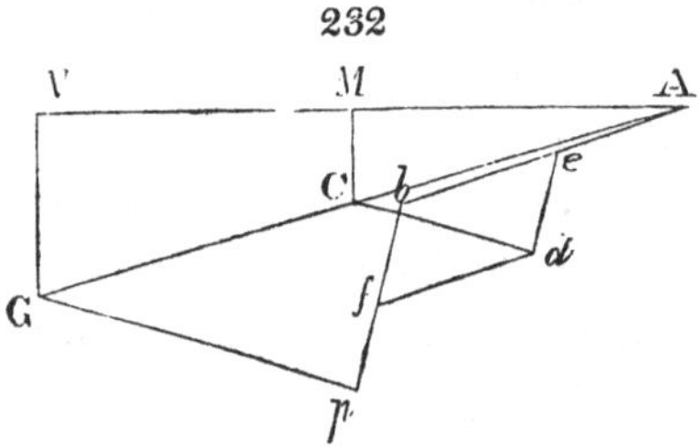

Join AC and produce it to G, making $AG = AC\,.\,\frac{Ab}{Ae}$, join Gp.

* From Prony, Arch. Hyd. Art. 1491.

Suppose Gp to be a new radius rod, moving round a fixed center G, it is clear in all positions of this arrangement that the lines Gp and Cd, bp and ed will remain parallel, on account of the fixed proportion of these lines respectively, therefore the point f would describe its straight line if fd were removed. But in that case the arrangement Ab, bp, pG considered separately forms a simple parallel motion of the first kind, and it appears that the more complex arrangement is equivalent to a simple one, occupying a greater space in the proportion of $AN : AM :: Ab : Ae$. Hence the convenience of the complex system.

450. There are various modifications of the latter arrangements, but the proportions of the rods may always be found in a similar manner to those already given. For example, in steam boats the beam is placed below the machinery, and the entire arrangement of the parallel motion inverted and otherwise altered to accommodate it to the necessity of compressing the entire machine into the smallest possible space.

Fig. 233 represents an arrangement of the parallel

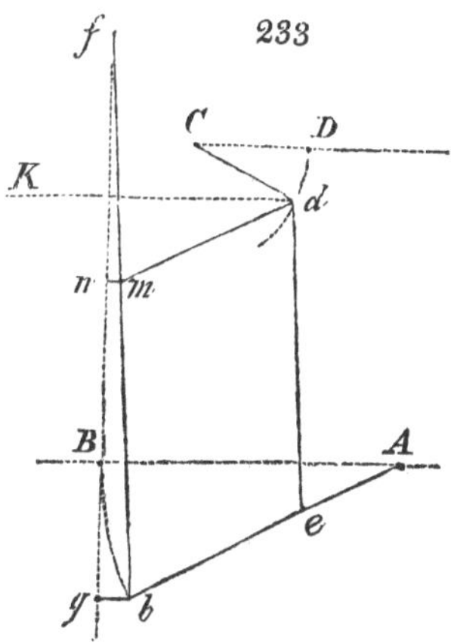

motion for steam boats, in which Ab is the beam, A its

center of motion; a short bridle rod, Cd, is employed, and the parallel rod dm is jointed to the main link bf below the parallel point f.

Let $Ae = R$, $eb = dm = R_{,}$, $Cd = r$, $DCd = \phi$, $BAb = \theta$. Draw AB, Kd horizontal, and fB vertical; then the point d is carried towards fB through a horizontal space

$$= Cd \text{ versin } DCd = 2r \,.\, \sin^2 \frac{\phi}{2}.$$

And the point m is carried horizontally to the left by this movement of d, and at the same time to the right through a space $= dm \times \text{versin } Kdm = 2R_{,} \sin^2 \frac{\theta}{2}$, since $dm = eb$ and $Kdm = BAb$.

The horizontal deviation of m from the vertical fB is therefore equal to $mn = 2R_{,} \sin^2 \frac{\theta}{2} - 2r \sin^2 \frac{\phi}{2}$.

Also the deviation of b from the vertical fB, is equal to $bg = Ab \times \text{versin } BAb = 2 \,.\, \overline{R + R_{,}} \,.\, \sin^2 \frac{\theta}{2}$,

and since f is the parallel point, we have

$$\frac{fm}{fb} = \frac{nm}{bg} = \frac{R_{,} \sin^2 \frac{\theta}{2} - r \sin^2 \frac{\phi}{2}}{\overline{R + R_{,}} \,.\, \sin^2 \frac{\theta}{2}}.$$

But in the system Cd, de, eA, we may assume

$$r \sin \frac{\phi}{2} = R \sin \frac{\theta}{2}, \quad \therefore \sin^2 \frac{\phi}{2} = \frac{R^2}{r^2} \sin^2 \frac{\theta}{2};$$

$\therefore$ putting $fm = x$, and $mb = l$, and arranging the terms,

we have $\dfrac{x}{l + x} = \dfrac{R_{,}r - R^2}{(R + R_{,})\, r}$, and $x = \dfrac{l}{R} \,.\, \dfrac{R_{,}r - R^2}{R + r}$.

If $R_{,}r = R^2$, $x = 0$ and the parallel point coincides with m, as in Art. 447. If $R_{,}r < R^2$, x becomes negative and the parallel point will fall between m and b.

451. Let an isosceles triangle *GFE*, fig. 234, be suspended by two equal radius rods *CE*, *AF*, moving on fixed centers *A* and *C*, and jointed to the two extremities of the base *FE* respectively.

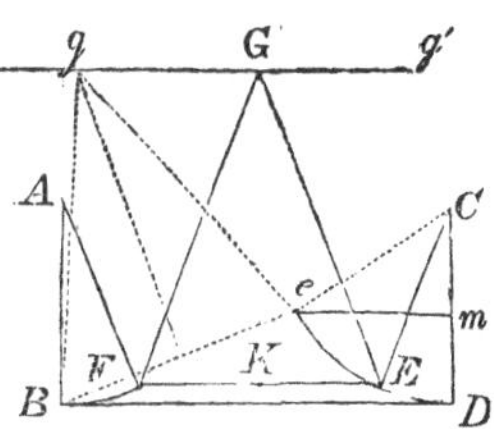

If now this triangle be swung from its central position, (that is, when the apex is equidistant from the points of suspension *A* and *C*), so as to carry its apex *G* to a little distance on either side, as for example to the position *g*, and to a similar one on the opposite side *g′*, then a describing point at *G* will draw a curve which will be found to vary very little from a right line whose direction is parallel to the base of the triangle when in its central position *GFE*, provided the proportions of the system be so arranged, that the three points *g G g′* are situated in a right line. This arrangement, which is the invention of Mr. Roberts of Manchester, furnishes a parallel motion which is in many cases more convenient than the former ones, especially if the path required be horizontal.

To investigate the proportions, draw the arcs *FB*, *DE*, make *AB*, *CD* perpendicular to *FE* and join *BD*. Let the extreme position be that in which the radius rod *AF* becomes perpendicular and coincident with *AB*, and the middle position that in which the base *FE* of the triangle is horizontal, and therefore parallel with *BD*. Then it remains to find such an altitude for the point *G*, that its vertical distance above *BD* may be the same in the middle and in the extreme position, in which case as the two extreme positions are symmetrical to the middle one, a right line parallel to *BD* will pass through the three positions of the apex *G*, as required.

Let $AB = CD = r$, $FE = b$, $BD = d$, $GK = h$,

$DCE = BAF = \theta$, $DCe = \phi$, $eBD = \psi$,

Then in the middle position, we have

$$2r \,.\, \sin\theta + b = d, \qquad (1)$$

in the extreme position,

$$b\cos\psi + r \,.\, \sin\phi = d, \qquad (2)$$

$$\text{and also } b \,.\, \sin\psi = r \,.\, \text{versin}\,\phi, \ (3).$$

Again, in the middle position, the altitude of G above BD is

$$h + r \,.\, \text{versin}\,\theta,$$

and in the extreme position the altitude of g above BD is

$$h \,.\, \cos\psi + \frac{b}{2}\sin\psi,$$

and these are equal by the conditions of the problem;

$$\therefore\ h + r \,.\, \text{versin}\,\theta = h \,.\, \cos\psi + \frac{b}{2}\sin\psi. \quad (4).$$

In these four equations we are at liberty to assume three of the quantities ϕ, ψ, θ, r, d, b, h, and the others may be determined, the most convenient is to assume values for r, d, and b. If $r = d = 1$, then the following table shews a few corresponding values of b and h.

b	h
$\frac{2}{3}$	3.95
.577	1.100
$\frac{1}{2}$	.943
.414	.654

But a convenient expression may be found by approximation, as follows: supposing that the angles of the system ϕ, θ and ψ are much smaller than those shewn in the figure;

for by (3) and (4) $\frac{h}{r} = \frac{\frac{1}{2}\text{ versin }\phi - \text{versin }\theta}{\text{versin }\psi}$,

in which if we assume

$$\text{versin }\phi = \frac{em^2}{2r^2},\ \text{versin }\theta = \frac{\left(\frac{em}{2}\right)^2}{2r^2} = \frac{1}{4}\cdot\frac{em^2}{2r^2},$$

$$\text{versin }\psi = \frac{mD^2}{2b^2},\ \text{where } mD = \frac{em^2}{2r},$$

$$\text{and } em = d - b,$$

$$\text{we finally obtain } \frac{h}{r} = \frac{b^2}{(d-b)^2}.$$

452. Let the lever AB, fig. 235, be jointed at the extremity A to a rod or frame EA moving round a fixed center E, and so long that the small arc Aa, through which the extremity of the lever A moves, may be taken for a right line in the direction of the line AF. CD is a bridle rod whose fixed center of motion C is in the line AF. Let $CD = r$, $AD = R$, $DB = R_{,}$, $DCA = \phi$, $DAC = \theta$, then, supposing as before for convenience that the machine is in a vertical plane and the line AF horizontal, the point D is carried horizontally to the right through a space $= r$ versin ϕ, and the point B receives this motion, and is also carried to the left horizontally by means of its inclination through a space $= R_{,}$ versin θ, and if these be equal, the horizontal distance of B from A will be the same as when the rods coincided with the horizontal line AF; therefore we must have

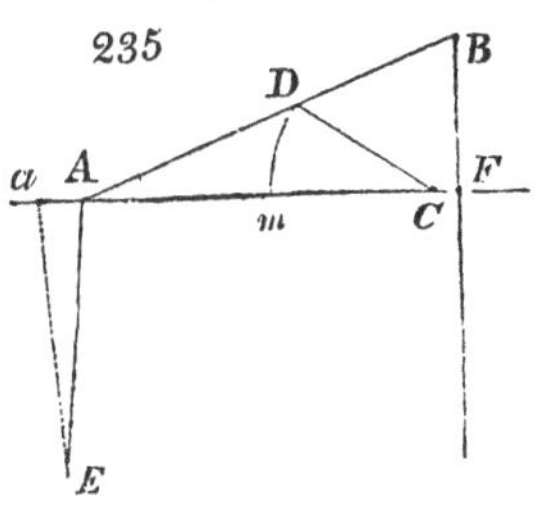

$$R_{,}\text{ versin }\theta = r\text{ versin }\phi, \qquad (1)$$

$$\text{also } Dm = R\sin\theta = r\,.\sin\phi \qquad (2).$$

From these two equations the value of $R_{,}$ may be obtained for any given values of R, r and θ; also,

$$\text{since } R_{,} \cdot \sin^2\frac{\theta}{2} = r\sin^2\frac{\phi}{2}, \quad \text{by (1)};$$

$$\text{and } R\sin\frac{\theta}{2} = r\,.\sin\frac{\phi}{2} \text{ very nearly, we obtain}$$

$$R_{,}r = R^2.$$

If the distances AD, DC, DB be equal, and the point A be made to travel in an exact straight line by sliding in a groove instead of the radial guide, then the parallel point will describe a true straight line perpendicular to AF, instead of the sinuous line which in all the other arrangements is substituted for it. For in this case the angle DAF is equal to DCA in all positions, and since $DB = DC$, a perpendicular from B upon AC will always pass through the same point C. In this respect this parallel motion has the advantage over all others.

If the friction of a sliding guide at A be considered objectionable, a small parallel motion of the first kind (Art. 443) may be substituted for it.

453. Toothed wheels are sometimes employed in parallel motions; their action is necessarily not so smooth as that of the link-work we have been considering, but on the other hand the rectilinear motion is strictly true, instead of being an approximation, as will appear by the two examples which follow.

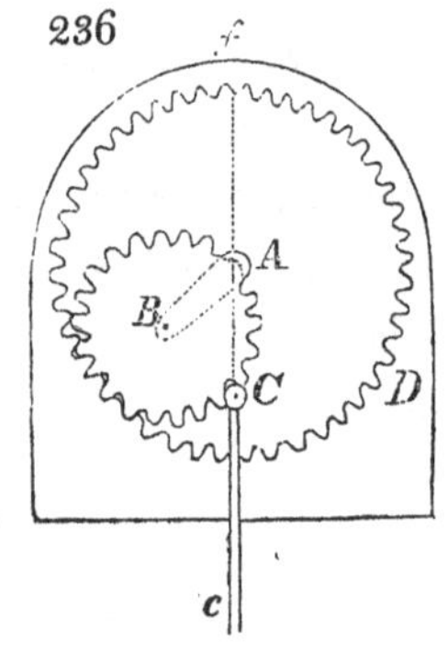

454. Ex. 1. In fig. 236 a fixed annular wheel D has an axis of motion A at the center of its pitch-line. An arm or crank AB revolves round this center of motion, and carries the center of a wheel B, whose pitch-line is exactly

of half the diameter of the annular wheel with whose teeth it geers. By the well known property of the hypocycloid any point C in the circumference of the pitch-line of B will describe a right line coinciding with a diameter of the annular pitch-circle. If then the extremity C of a rod Cc, be jointed to this wheel B by a pin exactly coinciding with the circumference of its pitch-circle, the rotation of the arm AB will cause C to describe an exact right line Cf, passing through the center A. This is termed White's parallel motion, from the name of its inventor*.

Since $AC = 2 \,.\, \cos BAC$, it is evident that the velocity ratio of C to BA is the same as in a common crank, and the motion produced in C equal to that which would be given by a crank with a radius equal to $2AB$, and an infinite link (Art. 328).

455. Ex. 2. Two equal toothed wheels, A and B, fig. 237, carry pins c and d at equal radial distances; and

237

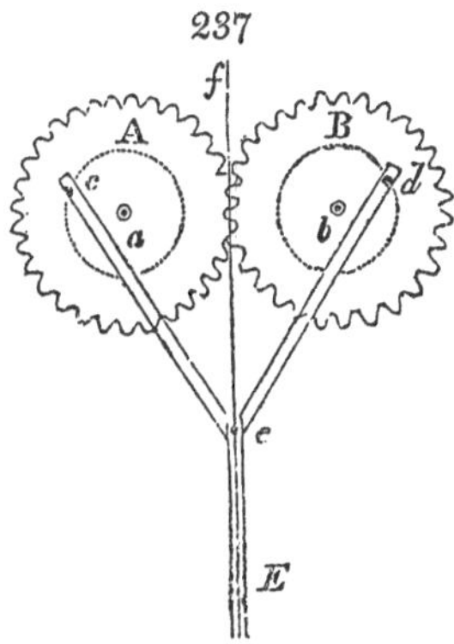

symmetrically placed with respect to the common tangent of the pitch-circles fe. If two equal links ce, de be jointed to these pins and to the extremity of a rod eE, the point e will plainly always remain in the common tangent, by virtue of

* Vide White's Century of Inventions.

the similar triangles formed by the rods, the tangent *fe*, and the line *cd*.

The velocity ratio of *e* to the wheels is not however the same as that produced by the common crank and link of fig. 168, Art. 328, for the path of *e* does not pass through the center of motion of the crank.

If however r be the radius of the crank *ac* or *bd*, R the radius of the pitch-circles of the wheels, l the length of the link *ce* or *ed*, and the angle $cab = \frac{\pi}{2} + \theta$, then it can be easily shewn that the distance of *e* from the line of centers *ab* is equal to $\sqrt{l^2 - (R \pm r \sin \theta)^2} \pm r \,.\, \cos \theta$.

PART THE THIRD.

ON ADJUSTMENTS.

CHAPTER I.

GENERAL PRINCIPLES.

456. In the elementary combinations which have occupied the two previous Parts of this subject, the angular velocity ratio and directional relation in any given combination are determined by the proportion and arrangement of the parts, and will either always remain the same, or their changes will recur in similar periods. But it is necessary in many machines that we should have the power of altering or adjusting these relations. These adjustments may be distributed under three heads.

(1.) To break off or resume at pleasure the communication of motion in any combination.

(2.) To reverse the direction of motion of the follower with respect to that of the driver; that is, to change their *directional relation.*

(3.) To alter the *velocity ratio* either by determinate or by gradual steps.

These changes may be either made by hand at any moment, or they may be effected by the machine itself, by means of a class of organs especially destined for that purpose; and which are in fact a kind of secondary moving powers to the machine.

457. The communication of motion may be broken off by detaching pieces that remain united during the action of the combination, and therefore move as one. Thus wheels and pullies are connected with their shafts for this purpose, by means of catches or bolts; and shafts are connected endlong with each other by couplings, or other contrivances which admit of being released or put in action at pleasure. Otherwise the communication may be broken off by disengaging the driver from the follower, which in the two kinds of contact action is effected by withdrawing the pieces from each other; in wrapping connections, by either slackening the belt or by slipping it off the pully; and in link-work, by disengaging the joints of the links.

458. But the whole of these contrivances as well as those by which the directional relation is changed, belong to constructive mechanism, and as they involve no calculations relating to the velocity ratio, which is the principal object of the present work, I shall not enter into any details respecting them, referring in the mean time to the Encyclopædias and other treatises on machinery, in which they are fully explained*. The case is different with respect to the third kind of adjustments in which the velocity ratio is the subject of alteration, and I shall therefore give examples of the principal methods of effecting this purpose.

The adjustments of the velocity ratio may consist either of (1) Determinate changes, which for the most part require the machine to be stopped, or of (2) Gradual changes, which do not require the machine to be stopped.

* Vide especially Buchanan's Essays on Mill-work by Rennie, in which these combinations are very fully treated of.

CHAPTER II.

TO ALTER THE VELOCITY RATIO BY DETERMINATE CHANGES.

---

459. Let there be two axes A, B, whose position in the machine is fixed; and let it be required to connect these by toothed wheels in such a manner that the velocity ratio may assume any one of a given set of values. The simplest method is to provide as many pairs of wheels as there are to be values, and let the sum of the pitch-radii of each pair equal the distance AB of the centers. Then to obtain any one of the required ratios, we have only to screw the proper pair of wheels to the ends of the axes. Sets of wheels for this purpose are commonly termed *Change-wheels.* It is generally convenient that all the change-wheels should be of the same pitch, and the numbers may be calculated as in the following example. Let the given set of values for the velocity ratio or the *change-ratios* be $\frac{1}{1}, \frac{2}{1}, \frac{3}{1}, \frac{4}{1}, \frac{3}{2}, \frac{5}{4}$. Then, since the pitch and distance of the centers are the same in every pair, the sum of their numbers of teeth must be the same; and this sum must also be divisible by the sum of the numerator and denominator of each of the above fractions, or by 2, 3, 4, 5, 9. The number required is therefore a multiple of $2^2 . 3^2 . 5 = 180$, and if 180 be taken as the least possible number, we have the following pairs of wheels, which manifestly fulfil the conditions:

Ratios.	Wheels.
1	90 ... 90
2	60 ... 120
3	45 ... 135
4	36 ... 144
$\frac{3}{2}$	72 ... 108
$\frac{5}{4}$	80 ... 100

460. To save the trouble of screwing and unscrewing the wheels, the entire set may remain fixed upon their respective axes, if arranged upon the principle of fig. 238.

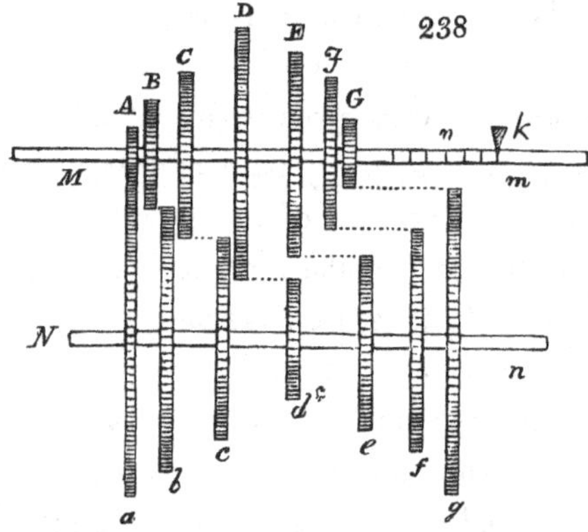

Mm, *Nn* are the two axes; *A*,*a* . *B*,*b* . *C*,*c* . &c.; the respective pairs of change-wheels and the sum of the radii of every such pair being equal to the distance of the axes, the teeth of any pair that are set opposite to each other will work. For this purpose the upper axis is capable of sliding endlong, and is retained in any required position by a bolt *k*, which enters into a groove *m* turned upon the axis. In the figure *A* and *a* are shewn in action, but if any other pair, as *D*, *d*, are required to work together, the bolt *k* must be removed, and the axis shifted endlong until *D* and *d* come into geer. The same motion will bring the groove n opposite to the bolt by which the shaft may be secured in this new position, and similarly for any other pair of wheels.

The wheels must be, however, so placed upon the shafts, that only one pair will come opposite to each other at the same time. To effect this, the wheels are arranged in the order of their magnitudes, placing the smallest at each end of the upper group, and the others in alternate order with the largest in the middle, and the wheels of the lower shaft in the reverse order, for a reason which will presently appear.

Let m be a quantity rather greater than the thickness of each wheel. Then, A and a being in contact, let the lateral distance of B from $b = m$, that from C to $c = 2m$, from D to $d = 3m$...... and that from the nth wheel to its fellow $= (n-1) \cdot m$.

But as every successive wheel B or C is too great to be pushed past the previous wheel a or b of the lower group, these upper wheels, to make the axis as short as possible, must each lie close to the previous wheel when the upper group is in its extreme position to the left; and therefore the smallest distance between the wheels of the upper set will be from A to $B = 0$, from B to $C = m$, from C to $D = 2m$, and so on; between the lowest set from a to $b = m$, from b to $c = 2m$... and so on; and if the wheels were each arranged in one conical group, as from A to D, and from a to d, the length of shaft required for n wheels would be the sum of the thickness of all the wheels + their distances, which, for the upper shaft, is equal to

$$[n + \{0 + 1 + 2 + \ldots (n-2)\}]m = \left\{(n-1) \cdot \frac{(n-2)}{2} + n\right\} m$$

and for the lower shaft equal to,

$$[n + \{1 + 2 + 3 + \ldots (n-1)\}]\, m = \frac{n+1}{2} \cdot nm.$$

By arranging the wheels in two conical groups, as in the figure, they occupy a much shorter length upon the shafts; for the central wheel D can be pushed past its own wheel d, and the same reasoning will then be true for the conical group $DEFG$ and $defg$.

Thus the length of shaft required for n wheels in two groups of $\frac{n}{2}$ each, will be for the lower shaft,

$$\frac{n}{2}\,\frac{\frac{n}{2}+1}{2}\,.\,m + \frac{n}{2}\,m,$$

(where $\frac{n}{2}m$ is the space between the two groups)

$$= n\,.\,\frac{n+6}{8}\,m,$$

which is much less than the former, and similarly for the upper shaft.

In our example, the wheels on the upper and lower shafts occupy spaces of $13m$ and $19m$ respectively, and if they had been arranged each in one conical group would have occupied spaces equal to $22m$ and $28m$.

Similar arrangements to this are adopted in cranes for raising weights, in which the choice of three or four velocity ratios is required between the handle and chain-barrel.

461. But it is often inconvenient to make the sum of the radii of change-wheels equal to the distance of the centers, and requires, moreover, as many different pairs of change-wheels as there are to be changes in the velocity ratio, unless indeed some of these ratios be merely the inverse of others. The more usual method therefore is, to screw a pair of wheels of the proper numbers to the end of the axes, without

regard to their radii, and afterwards to connect them by an idle-wheel. Art. 223.

Thus let *a* and *b*, fig. 239, be the axes upon which a pair of change-wheels *A* and *B* have been fitted.

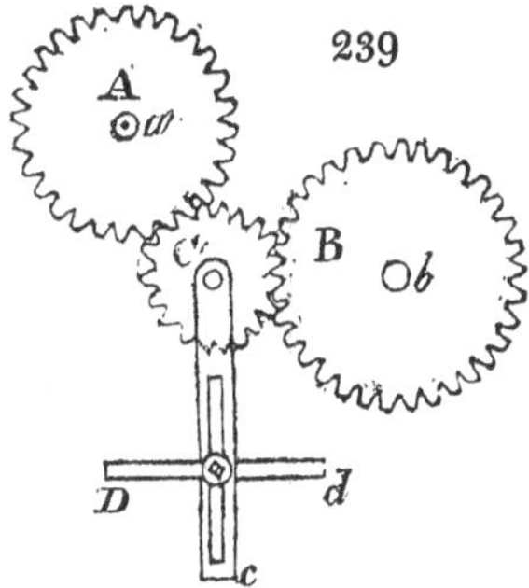

C is the idle-wheel which may revolve upon a pin or stud fitted to the end of a piece *Cc*, which has a long slit at its extremity. A slit *Dd* in the transverse direction is formed in the frame of the machine, and the piece *Cc* which carries the idle-wheel is fixed in its place by a bolt passing through the two slits at their intersection.

By this method of fixing the idle-wheel it admits of being shifted about so as to be put in geer with the two change-wheels whatever be their diameters.

There are various other methods of shifting and fixing the variable center of the idle-wheel, but the effect is the same in all. If it be required also to have the power of changing the directional relation, another piece like *Cc* must be provided, upon which two idle-wheels in geer are mounted, and this piece must be brought into such a position that one of these wheels shall geer with *B* and the other with *A*; *A* and *B* will therefore turn in opposite directions, whereas in fig. 239 they turn in the same direction.

The number of change-wheels is greatly reduced in this manner, because they admit of being combined in any pairs; thus, in the example (Art. 459), six change-wheels will be sufficient instead of twelve, thus:

Ratios.	Wheels.
1	24 ... 24
2	24 ... 48
3	24 ... 72
4	24 ... 96
$\frac{3}{2}$	48 ... 72
$\frac{5}{4}$	48 ... 60

462. *On Speed Pullies.* Let there be two parallel axes *Aa*, *Bb*, fig. 240, upon each of which is fixed a group of pullies adapted for belts or bands, and of different diameters. A ready mode is thus provided of changing the angular velocity ratio of the shafts by merely shifting the belt from one pair of pullies to another. Such groups of pullies are termed *Speed Pullies.* The diameters of every pair of opposite pullies ought to be so adjusted that the belt shall be equally tight upon any pair. If the belt be crossed, it is easy to shew that this object will be attained by making the sum of the diameters of every pair of opposite pullies the same throughout the set. For let *DK*, *FG* be the radii of any pair, make *GK* a common tangent to the pullies, draw *FE* parallel to *GK* and describe a circle with radius $DE = DK + FG$.

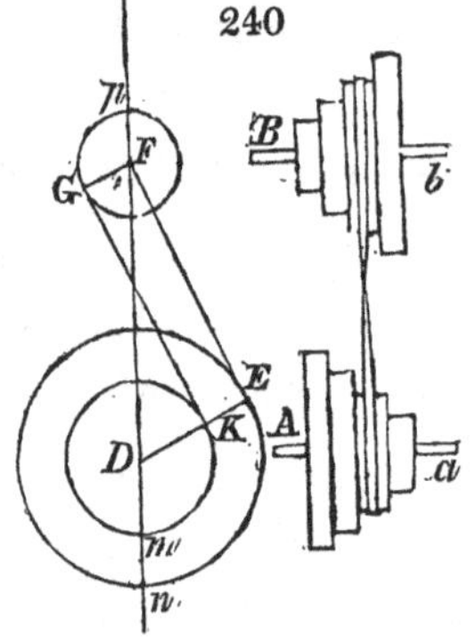

$$\text{Then } \tfrac{1}{2} \text{ length of belt} = mK + KG + Gp,$$

$$\text{and } mK + Gp = Dm \,.\, mDK + FG \,.\, GFp$$

$$= DE \times mDK \text{ for } mDK = GFp;$$

$$\therefore\ \tfrac{1}{2} \text{ length} = nE + EF,$$

which is constant for any pair of pullies of which the sum of the radii equals DE.

463. In any group of speed-pullies if D be the diameter of any follower, and K the constant sum of the diameters, $K-D$ will be the diameter of its driver. And if L, l be the synchronal rotations of the driver and follower respectively,

$$\frac{l}{L} = \frac{K - D}{D} = \frac{K}{D} - 1,$$

$$\text{and } D = \frac{KL}{L + l},$$

in which equation putting for L and l the required series of values, the corresponding diameters of the speed-pullies may be obtained.

464. To save founders' patterns it is usual in practice to make the two groups of speed-pullies exactly alike, placing the small end of one opposite to the large end of the other.

A regular geometrical series of values of $\frac{L}{l}$ may be obtained for such a pair of similar pullies, as follows: Let r be the common ratio of this series, n the number of terms, then the extreme terms of the series must evidently be the reciprocals of each other, therefore the series will be (putting $m = \frac{n-1}{2}$ for convenience) of the form,

$$\frac{1}{r^m}\ \frac{1}{r^{m-1}}\ \frac{1}{r^{m-2}} \ldots\ldots r^{m-2},\ r^{m-1},\ r^m.$$

But if K be the constant sum of the diameters, and $D_1 D_2 \ldots$ the diameters of the pullies in order, the same series will be

$$\frac{D_1}{K-D_1},\ \frac{D_2}{K-D_2},\ \ldots\ldots \frac{K-D_2}{D_2},\ \frac{K-D_1}{D_1},$$

and comparing the corresponding terms, we have

$$\frac{D_1}{K-D_1}=\frac{1}{r^m};\quad \therefore\ D_1=\frac{K}{1+r^m},\ \text{similarly } D_2=\frac{K}{1+r^{m-1}},$$

and so on.

465. Ex. 1. To find the diameters of a set of speed-pullies that shall give four values for $\frac{l}{L}$, with a common ratio of 1.38; the sum of the diameters of the corresponding pullies being 25 inches.

$$\text{Here } K=25,\ r=1.38,\ n=4,\ m=\frac{3}{2};$$

$$\therefore\ D_1=\frac{250}{26}=9.6,\ D_2=\frac{250}{22}=11.4,$$

$$D_3=K-D_2=13.6;\ \text{and } D_4=K-D_1=15.4,$$

are the diameters in inches.

Ex. 2. Let there be a set of six speed-pullies in each group, of which the diameters of the extremes are 13in. and 4in.: to find the intermediate diameters.

The first and last terms of the geometrical series of six velocity ratios is $\frac{4}{13}$ and $\frac{13}{4}$, hence the common ratio being found by logarithms as usual, gives $r = 1.61$.

$$\text{Also } K=17,\ m=\frac{5}{2};$$

whence the successive diameters are 4, 5.6, 7.5, 9.5, 11.4, 13, in inches.

466. If a great number of changes of velocity be required either in the case of speed-pullies or toothed wheels, a *train* of axes must be employed, with the power of introducing a given number of changes between each, in which case the total number of changes in the system will be the continual product of the numbers of changes that can take place between each pair. Considering only a set of four shafts for the sake of simplicity, let A_1, A_2, A_3, A_4, be the angular velocities of the axes in order, and let the series of changes in the value of $\frac{A_1}{A_2}$ form a geometrical series whose common ratio is r, and first term a; $\therefore \frac{A_1}{A_2} = ar^{n-1}$ is the nth term of this series. Similarly, let the mth term of the series of values of $\frac{A_2}{A_3} = bs^{m-1}$, and the kth term of the series of values of $\frac{A_3}{A_4} = ct^{k-1}$. $\therefore$ Angular velocity ratio of the extreme axes of the train when the nth, m^{th}, and k^{th} values of the respective ratios are employed

$$= \frac{A_1}{A_4} = abc \,.\, r^{n-1} \,.\, s^{m-1} \,.\, t^{k-1} = Cr^{n-1} \,.\, s^{m-1} \,.\, t^{k-1} \text{ suppose.}$$

Let the number of changes or terms of which each of these series consists be m, n and k respectively, then may the entire set of changes in the system be arranged in a continuous geometrical series with a common ratio t, as in the margin; provided we have

$$\begin{array}{l} C \\ Ct \\ Ct^2 \\ \vdots \\ Ct^{k-1} \\ Cs \\ Cst \\ Cst^2 \\ \vdots \\ Cst^{k-1} \\ \vdots \\ Cs^2t^{k-1} \\ \vdots \\ Cs^{m-1}t^{k-1} \\ Cr \\ Crt \\ \vdots \\ Cr^{n-1}s^{m-1}t^{k-1} \end{array}$$

$$\frac{Cs}{Ct^{k-1}} = t, \quad \therefore s = t^k.$$

And also $\frac{Cr}{Cs^{m-1}t^{k-1}} = t$;

$$\therefore r = s^{m-1}t^k = s^m = t^{km}.$$

If however we had counterchanged the values by making $\frac{A_1}{A_2} = c't^{k-1} \; \frac{A_2}{A_3} = ar^{n-1}$, and so on, the same value would have been obtained for $\frac{A_1}{A_4}$. It appears therefore that to form a regular geometrical series of changes whose velocity ratio shall be t, the separate series of change-values of the velocity ratios $\frac{A_1}{A_2}$, $\frac{A_2}{A_3}$ &c., must be so arranged that the common ratio of some one of these series must be t, and if there be k changes or terms in this series, then the common ratio of a second must be t^k; also if this have m changes, the common ratio of a third set must be t^{km}, and so on.

467. Ex. 1. Change-wheels are employed in lathes for cutting screws of any required pitch, and also in self-acting lathes. The diagram, fig. 241, represents the general arrangement of this mechanism.

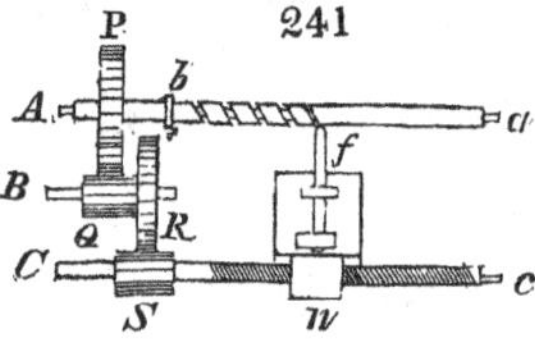

Ab is the spindle or mandrel of the lathe, to which is united in the usual way a cylindrical rod *ba* upon which the screw is to be cut. *Cc* is a long screw revolving in bearings fixed to the frame of the lathe, and giving motion by means of the nut *n* to a sliding table or *saddle* upon which is clamped the pointed tool *f*, which is intended to cut the screw*.

Every revolution of the screw *Cc* will therefore advance the tool through the space of one pitch, and if the spindle *Aa* revolve with the same velocity as the screw, the tool will trace upon the surface of *ba* a screw exactly of the same

* This construction of a screw-cutting engine was first employed, I believe, by Ramsden, and is at present universally followed. Vide Desc. of the Engine for dividing Math. Inst. by Ramsden.

pitch as Cc. But if Aa revolve with a less velocity than the screw, ba will have a greater pitch.

If Aa and Cc be connected by a set of change-wheels P, S, as in fig. 239, we can, by properly choosing the numbers of these wheels, obtain any desired pitch for the screw ba.

B is an intermediate axis supported by a slit piece as in fig. 239, and either carrying an idle-wheel or two additional change-wheels Q and R. The pitch of screws is commonly defined by stating the number of threads in an inch. Let the screw Cc have n threads in the inch. Then one turn of Cc advances the tool through the space of $\frac{\text{inch}}{n}$, and one turn of Aa advances the tool through the space which corresponds to $\frac{PR}{QS}$ turns of Cc, that is, through $\frac{PR}{QSn}$ inches. The pitch of the screw Aa is therefore $\frac{QSn}{PR}$ threads in the inch. Thus by providing the proper change-wheels, a screw of any required pitch can be cut. The pitches usually cut upon these lathes extend from about four to fifty threads in the inch, and a set of twenty change-wheels will be generally sufficient to supply all the values required for $\frac{QS}{PR}$. These should be arranged in a table, and the wheels corresponding to each written opposite to them, to save the trouble of computation during the work.

468. If the apparatus, fig. 241, is used for turning cylinders instead of for cutting screws, the arrangement will not essentially differ, for the motion by which a tool traces a cylinder is precisely the same as when it cuts a screw, only that the spiral thread is much closer. In a lathe for turning, the number of cuts will be from 50 to 1000 in an inch.

In computing the change-wheels for this purpose, we may employ the principle of Art. 466, as in the following Example.

469. Ex. 2. Let it be required to compute a set of change-wheels for a self-acting turning lathe, that shall have a choice of twelve different pitches for the cuts, varying from about 50 to 1000 in the inch.

The motion to be produced in the tool f is very slow, and an endless screw may be therefore substituted for the wheel P, and as this will place the axis B at right angles to Cc, the wheels R and S must be bevil wheels.

Let the screw Cc have 9 threads to the inch, therefore $n = 9$, and $P = 1$, being an endless screw, therefore the number of cuts in the inch $= 9 \cdot Q \cdot \frac{S}{R}$.

This quantity by the conditions of the problem is to have twelve values, forming a geometrical series of which the first and last terms are 50 and 1000, and therefore the common

$$\text{ratio} = t = \left(\frac{1000}{50}\right)^{\frac{1}{11}} = 20^{\frac{1}{11}} = 1.313 \text{ by logarithms.}$$

By Art. 466, it appears that if we give to Q four values, and to $\frac{S}{R}$ three values, these sets must each form a geometrical series, of which if the common ratio of the first $= t = 1.313$, that of the second must $= t^4 = 2.972$, $= 3$ very nearly.

Let the intermediate change of $\frac{S}{R}$ be made by employing two equal wheels, then the three values of $\frac{S}{R}$ will stand thus, $\frac{1}{3}$, 1, 3, and the same pair of wheels will serve for the two extreme values by merely reversing their positions as driver and follower; thus $\frac{20}{60}$, $\frac{40}{40}$, $\frac{60}{20}$, may be the three values of $\frac{S}{R}$, which are obtained by four wheels only.

Geometrical Series.	Q	$\frac{S}{R}$	Cuts in the Inch.
1000	37		999
761.6	28	$\frac{60}{20}$	756
580.	21		567
441.7	16		432
336.4	37		333
256.2	28	$\frac{40}{40}$	252
195.1	21		189
148.6	16		144
113.2	37		111
86.2	28	$\frac{20}{60}$	84
65.6	21		63
50	16		48

The geometrical series of values of $9 \, . \, Q \, . \, \frac{S}{R}$ being obtained, as in the first column of the table, we have for the four middle terms $\frac{S}{R} = 1$, and therefore the values of Q, that is, the numbers of teeth of the endless screw-wheels will be obtained by dividing these terms by nine and taking the nearest whole numbers, by which we get 37, 28, 21, 16*. The difference between the last column of the table and the first is occasioned by the necessary substitution of whole numbers for decimals in the teeth of the wheels.

This system requires eight wheels for the twelve changes, but by a slightly different arrangement seven wheels may be made to answer the same purpose.

* These numbers of teeth are the same as those of a lathe by Mr. Clements, Trans. Soc. Arts. Vol. 46.

Let three values be given to Q and four to $\frac{S}{R}$, then the common ratio of the values of Q being as before $t = 1.313$, that of the values of $\frac{S}{R}$ will now be $t^3 = 2.26$, and these values may be obtained by four wheels thus,

$$\frac{20}{68}, \frac{32}{48}, \frac{48}{32}, \frac{68}{20}.$$

Let the screw Cc have ten threads in the inch, then we easily find the numbers for the endless screw-wheel Q to be 29, 22, 17, and the table for this second system will stand as follows, employing only seven wheels, namely, two pair of bevil-wheels, and three screw-wheels.

Geometrical Series.	Q	$\frac{S}{R}$	Cuts in the Inch.
1000	29		986
761.6	22	$\frac{68}{20}$	748
580	17		578
441.7	29		435
336.4	22	$\frac{48}{32}$	330
256.2	17		255
195.1	29		193
148.6	22	$\frac{32}{48}$	147
113.2	17		113
86.2	29		85
65.6	22	$\frac{20}{68}$	65
50	17		50

470. Ex. 3. In large engineers' lathes for turning metal the motion is derived from a shaft which revolves uniformly under the action of a steam-engine; but it is

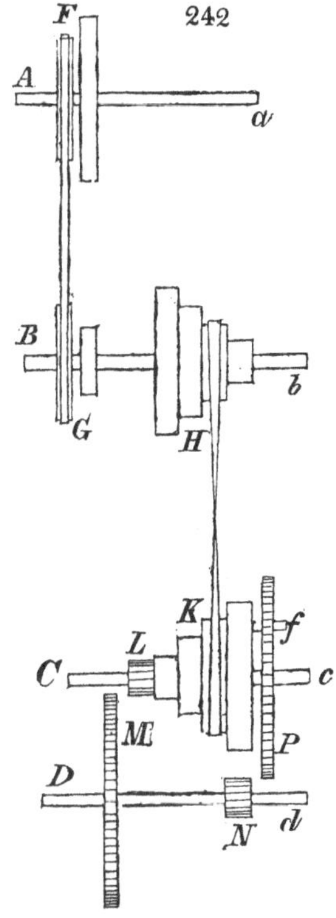

necessary to have the power of changing the velocity of the mandrel of the lathe, to accommodate the different diameters of the work, or the material of which it is composed. The usual arrangement for this purpose is shewn in the diagram fig. 242. *Aa* is the shaft which is driven uniformly by the steam-engine, *Bb* a second shaft termed the counter-shaft. Two pullies are fixed at *F* and two others opposite to them at *G*, and an endless band upon either pair will thus enable *Aa* to drive *Bb*. *Cc* is the mandrel of the lathe, upon which is fixed a toothed wheel *P*: a group of four or more speed-pullies *K* runs loose upon the mandrel, but may be locked fast to the wheel *P*, at pleasure, by a bolt *f*.

Opposite to *K* a similar group of speed-pullies is fixed at *H* to the counter-shaft *Bb*, so that if *K* be locked fast to the mandrel motion is given to the latter from the counter-shaft, by means of an endless band placed upon any pair of the speed-pullies. But if the pullies *K* be loosed from the wheel *P* by withdrawing the bolt *f*, their motion is conveyed to the mandrel by means of a pinion *L* which is attached to the end of the speed-pullies. In this case the spindle *Dd* is pushed endlong through a small space, so as to bring its toothed-wheel *M* into geer with *L*, and at the same time its pinion *N* into geer with *P*, so that the mandrel and its wheel *P* now derive their motion from the shaft *Dd* which is turned by the speed-pullies. In this latter arrangement the motion of the mandrel Cc is very much slower than that of the speed-pullies.

In this system then we have two changes between Aa and Bb, or two values of $\frac{F}{G}$*; four between Bb and the speed-pullies K, or four values of $\frac{H}{K}$; and two changes between the speed-pullies K and the mandrel; that is unity and $\frac{LN}{MP}$; making the total number of changes of the velocity ratio between Aa and Cc equal to $2 \times 4 \times 2 = 16$; and we may arrange them (by Art. 446) in a geometrical series whose common ratio is t. Thus let the common ratio of the series of four values of $\frac{H}{K} = t$, and that of the two values of $\frac{F}{G} = t^4$, then will that of $\frac{MP}{LN} = t^8$.

For example, let the shaft Aa revolve at the rate of sixty turns in a minute, and let it be required that the mandrel Cc shall revolve from 2 to 270 in a minute. A geometrical series of sixteen terms of which 2 and 270 are the extremes, would have a common ratio of

$$1.38 = t; \quad \therefore\ t^4 = 3.7, \text{ and } t^8 = 13.68.$$

The diameters of the speed-pullies with the ratio of 1.38 have been already obtained in Ex. 1, Art. 465, and are 9.6, 11.4, 13.6, 15.4, and as the quick ratio between the speed-pullies and mandrel is unity, we have, when the mandrel revolves at its extreme ratio of 270 in a minute,

$$\frac{270}{60} = \frac{15.4}{9.6} \times \frac{F}{G};$$

$$\text{whence } \frac{F}{G} = 2.8 \text{ is the quick value of } \frac{F}{G},$$

$$\text{and its second value} = \frac{2.8}{t^4} = \frac{2.8}{3.7} = .75.$$

If the diameters of the pullies at F be 15^{in} and 28^{in}, those at G must be 20^{in} and 10^{in}.

* The letters of reference opposite to each group of change-wheels are here used to represent the pair which is in action.

Again, to find the numbers of the train of toothed wheels, we have

$$\frac{MP}{LN} = t^8 = 13.68 = \frac{1368}{100} = \frac{2 \cdot 3^2 \cdot 19}{5^2}.$$

Now the pinions L and N ought not to have less than twelve leaves, and it appears from this fraction that they must be multiples of five, we may therefore give them fifteen leaves each; whence the convenient train

$$\frac{MP}{LN} = \frac{54 \times 57}{15 \times 15}.$$

The following table shews the result of these arrangements.

Geometric Series of Turns per min. of *Cc.*	Values of *H.*	*K.*	Values of *F.*	*G.*	
2	9.6	15.4			
2.8	11.4	13.6	15	20	
3.8	13.6	11.4			train
5.3	15.4	9.6			$\frac{54 \times 57}{15 \times 15}$
7.4	9.6	15.4			employed.
10.3	11.4	13.6			
14.2	13.6	11.4	28	10	
19.7	15.4	9.6			
27.4	9.6	15.4			
37.9	11.4	13.6	15	20	
52.6	13.6	11.4			
73.	15.4	9.6			pullies *K* bolted to mandrel.
101.2	9.6	15.4			
140.4	11.4	13.6			
194.7	13.6	11.4	28	10	
270.	15.4	9.6			

471. In adjusting trains upon these principles it must be remarked, that for a given series of velocity ratios between the extreme axes, the total number of separate changes will be the least when the number of changes allotted to the component series are equal, or $m = n = k$ (Art. 466). But the nature of the mechanism will not always allow of this with convenience. For example, since the ratios of the component geometrical series are necessarily each greater than the previous one in order, as t, t^k, t^{km}, &c....; it appears that the differences of value in the radii of the pullies or wheels of the first set is much less than in those of the succeeding ones, and therefore it may be better to assign a greater number of change values to that series whose common ratio is the smallest, or t; although by so doing the last ratio t^{km} is increased, because a group of speed-pullies will always readily supply a series of values provided their common ratio is not too great. Indeed, the values of the separate common ratios would be diminished by assigning a greater number of changes to that series whose common ratio is t^{km}; that is, by giving a higher value to n which does not enter into the common ratios, than to k and m which do; thus in the last example, the respective values of k, m, n, are 4, 2, 2; if we take for these, 2, 2, 4; we obtain $t = 1.38$, $t^2 = 1.904$, $t^4 = 3.7$, which avoids the great common ratio 13.68, but here the ratio 3.7 is too great for a set of four speed-pullies.

Again, if the respective values of k, m, n were made 3, 3, 2, the number of component changes would be the same as before, that is, $3 + 3 + 2 = 8$, but the total number of changes would be increased to $3 \times 3 \times 2 = 18$, and the common ratios would be $t = 1.33$, $t^3 = 2.37$, $t^9 = 13.42$, so that by putting three pair of speed-pullies at F, G, and three at H, K, with the common ratios of 2.37 and 1.33 two more changes are added to the system without increasing the number of speed-pullies, and the great ratio 13.42 rather lessened.

However, it is plain that the nature of the mechanism that admits of being conveniently employed and the amount of changes required must always be taken into account in every particular case, and a number of different trains calculated to choose from. When change-wheels are employed, as in Art. 459, their number may sometimes be reduced by computing their teeth upon the principles of Art. 464, which plainly apply as well to tooth-numbers as to the diameters of speed-pullies. Thus every pair of the series is used twice, since every two terms equidistant from the ends are the inverse of each other.

472. In link-work adjustments are very simply made by drilling holes in the arms and shifting the joint-pins from one to another, or by more elaborate constructive devices for altering the efficient lengths of the arms of the links; the details of which do not fall within the plan of our present work.

CHAPTER III.

TO ALTER THE VELOCITY RATIO BY GRADUAL CHANGES.

473. In the methods of the last Chapter it is obviously necessary that the machines should be stopped in order to effect the necessary changes of the wheels, or in the position of the bolts, and so on; and besides, the series of changes themselves are not continuous, and we have only the choice of a few given intermediate ratios between the extremes. We have now to consider how the velocity ratio may be altered by gradual changes, so as to enable us to take any value for it between the extremes. The same constructions will generally enable the changes to be made without interrupting the motions of the machine.

474. Let *Aa*, *Bb*, fig. 243, be parallel axes, *C*, *D*

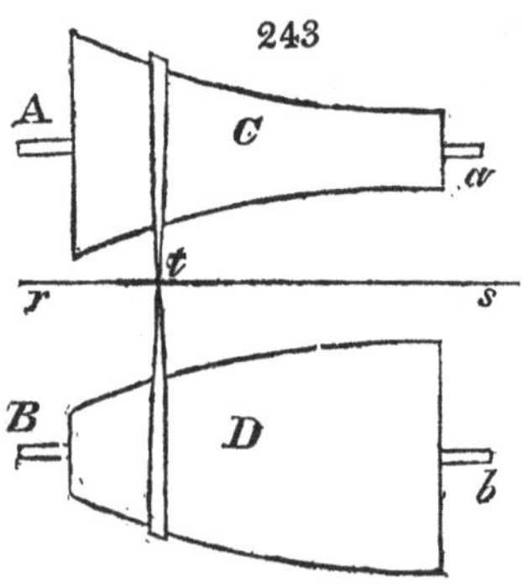

solids of revolution or long pullies connected by an endless strap. If this strap be crossed and the sum of every opposite pair of diameters of these solids be constant, the strap will be tight in any position upon them. A bar rs slides in the direction of its own length, and is provided at t with a loop or with friction-rollers, between which the belt passes, and which serves to retain it in its place. In Art. 184 it is shewn that a belt may be guided *by its advancing side* to any point of the surface of a revolving cylinder; and this guide-loop embracing the sides of the belt which are advancing to the two pullies is sufficient to retain them in any position upon their surfaces, provided the tangents to the generating curves of the solids do not make too great an angle with the axis. If the bar were removed, the two ends of the belt would be drawn each towards the large end of its pully, by Art. 181; but the loop is sufficient to prevent this action. By sliding the bar and belt to different points the velocity ratio will be gradually changed as the acting diameters of the driver and follower are thus both gradually altered.

475. The solids are easily formed to suit the condition of the constancy of their added diameters; for draw AM, ab, fig. 244,

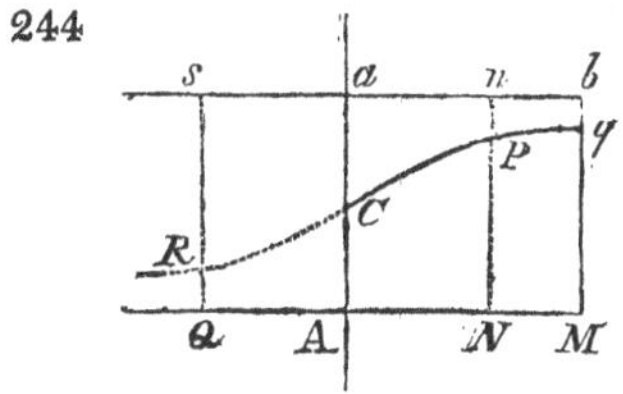

parallel and at a distance equal to the given sum of the radii, and let CPq be the generating curve of one pully round AM, then will the same curve generate the other pully by revolving round ab.

476. Let $AN = x$, $NP = y$, $nP = y_{\prime}$, A and a be angular velocities of the axes AM, ab, respectively,

$$\frac{A}{a} = \frac{y_{\prime}}{y}.$$

Now if the strap is to remain equally tight in every position, we must have $y + y_{,} = c$;

$$\therefore \frac{A}{a} = \frac{c - y}{y}.$$

If the solids be cones, of which $AM = l$, and $Mq = r$,

$$\text{we have } y = \frac{xr}{l}; \ \therefore \frac{A}{a} = \frac{c - \frac{r}{l} \cdot x}{\frac{r}{l} x} = \frac{\frac{lc}{r} - x}{x}.$$

If equal shifts of the belt between A and M are to produce equal differences in the velocity ratios, we have

$$\frac{A}{a} \propto x \propto \frac{c - y}{y}.$$

If equal shifts of the belt are to produce a geometrical series of velocity ratios, then

$$\frac{NP}{nP}, \text{ or } \frac{y}{c - y} = g^x,$$

and when $x = 0$, $NP = nP$; therefore the origin of x is at the point A, if $AC = aC$,

$$\text{and } \frac{c}{y} = g^{-x} + 1; \ \therefore y = \frac{c}{g^{-x} + 1}$$

is equation to curve.

$$\text{Also, } c - y = c - \frac{c}{g^{-x} + 1} = \frac{c}{g^x + 1};$$

which shews that if we set off from the point A equal abscissæ AN, AQ, in opposite directions the ordinates NP, sR will be equal.

477. But in practice it is more usual to make the solid pullies into cones, because the strap is apt to slip when the inclination is great. In this case the desired succession

of velocity ratios is obtained by making the shifts of the belt unequal.

When cones are employed,

$$\frac{A}{a} = \frac{\frac{lc}{r} - x}{x}, \text{ and } x = \frac{lc}{r} \times \frac{1}{1 + \frac{A}{a}},$$

from which the shifts or values of x can be computed for any required succession of values in $\frac{A}{a}$.

Sometimes a cone and cylinder are employed for the two solids, but in that case a stretching pully is required for the belt, because the sum of the corresponding diameters is no longer constant. If the cone be the driver the velocity ratio $\frac{a}{A}$ will vary directly as the distance of the belt from the apex of the cone.

478. Variable velocity ratios are also obtained from wrapping connectors by means of pullies so contrived as to expand and contract their acting diameters, the structure of which belonging to constructive mechanism, may be found in Rees' Cyclopædia; they are termed Expanding Riggers.

479. The *disk and roller* is often used for the purpose of obtaining an adjustable velocity ratio by rolling contact.

245

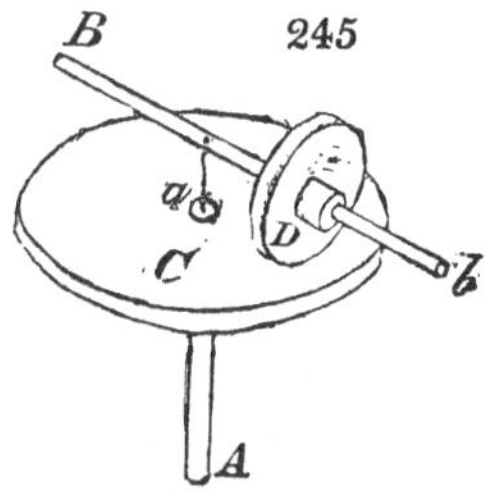

Aa the driving axis, to which is fixed a plain disk C. Bb the following axis whose direction meets that of Aa. A plain roller D, whose edge is covered with a narrow belt of soft leather, is mounted upon the axis Bb, so that it can be made to slide at pleasure to different distances from the point of intersection of the axes, but yet is prevented from turning with respect to Bb. This roller and its axis will therefore receive from the disk a rotation by rolling contact; and if r be the radius of the roller, R the adjustable radius of its point of contact with the disk, A and a the respective angular velocities of Aa and Bb, we have

$$\frac{a}{A} = \frac{R}{r} \text{ varies directly as } R.$$

But the rolling contact of the surfaces is imperfect, for perfect contact in the case of intersecting axes can only take place between cones whose apex coincides with the point or intersection. The following combination is more perfect in its action, but not so simple in construction.

480. Let AB, fig. 246, be the axis of the driver, which is a solid of revolution whose generating curve is Nn. The

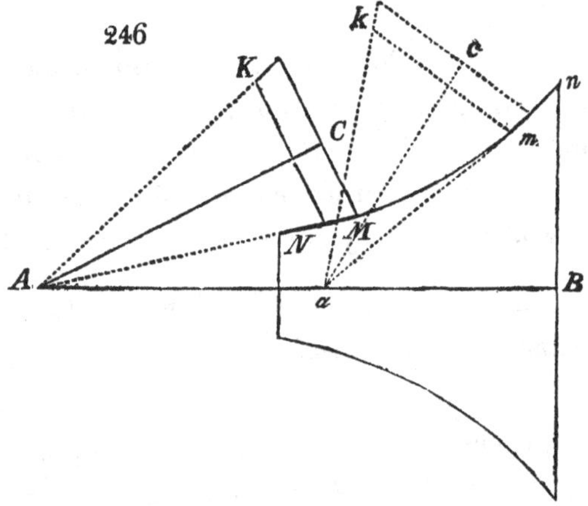

follower is a conical frustum KM, whose axis AC must be mounted in a frame in such a manner that the apex A of the cone may travel in a line Aa coinciding with the axis of the driver, and that the axis AC shall have the power of turning in position about the point A, so as to enable the frustum to rest upon the surface of the solid pully in every position of

AC, and thus to receive motion from it by rolling contact. Thus km is a position of the frustum in which it touches the solid at m, and its apex has moved from A to a, still remaining in the line AaB. If now the line AM touch the generating curve Nn in all these positions of AC, the portion of the solid in contact with the frustum is so small that it will nearly coincide with the corresponding frustum of a cone whose apex would be at A, and therefore coincide with that of the follower. The contact action therefore will in this case be complete.

But AM the tangent of Nn is thus shewn to be of a constant length, Nn is therefore the equitangential curve or tractory (Peacock's Ex. p. 174), to find the equation to which, we have, if AB be the axis of x,

$$\tan = \frac{y\sqrt{dx^2 + dy^2}}{dy} = t \text{ a constant.}$$

$$\therefore\ dx = \frac{dy}{y}\sqrt{t^2 - y^2} \text{ is equation to curve;}$$

which integrated gives

$$x = \sqrt{t^2 - y^2} + \frac{t}{2}\log\frac{t - \sqrt{t^2 - y^2}}{t + \sqrt{t^2 - y^2}};$$

whence from assumed values of y the curve may be constructed by points.

$$\sqrt{t^2 - y^2} \text{ is the subtangent} = s \text{ suppose;}$$

$$\therefore\ x = s - \frac{t}{2}\log\frac{t + s}{t - s}.$$

y	s	x
.9	4.72	6.75
1.	4.70	6.29
1.1	4·68	5.80
1.2	4.65	5.29
1.3	4.62	4.88
1.4	4.59	4.53
1.5	4.56	4.23
1.6	4.53	3.97
1.8	4.46	3.47
2.0	4.37	2.97
2.2	4.27	2.54
2.4	4.16	2.17
2.6	4.04	1.85
2.8	3.90	1.54
3	3.76	1.30

In the above table values of y are taken from 3 inches to 9, and the constant tangent $t = 4.8$ inches. From this the curve may be easily constructed by points.

481. The solid cam, (Art. 363) may be used to obtain adjustable motion, in which case the screw a and its nut must be removed, and the cam may then be shifted at pleasure so as to bring any section of it into action upon the follower Dd; and also this section may be allowed to continue its action as long as we please; thus we may, by properly forming the successive sections of the solid, retain at pleasure the law of motion that belongs to any one of them, or gradually change it into that which is appropriated to any other section, by shifting the cam so as to bring that section under the follower.

482. In link-work gradual changes of the velocity ratio are effected by fixing the pins upon the arms in slits or sliding pieces, that thus allow of gradual changes in the effective lengths of these arms upon which the velocity ratio depends. This may be managed in various ways. I shall conclude this Part with a piece of link-work by which such changes may be effected without the use of these adjustable pins.

483. *A*, fig. 247, is the fixed center of motion of a crank or excentric *Am*, which by means of a link *mb* communicates in the usual way a small reciprocating motion to the arm *Bb*, whose center of motion is *B*. The end of *b* is also joined by a link *bc* to an arm *Cc*, therefore the reciprocation of *b* is communicated to *c*. But the center of motion of the arm *B* is itself mounted on a shifting arm whose center is near *b*; and the radial distance of *B* from this center is made equal to the link *bc*; thus the position of the center *B* can be shifted to any point of the arc *Bc*, or even be brought to coincide with *c*. In all these different positions the quantity of motion which *b* receives from *Am* will be nearly the same, but the arc described by *c* will vary; for when the center is at *B*, it will give to *c* very nearly its own motion; but when *B* is moved to *c* it will communicate no motion at all to *c*, for the link *bc* will then coincide with *Bb*, and will vibrate as one piece with it round the point *c*. In any intermediate positions of *Bb*, as at *Db*, the velocity of *c* and the extent of its excursion will vary nearly as *Dn*, the perpendicular upon *bc*, which vanishes when *D* comes to *c*.

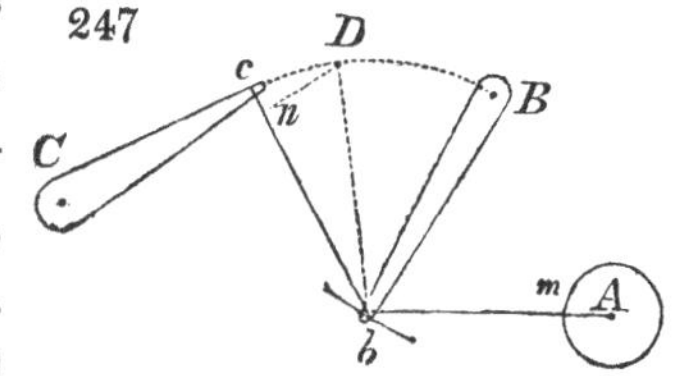

As the travelling of the center *B* does not stop the motion of the system, this combination affords a ready method of adjusting the relative velocity in link-work, or of entirely cutting off the motion of the follower *Cc* without stopping the motion of the driver *Am*.

THE END.

Printed in the United States
By Bookmasters